D1164067

High-Level Synthesis

Philippe Coussy • Adam Morawiec
Editors

High-Level Synthesis

From Algorithm to Digital Circuit

Philippe Coussy
Université Européenne
de Bretagne - UBS
Laboratoire Lab-STICC
Centre de Recherche
BP 92116
56321 Lorient Cedex
France
philippe.coussy@univ-ubs.fr

Adam Morawiec
European Electronic Chips & Systems
design Initiative (ECSI)
2 av. de Vignate
38610 Grieres
France
adam.morawiec@ecsi.org

ISBN 978-1-4020-8587-1 e-ISBN 978-1-4020-8588-8

Library of Congress Control Number: 2008928131

Cover illustration: Cover design by Martine Piazza, Adam Morawiec and Philippe Coussy

Printed on acid-free paper.

9 8 7 6 5 4 3 2 1

springer.com

Foreword

High-level synthesis – also called behavioral and architectural-level synthesis – is a key design technology to realize systems on chip/package of various kinds, whether single or multi-processors, homogeneous or heterogeneous, for the embedded systems market or not. Actually, as technology progresses and systems become increasingly complex, the use of high-level abstractions and synthesis methods becomes more and more a necessity. Indeed, the productivity of designers increases with the abstraction level, as demonstrated by practices in both the software and hardware domains. The use of high-level models allows designers with systems, rather than circuit, background to be productive, thus matching the trend of industry which is delivering an increasingly larger number of integrated systems as compared to integrated circuits.

The potentials of high-level synthesis relate to leaving implementation details to the design algorithms and tools, including the ability to determine the precise timing of operations, data transfers, and storage. High-level optimization, coupled with high-level synthesis, can provide designers with the optimal concurrency structure for a data flow and corresponding technological constraints, thus providing the balancing act in the trade-off between latency and resource usage. For complex systems, the design space exploration, i.e., the systematic search for the Pareto-optimal points, can only be done by automated high-level synthesis and optimization tools.

Nevertheless, high-level synthesis has been showing a long gestation period. Despite early results in the 1980s, it is still not common practice in hardware design. The slow acceptance-rate of this important technology has been attributed to a few factors such as designers' desire to micromanage integrated systems by controlling their internal timing and the lack of a universal standard front-end language. The former issue is typical of novel technologies: as systems grow in size it will be necessary for designers to show a broader system vision and fewer concerns on internal timing. In other words, this problem will naturally disappear.

The Babel of high-level modeling languages has been a significant obstacle to the development of this technology. When high-level synthesis was introduced in the 1980s, the designer community embraced Verilog and VHDL as specification languages, due to their ability to perform efficient simulation. Nevertheless, such languages were conceived without an intrinsic hardware semantics, making synthesis more cumbersome.

C-based hardware description languages (CHDLs) surfaced in the 1980s as well, such as HardwareC and its hardware compiler Hercules. The limitations of HardwareC and similar CHDLs are rooted in the modification of the C language semantics to support hardware constructs, thus making each CHDL a different dialect of C. The introduction of SystemC in the 1990s solved the problem by not modifying the software programming language (in this case C++) and by introducing a class library with a well-defined hardware semantics. It is regrettable that the initial enthusiasm was mitigated by the limited support of high-level synthesis for SystemC.

The turn of the century was characterized by a renewed interest in CHDLs and in high-level synthesis from CHDLs. New companies carried the torch of educating designers with new models and tools for design. Today, there are several offers in high-level synthesis tools that provide effective solutions in silicon. Moreover, some of the technical roadblocks to high-level synthesis have been overcome. Synthesis of C-based models with pointers and memory allocators was demonstrated and patented by Stanford jointly with NEC, thus removing the last hard technical difficulty to synthesize full C-based models.

At present, the potentials of high-level synthesis are still very good, even though the designers' community has not yet converged on a single modeling language that would lower the entry barrier of tools into the marketplace. This book presents an excellent collection of contributions addressing different aspects of high-level synthesis from both industry and academia. This book should be on each designer's and CAD developer's shelf, as well as on those of project managers who will soon embrace high-level design and synthesis for all aspects of digital system design.

EPF Lausanne, 2008 *Giovanni De Micheli*

Contents

Contributors

Shail Aditya
Synfora, Inc., 2465 Latham Street, Suite #300, Mountain View, CA 94040, USA, shail.aditya@synfora.com

Ivan Augé
UPMC-LIP6/SoC, Équipe ASIM/LIP6, Université Pierre et Marie Curie, Paris, France, Ivan.Auge@lip6.fr

Thomas Bollaert
Mentor Graphics, 13/15 rue Jeanne Braconnier, 92360 Meudon-la-Foret, France, Thomas_Bollaert@mentor.com

Pierre Bomel
European University of Brittany – UBS, Lab-STICC, BP 92116, 56321 Lorient Cedex, France, pierre.bomel@univ-ubs.fr

Christos-Savvas Bouganis
Department of Electrical and Electronic Engineering, Imperial College London, South Kensington Campus, London SW7 2AZ, UK, christos-savvas.bouganis@imperial.ac.uk

Forrest Brewer
Electrical and Computer Engineering, University of California, Santa Barbara, CA 93106-9560, USA, forrest@ece.ucsb.edu

Cyrille Chavet
European University of Brittany – UBS, Lab-STICC, BP 92116, 56321 Lorient Cedex, France, chavet@univ-ubs.fr

Jason Cong
AutoESL Design Technolgoies, Inc., 12100 Wilshire Blvd, Los Angeles, CA 90025, USA

and

UCLA Computer Science Department, Los Angeles, CA 90095-1596, USA, cong@autoesl.com, cong@cs.ucla.edu

George A. Constantinides
Department of Electrical and Electronic Engineering, Imperial College London, South Kensington Campus, London SW7 2AZ, UK, george.constantinides@ieee.org

Philippe Coussy
European University of Brittany – UBS, Lab-STICC, BP 92116, 56321 Lorient Cedex, France, philippe.coussy@univ-ubs.fr

Steven Derrien
Irisa, universit'e de Rennes 1, Campus de beaulieu, 35042 Rennes Cedex, France, steven.derrien@irisa.fr

Robert P. Dick
Department of Electrical Engineering and Computer Science, Northwestern University, Evanston, IL, USA, dickrp@northwestern.edu

Yiping Fan
AutoESL Design Technolgoies, Inc., 12100 Wilshire Blvd, Los Angeles, CA 90025, USA, fanyp@autoesl.com

Wenrui Gong
Department of Electrical and Computer Engineering, University of California, Santa Barbara, CA 93106, USA, gong@ece.ucsb.edu

Alexandre Gouraud
France Telecom R&D, 38-40 rue du General Leclerc, 92794 Issy Moulineaux Cedex 9, France, alexandre.gouraud@orange-ftgroup.com

Rajesh Gupta
Computer Science and Engineering, University of California, San Diego, 9500 Gilman Drive, La Jolla, CA 92093-0404, USA, rgupta@ucsd.edu

Guoling Han
AutoESL Design Technolgoies, Inc., 12100 Wilshire Blvd, Los Angeles, CA 90025, USA, leohgl@autoesl.com

Dominique Heller
European University of Brittany – UBS, Lab-STICC, BP 92116, 56321 Lorient Cedex, France, dominique.heller@univ-ubs.fr

Román Hermida
Facultad de Informática, Universidad Complutense de Madrid, c/Prof. José García Santesmases s/n, 28040 Madrid, Spain, rhermida@dacya.ucm.es

Niraj K. Jha
Department of Electrical and Engineering, Princeton University, Princeton, NJ 08544, USA, jha@princeton.edu

Wei Jiang
AutoESL Design Technolgoies, Inc., 12100 Wilshire Blvd, Los Angeles, CA 90025, USA, wjiang@autoesl.com

Ryan Kastner
Department of Electrical and Computer Engineering, University of California, Santa Barbara, CA 93106, USA, kastner@ucsd.edu

Vinod Kathail
Synfora, Inc., 2465 Latham Street, Suite # 300, Mountain View, CA 94040, USA, vinod.kathail@synfora.com

Hyukmin Kwon
Samsung Electronics Co., Suwon, Kyunggi Province, South Korea, hm25.kwon@samsung.com

Eric Martin
European University of Brittany – UBS, Lab-STICC, BP 92116, 56321 Lorient Cedex, France, eric.martin@univ-ubs.fr

José Manuel Mendías
Facultad de Informática, Universidad Complutense de Madrid, c/Prof. José García Santesmases s/n, 28040 Madrid, Spain, mendias@dacya.ucm.es

Michael Meredith
VP Technical Marketing, Forte Design Systems, San Jose, CA 95112, USA, mmeredith@ForteDS.com

María Carmen Molina
Facultad de Informática, Universidad Complutense de Madrid, c/Prof. José García Santesmases s/n, 28040 Madrid, Spain, cmolinap@dacya.ucm.es

Rishiyur S. Nikhil
Bluespec, Inc., 14 Spring Street, Waltham, MA 02451, USA, nikhil@bluespec.com

Frédéric Pétrot
INPG-TIMA/SLS, 46 Avenue Félix Viallet, 38031 Grenoble Cedex, France, Frederic.Petrot@imag.fr

Patrice Quinton
ENS de Cachan, antenne de Bretagne, Campus de Ker Lann, 35 170 Bruz Cedex, France, patrice.quinton@irisa.fr

Sanjay Rajopadhye
Department of Computer Science, Colorado State University, 601 S. Howes St. USC Bldg., Fort Collins, CO 80523-1873, USA, Sanjay.Rajopadhye@colostate.edu

Tanguy Risset
CITI – INSA Lyon, 20 avenue Albert Einstein, 69621, Villeurbanne, France, tanguy.risset@insa-lyon.fr

Rafael Ruiz-Sautua
Facultad de Informática, Universidad Complutense de Madrid, c/Prof. José García Santesmases s/n, 28040 Madrid, Spain, rsautua@fdi.ucm.es

Benjamin Carrion Schafer
EDA R&D Center, Central Research Laboratories, NEC Corp., Kawasaki, Japan, schaferb@bq.jp.nec.com

Eric Senn
European University of Brittany – UBS, Lab-STICC, BP 92116, 56321 Lorient Cedex, France, eric.senn@univ-ubs.fr

Li Shang
Department of Electrical and Computer Engineering, Queen's University, Kingston, ON, Canada K7L 3N6, li.shang@queensu.ca

Pascal Urard
STMicroelectronics, Crolles, France, pascal.urard@st.com

Kazutoshi Wakabayashi
EDA R&D Center, Central Research Laboratories, NEC Corp., Kawasaki, Japan, wakaba@bl.jp.nec.com

Gang Wang
Technology Innovation Architect, Intuit, Inc., 7535 Torrey Santa Fe Road, San Diego, CA 92129, USA, Gang_Wang@intuit.com

Changqi Yang
AutoESL Design Technolgoies, Inc., 12100 Wilshire Blvd, Los Angeles, CA 90025, USA, charles@autoesl.com

Joonhwan Yi
Samsung Electronics Co., Suwon, Kyunggi Province, South Korea, joonhwan.yi@samsung.com, joonhwan.yi@gmail.com

Zhiru Zhang
AutoESL Design Technolgoies, Inc., 12100 Wilshire Blvd, Los Angeles, CA 90025, USA, zhiruz@autoesl.com

List of Web sites

Chapter 2

Microelectronic Embedded Systrems Laboratory at UCSD hosts a number of projects related to system level design, synthesis and verification. Our recent projects include the SPARK parallelizing synthesis framework, SATYA verification framework. Earlier work from the laboratory formed the technical basis for the SystemC initiative.
http://mesl.ucsd.edu/

Chapter 3

Catapult Synthesis product information page
The home page for Catapult Synthesis on www.mentor.com, with links to product datasheets, free software evaluation, technical publications, success stories, testimonials and related ESL product information.
http://www.mentor.com/products/esl/high_level_synthesis/

Algorithmic C datatypes download page
The Algorithmic C arbitrary-length bit-accurate integer and fixed-point data types allow designers to easily model bit-accurate behavior in their designs. The data types were designed to approach the speed of plain C integers. It is no longer necessary to compromise on bit-accuracy for the sake of speed or to explicitly code fixed-point behavior using integers in combination with shifts and bit masking.
http://www.mentor.com/products/esl/high_level_synthesis/ac_datatypes

Chapter 4

Synfora, Inc. is the premier provider of PICO family of algorithmic synthesis tools to design complex application engines for SoCs and FPGAs. Synfora's technology helps to reduce design costs, dramatically speed IP development and verification, and reduce time-to-market. For the latest information on Synfora and PICO products, please visit http://www.synfora.com

Chapter 5

More information on Cynthesizer from Forte Design Systems can be found at http://www.ForteDS.com

Chapter 6

More information on AutoPilotTM from AutoESL Design Technologies can be found at http://www.autoesl.com and http://cadlab.cs.ucla.edu/soc/

Chapter 7

Home Page for CyberWorkBench from NEC
http://www.cyberworkbench.com

Chapter 8

More information on Bluespec can be found at http://www.bluespec.com
Documentation, training materials, discussion forums, inquiries about Bluespec SystemVerilog. http://csg.csail.mit.edu/oshd/
Open source hardware designs done by MIT and Nokia in Bluespec SystemVerilog for H.264 decoder (baseline profile), OFDM transmitter and receiver, 802.11a transmitter, and more.

Chapter 9

GAUT is an open source project at UEB-Lab-STICC. The software for this project is freely available for download. It is provided with a graphical user interface, a quick start guide, a user manual and several design examples. GAUT is currently supported on Linux and Windows. GAUT has already been downloaded more than 200 times by people from industry and academia in 36 different countries. For more information, please visit:
http://web.univ-ubs.fr/gaut/

Chapter 10

More information can be found on UGH from at UPMC-LIP6/SoC and INPG-TIMA/SLS at http://www-asim.lip6.fr/recherche/disydent/
This web site contains introduction text, source code and tutorials (through CVS) of the opensource Dysident framework that includes the UGH HLS tool.

Chapter 11

More information on Chapter 11 can be found at
http://cas.ee.ic.ac.uk/

Chapter 12

More information on MMAlpha can be found at
http://www.irisa.fr/cosi/ALPHA/

Chapter 13

More information Chapter 13 can be found on at
http://www.cse.ucsd.edu/∼ kastner/research/aco/

Chapter 14

More information on Chapter 14 can be found at
http://atc.dacya.ucm.es/

Chapter 15

More information on Chapter 15 can be found at
http://www.princeton.edu/∼jha

Chapter 1
User Needs

Pascal Urard, Joonhwan Yi, Hyukmin Kwon, and Alexandre Gouraud

Abstract One can see successful adoption in industry of innovative technologies mainly in the cases where they provide acceptable solution to very concrete problems that this industry is facing. High-level synthesis promises to be one of the solutions to cope with the significant increase in the demand for design productivity beyond the state-of-the-art methods and flows. It also offers an unparalleled possibility to explore the design space in an efficient way by dealing with higher abstraction levels and fast implementation ways to prove the feasibility of algorithms and enables optimisation of performances. Beyond the productivity improvement, which is of course very pertinent in the design practice, the system and SoC companies are more and more concerned with their overall capability to design highly complex systems providing sophisticated functions and services. High-level synthesis may considerably contribute to maintain such a design capability in the context of continuously increasing chip manufacturing capacities and ever growing customer demand for function-rich products.

In this chapter three leading industrial users present their expectations with regard to the high-level synthesis technology and the results of their experiments in practical application of currently available HLS tools and flows. The users also draw conclusions on the future directions in which they wish to see the high-level synthesis evolves like multi-clock domain support, block interface synthesis, joint optimisation of the datapath and control logic, integration of automated testing to the generated hardware or efficient taking into account of the target implementation technology for ASICs and FPGAs in the synthesis process.

Pascal Urard
STMicroelectronics

Joonhwan Yi and Hyukmin Kwon
Telecommunication R&D, Samsung Electronics Co., South Korea

Alexandre Gouraud
France Telecom R&D

P. Coussy and A. Morawiec (eds.) *High-Level Synthesis*.

Keywords: High-level synthesis, Productivity, ESL, ASIC, SoC, FPGA, RTL, ANSI C, C++, SystemC, VHDL, Verilog, Design, Verification, IP, TLM, Design space exploration, Memory, Parallelism, Simulation, Prototyping

1.1 System Level Design Evolution and Needs for an IDM Point of View: STMicroelectronics[1]

Pascal Urard, STMicroelectronics

The complexity of digital integrated circuits has always increased from a technology node to another. The designers often had to adapt to the challenge of providing commercially acceptable solution with a reasonable effort. Many evolutions (and sometimes revolutions) occurred in the past: back-end automation or logical synthesis were part of those, enabling new area of innovation. Thanks to the increasing integration factor offered by technology nodes, the complexity in latest SoC has reached tens of millions of gates. Starting with 90 nm and bellow, RTL design flow (Fig. 1.1) now shows its limits.

The gap between the productivity per designer and per year and the increasing complexity of the SoC, even taking into account some really conservative number of gates per technology node, lead to an explosion of the manpower for SoCs in the coming technology node (Fig. 1.2).

There is a tremendous need for productivity improvement at design level. This creates an outstanding opportunity for new design techniques to be adopted: designers, facing this challenge, are hunger to progress and open to raise the level of abstraction of the golden reference model they trust.

A new step is needed in productivity. Part of this step could be offered by ESLD: Electronics System Level Design. This includes HW/SW co-design and High-Level Synthesis (HLS).

HW/SW co-design deployment has occurred few years ago, thanks to SystemC and TLM coding. HLS however is new and just starting to be deployed. Figure 1.3 shows the basis of STMicroelectronics C-level design methodology. A bit-accurate reference model is described at functional level in C/C++ using SystemC or equivalent datatypes. In the ideal case, this C-level description has to be extensively validated using a C-level testbench, in the functional environment, in order to become the golden model of the implementation flow. This is facilitated by the simulation speed of this C model, usually faster than other kinds of description. Then, taking into account technology constraints, the HLS tool produces an RTL representation, compatible with RTL-to-GDS2 flow. Verification between C-level model and RTL is done either thanks to sequential equivalence checking tools, or by extensive simulations. Started in 2001 with selected CAD-vendors, the research on new flows

[1] (C) Pascal Urard, STMicroelectronics Nov. 2006. Extracted for P. Urard presentation at ICCAD, Nov. 2006, San José, California, USA.

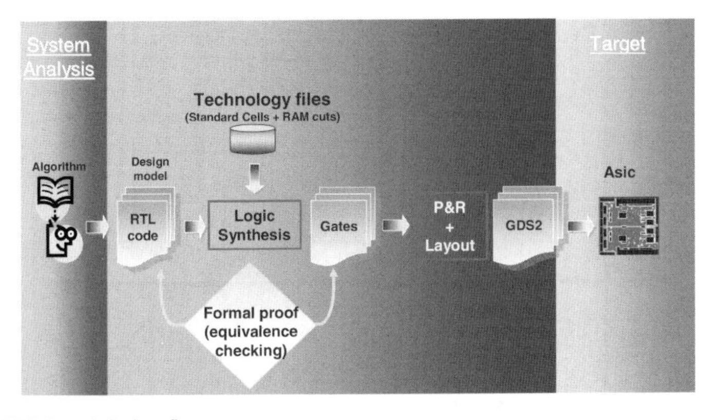

Fig. 1.1 RTL Level design flow

1991	1994	1006	1998	2000	2002	2004	2006	2008	2010
0.7	0.5	0.35	0.25	0.18	0.13	90	65	45	32
1k	5k	15k	30k	45k	80k	150k	300k	600k	1.2M
#Gates / Die (50mm2) conservative numbers									
50K	250K	750k	1.5M	2.2M	4M	7.5M	15M	30M	60M
#Gates per Designer per year									
4k	6k	9k	40k	56k	91k	125k	200k	200k	200k
Men / Years per 50 mm2 Die									
~10	~40	~80	~40	~40	~43	~60	~75	~150	~300

→ **It is urgent to win some productivity**

Fig. 1.2 Design challenges for 65 nm and below

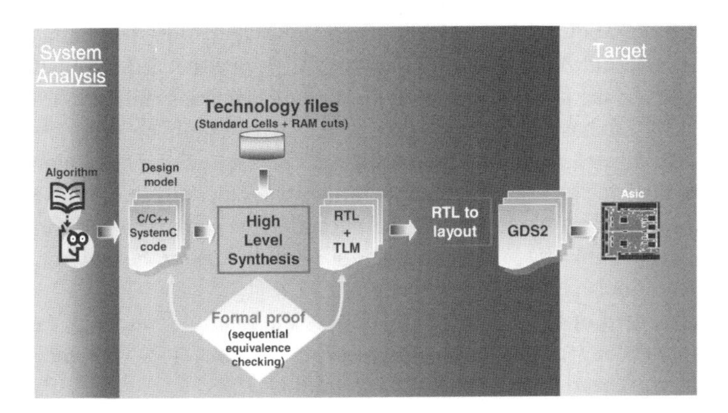

Fig. 1.3 High level synthesis flow

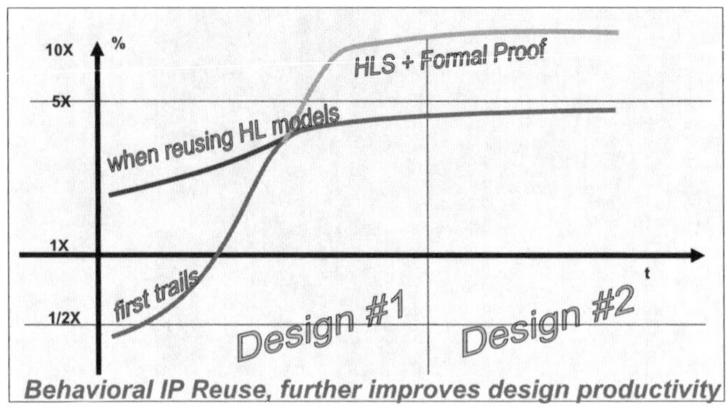

Fig. 1.4 Learning curve

has allowed some deployment of HLS tools within STMicroelectronics starting in 2004, with early division adopters. We clearly see in 2007 an acceleration of the demand from designers. Those designers report to win a factor $\times 5$ to $\times 10$ in terms of productivity when using C-level design methodology depending on the way they reuse in design their IPs (Fig. 1.4). More promising: designers that moved to C-level design usually don't want to come back to RTL level to create their IPs...

Side benefit of these C-level design automation, the IP reuse of signal processing IP is now becoming reality. The flow automation allows to have C-IPs quite independent of implementation constraints (technology, throughput, parameters), described at functional level, easy to modify to cope with new specification and easy to re-synthesize. Another benefit: the size of the manual description (C instead of RTL) is reduced by roughly a factor 10. This reduces the time to modification (ECO) as well as the number of functional bugs manually introduced in the final silicon.

The link with Transactional Level Modelling (TLM) platform has to be enhanced. Prior to HLS flow, both TLM and RTL descriptions where done manually (Fig. 1.5).

HLS tools would be able to produce the TLM view needed for platform validation. However, the slowing-down of TLM standardization did not allow in 2006 neither H1-2007 to have a common agreement of what should be TLM 2.0 interface. This lack of standardization has penalized the convergence of TLM platform flow and C-level HW design flow. Designer community would benefit of such a common agreement between major players of the SystemC TLM community. More and more, we need CAD community to think in terms of flows in their global environment, and not in terms of tools alone.

Another benefit of HLS tools automation is the micro-architecture exploration. Figure 1.6 basically describes a change of paradigm: clock frequency can be partially de-correlated from throughput constraints.

This means that, focusing on the functional constraints (throughput/latency), designer can explore several solutions fulfilling the specifications, but using various clock frequencies. Thanks to released clock constraint, the low-speed design will not have the area penalty of the high-speed solution. Combining this exploration

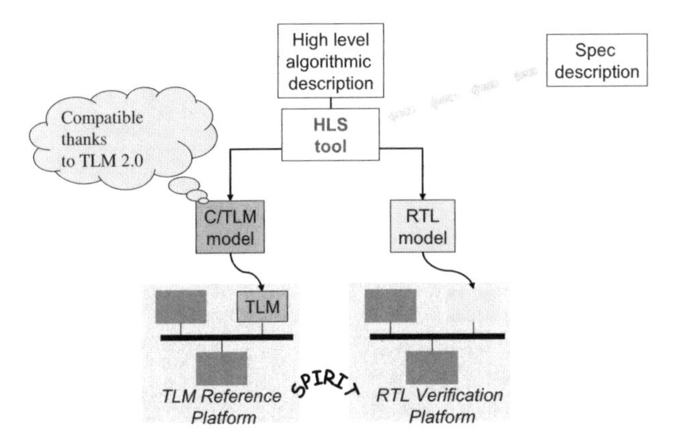

Fig. 1.5 Convergence of TLM and design flows

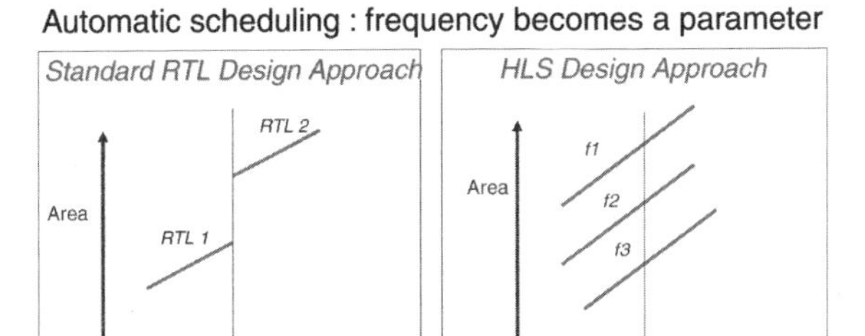

Fig. 1.6 One benefit of automation: exploration

to memory partitioning and management exploration leads to some very interesting solutions. As an example, Fig. 1.7 shows that combining double buffering of an LDPC encoder to a division by 2 of the clock speed, produces a $\times 0.63$ lower power solution for a 27% area penalty. The time-to-solution is dramatically reduced thanks to automation. The designer can then take the most appropriated solution depending on application constraints (area/power). Currently, power is estimated at RTL level, on automatically produced RTL, thanks to some specialized tools. Experience shows that power savings can be greatly improved at architectural level, compared to back-end design level.

There is currently no real power-driven synthesis solution known to us. This is one of the major needs we have for the future. Power driven synthesis will have to be much more than purely based on signals activity monitoring in the SoC buses. It will need also to take into account leakage current, memory consumption and will have to be compliant with multi-power-modes solutions (voltage and frequency scaling). There are many parameters to take into account to determine a power optimized solution, the ideal tool would have to take care of all these parameters in order to

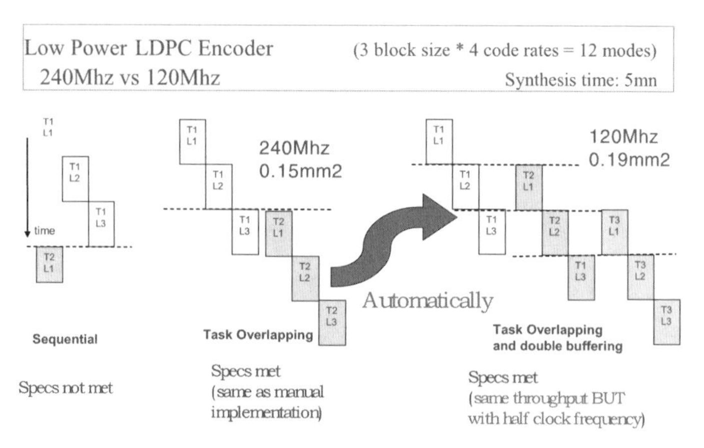

Fig. 1.7 HLS architecture explorations

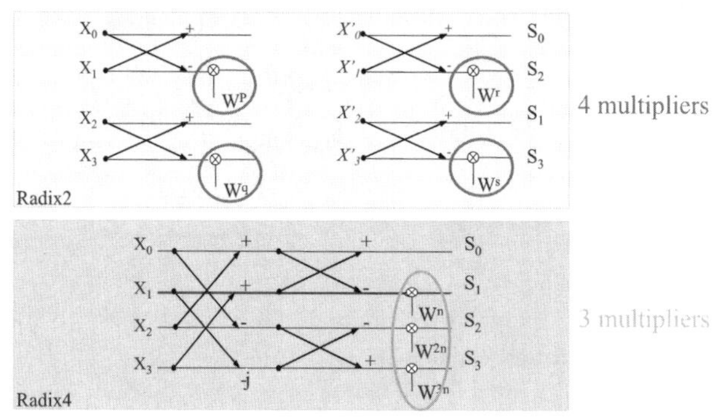

Fig. 1.8 Medium term need: arithmetic optimizations

allow the designer to keep a high level of abstraction and to focus on functionality. For sure this would have to be based on some pre-characterization of the HW.

Now HLS is being deployed, new needs are coming out for more automation and more optimization. Deep arithmetic reordering is one of those needs. The current generation of tools is effectively limited in terms of arithmetic reordering. As an example: how to go from a radix2 FFT to a radix4 FFT without re-writing the algorithm? Figure 1.8 shows one direction new tools need to explore. Taylor Expansion Diagrams seems promising in this domain, but up to now, no industrial EDA tool has shown up.

Finally after a few years spent in the C-level domain, it appears that some of the most limiting factors to exploration as well as optimization are memory accesses. If designer chose to represent memory elements by RAMs (instead of Dflip-flop), then the memory access order needs to be explicit in the input C code, as soon as this is not a trivial order. Moreover, in case of partial unroll of some FOR loops dealing with data stored in a memory, the access order has to be re-calculated and C-code has to be rewritten to get a functional design. This can be resumed to a problem of memory precedence optimization. The current generation of HLS tools have a very low level of exploration of memory precedence, when they have some: some tool simply ignore it, creating non-functional designs! In order to illustrate this problem, let take an in-place FFT radix2 example. We can simplify this FFT to a bunch of butterflies, a memory (RAM) having the same width than the whole butterflies, and an interconnect. In a first trial, with a standard C-code, let flatten all butterflies (full unroll): we have a working solution shown in Fig. 1.9.

Keep in mind that during third stage, we store the memory the $C_0 = K.B_0 + B_4$ calculation. Let now try to not completely unroll butterflies but allocate half of them (partial unroll). Memory will have the same number of memory elements, but twice deeper, and twice narrower. Calculation stages are shown in Fig. 1.10.

We can see that the third stage has a problem: C_0 cannot be calculated in a single clock cycle as B_0 and B_4 are stored at two different addresses of the memory. With current tools generation, when B_0 is not buffered, then RTL is not-functional

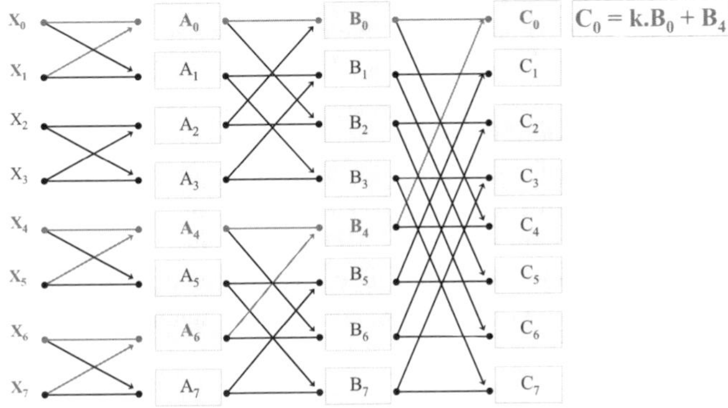

Fig. 1.9 Medium term need: memory access problem

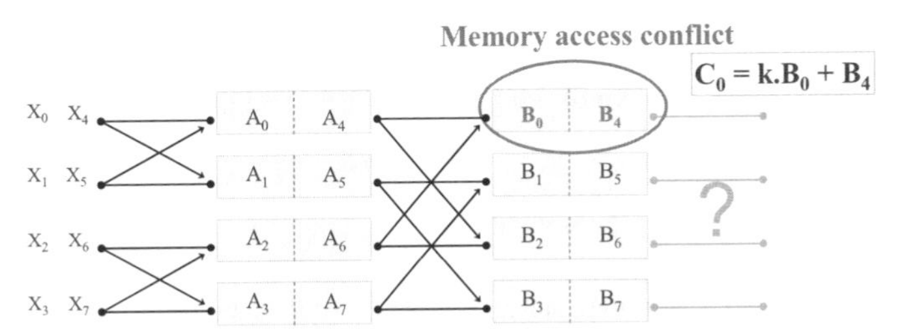

Fig. 1.10 Medium term need: memory access problem

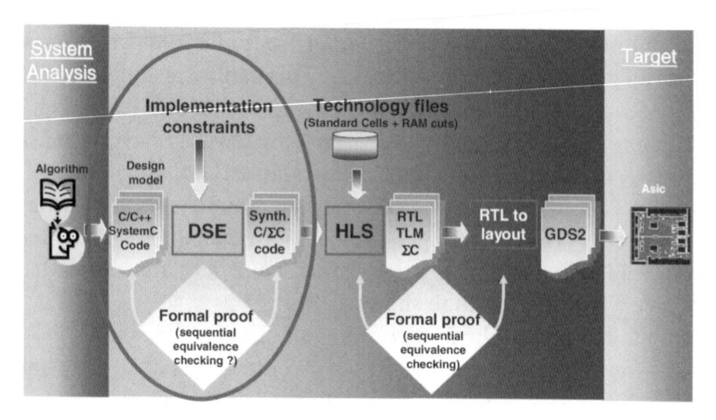

Fig. 1.11 HLS flow: future enhancements at design space exploration level

because tools have weak check of memory precedence. HLS designers would need a tool that re-calculate memory accesses given the unroll factors and interface accesses. This would ease a lot the Design Space Exploration (DSE) work, leading to find much optimized solutions. This could also be part of higher level optimizations tools: DSE tools (Fig. 1.11).

Capacity of HLS tools is another parameter to be enhanced, even if tools have done enormous progresses those last years. The well known Moore's law exists and even tools have to follow the semi-conductor industry integration capacity.

As a conclusion, let underline that HLS tools are working, are used in production flows on advanced production chips. However, some needs still exist: enhancement of capacity, enhancement of arithmetic optimizations, or automation of memory allocation taking into account micro-architecture. We saw in the past many stand-alone solutions for system-level flows, industry now needs academias and CAD vendors to think in terms of C-level flows, not anymore stand-alone tools.

1.2 Samsung's Viewpoints for High-Level Synthesis

Joonhwan Yi and Hyukmin Kwon, Telecommunication R&D, Samsung Electronics Co.

High-level synthesis technology and its automation tools have been in the market for many years. However the technology is not mature enough for industry to widely accept it as an implementation solution. Here, our viewpoints regarding high-level synthesis are presented.

The languages that a high-level synthesis tool takes as an input often characterize the capabilities of the tool. Most high-level synthesis languages are C-variant including SystemC [1]. Some tools take C/C++ codes as inputs and some take SystemC as inputs. These languages differ from each other in several aspects, see

Table 1.1 The differences between C/C++ and SystemC as a high-level synthesis language

	ANSI C/C++	SystemC
Synthesizable code	Untimed C/C++	Untimed/timed SystemC
Abstraction level	Very high	High
Concurrency	Proprietary support	Standard support
Bit accuracy	Proprietary support	Standard support
Specific timing model	Very hard	Standard support
Complex interface design	Impossible	Standard support, but hard
Ease of use	Easy	Medium

Table 1.1. Based on our experience, C/C++ is good at describing hardware behavior in a higher level than SystemC. On the other hand, SystemC is good at describing hardware behavior in a bit-accurate and/or timing-specific fashion than C/C++. High-level synthesis tools for C/C++ usually provide proprietary data types or directives because C/C++ has no standard syntax for describing timing. Of course, the degree of detail in describing timing by the proprietary mean is somewhat limited comparing to SystemC. So, there exists a trade-off between two languages. A hardware block can be decomposed into block body and its interface. Block body describes the behavior of the block and its interface defines the way of communication with the outer world of the block. A higher level description is preferred for a block body while a bit-accurate and timing-specific detail description needs to be possible for a block interface. Thus, a high-level synthesis tool needs to provide ways to describe both block bodies and block interfaces properly.

Generally speaking, high-level synthesis tools need to support common syntaxes and commands of C/C++/SystemC that are usually used to describe the hardware behavior at the algorithm level. They include arrays, loops, dynamic memories, pointers, C++ classes, C++ templates, and so on. Current high-level synthesis tools can synthesize some of them but not all. Some of these commands or syntaxes may not be directly synthesizable.

Although high-level synthesis intends to automatically convert an algorithm level specification of a hardware behavior to a register-transfer level (RTL) description that implements the behavior, it requires many code changes and additional inputs from designers [2]. One of the most difficult problems for our high-level synthesis engineers is that the code changes and additional information needed for desired RTL designs are not clearly defined yet. Behaviorally identical two high-level codes usually result in very different RTL designs with current high-level synthesis tools. Recall that RTL designs also impose many coding rules for logic synthesis and lint tools exist for checking those rules. Likewise, a set of well defined C/C++/SystemC coding rules for high-level synthesis should exist. So far, this problem is handled by a brute-force way and well-skilled engineers are needed for better quality of results.

One of the most notable limitations of the current high-level synthesis tools is not to support multiple clock domain designs. It is very common in modern hardware designs to have multiple clock domains. Currently, blocks with different clock domains should be synthesized separately and then integrated manually. Our high-level synthesis engineers experienced significant difficulties in integrating synthesized RTL blocks too. A block interface of an algorithm level description is usually not detailed enough to synthesize it without additional information. Also, integration of the synthesized block interface and the synthesized block body is done manually. Interface synthesis [4] is an interesting and important area for high-level synthesis.

Co-optimization of datapath and control logic is also a challenging problem. Some tools optimize datapath and others do control logic well. But, to our knowledge, no tool can optimize both datapath and control logic at the same time. Because a high-level description of hardware often omits control signals such as valid, ready, reset, test, and so on, it is not easy to automatically synthesize them. Some additional information may need to be provided. In addition, if possible, we want to define the timing relations between datapath signals and control signals.

High-level synthesis should take into account target process technology for RTL synthesis. The target library can be an application specific integrated circuit (ASIC) or a field programmable logic array (FPGA) library. Depending on the target technology and target clock frequency, RTL design should be changed properly. The understanding of the target technology is helpful to accurately estimate the area and timing behavior of resultant RTL designs too. A quick and accurate estimation of the results is also useful because users can quickly measure the effects of high-level codes and other additional inputs including micro architectural and timing information.

The verification of a generated RTL design against its input is another essential capability of high-level synthesis technology. This can be accomplished either by a sequential equivalence checking [3] or by a simulation-based method. If the sequential equivalence checking method can be used, the long verification time of RTL designs can be alleviated too. This is because once an algorithm level design D_h and its generated RTL design D_{RTL} are formally verified, fast algorithm level design verification will be sufficient to verify D_{RTL}. Sequential equivalence checking requires a complete timing specification or timing relation between D_h and D_{RTL}. Unless D_{RTL} is automatically generated from D_h, it is impractical to manually elaborate the complete timing relation for large designs.

Seamless integration to downstream design flow tools is also very important because the synthesized RTL designs are usually hard to understand by human. First of all, design for testability (DFT) of the generated RTL designs should be taken into account in high-level synthesis. Otherwise, the generated RTL designs cannot be tested and thus cannot be implemented. Secondly, automatic design constraint generation is necessary for gate-level synthesis and timing analysis. A high-level synthesis tool should learn all the timing behavior of the generated RTL designs such as information of false paths and multi-cycle paths. On the other hand, designers have no information about them.

We think high-level synthesis is one of the most important enabling technologies that fill the gap between the integration capacity of modern semiconductor processes and the design productivity of human. Although high-level synthesis is suffering from several problems mentioned above, we believe these problems will be overcome soon and high-level synthesis will prevail in commercial design flows in a near future.

1.3 High Level Design Use and Needs in a Research Context

Alexandre Gouraud, France Telecom R&D

Implementing algorithms onto electronic circuits is a tedious task that involves scheduling of the operations. Whereas algorithms can theoretically be described by sequential operations, their implementations need better than sequential scheduling to take advantage of parallelism and improve latency. It brings signaling into the design to coordinate operations and manage concurrency problems. These problems have not been solved in processors that do not use parallelism at algorithm level but only at instruction level. In these cases, parallelism is not fully exploited. The frequency race driven by processor vendors shadowed the problem replacing operators' parallelism by faster sequential operators. However, parallelism remains possible and it will obviously bring tremendous gains in algorithms latencies. HLS design is a kind of answer to this hole, and opens a wide door to designers.

In research laboratories, innovative algorithms are generally more complex than in market algorithms. Rough approximations of their complexity are often the first way to rule out candidates to implementation even though intrinsic (and somehow often hidden) complexity might be acceptable. The duration of the implementation constrains the space of solutions to a small set of propositions, and is thus a bottleneck to exploration. HLS design tools bring to researchers a means to test much more algorithms by speeding up drastically the implementation phase. The feasibility of algorithms is then easily proved, and algorithms are faster characterized in term of area, latency, memory and speed.

Whereas implementation on circuits was originally the reserved domain of specialists, HLS design tools break barriers and bring the discipline handy to non-hardware engineers. In signal processing, for instance, it allows faster implementation of algorithms on FPGA to study their behavior in more realistic environment. It also increases the exploration's space by speeding up simulations.

Talking more specifically about the tools themselves, the whole stake is to deduce the best operations' scheduling from the algorithm description, and eventually from the user's constraints. A trade-off has to be found between user's intervention and automatic deduction of the scheduling in such a way that best solutions are not excluded by the tool and complicated user intervention is not needed.

In particular, state machine and scheduling signals are typical elements that the user should not have to worry about. The tool shall provide a way to show operations' scheduling, and eventually a direct or indirect way to influence it. The user shall neither have to worry about the way scheduling is implemented nor how effective this implementation is. This shall be the tool's job.

Another interesting functionality is the bit-true compatibility with the original model/description. This guarantee spares a significant part of the costly time spent to test the synthesized design, especially when designs are big and split into smaller pieces. Whereas each small piece of code needed its own test bench, using HLS tools allows work on one bigger block. Only one test bench of the global entity is implemented which simplifies the work.

Models are generally complex, and their writing is always a meticulous task. If one can avoid their duplication with a different language, it is time saving. This raises the question whether architectural and timing constraints should be included inside the original model or not. There is no clear answer yet, and tools propose various interfaces described in this book. From a user's perspective, it is important to keep the original un-timed model stable. The less it is modified, the better it is manageable in the development flow. Aside from this, evolutions of the architecture along the exploration process shall be logged using any file versioning system to allow easy backward substitution and comparisons.

To conclude this introduction, it is important to point out that introduction of HLS tools should move issues to other fields like dimensioning of variables where tools are not yet available but the engineer's brains.

References

1. T. Grotker et al., *System design with SystemC*, Kluwer, Norwell, MA, 2002
2. B. Bailey et al., *ESL design and verification*, Morgan Kaufmann, San Mateo, 2007
3. Calypto design systems, available at http://www.calypto.com/products/index.html
4. A. Rajawat, M. Balakrishnan, A. Kumar, Interface synthesis: issues and approaches, *Int. Conf. on VLSI Design*, pp. 92–97, 2000

Chapter 2
High-Level Synthesis: A Retrospective

Rajesh Gupta and Forrest Brewer

Abstract High-level Synthesis or HLS represented an ambitious attempt by the community to provide capabilities for "algorithms to gates" for a period of almost three decades. The technical challenge in realizing this goal drew researchers from various areas ranging from parallel programming, digital signal processing, and logic synthesis to expert systems. This article takes a journey through the years of research in this domain with a narrative view of the lessons learnt and their implication for future research. As with any retrospective, it is written from a purely personal perspective of our research efforts in the domain, though we have made a reasonable attempt to document important technical developments in the history of high-level synthesis.

Keywords: High-level synthesis, Scheduling, Resource allocation and binding, Hardware modeling, Behavioral synthesis, Architectural synthesis

2.1 Introduction

Modern integrated circuits have come to be characterized by the scaling of Moore's law which essentially dictates a continued doubling in the capacity of cost-efficient ICs every so many months (every 18 months in recent trends). Indeed, *capacity* and *cost* are two major drivers of the microelectronics based systems on a chip (or SOC). A pad limited die of 200 pins on a 130 nm process node is about 50 square millimeters in area and comes to about $5 or less in manufacturing and packaging costs per part given typical yield on large volumes of 100,000 units or more. That is area sufficient to implement a large number of typical SOC designs without pushing the envelope on die size or testing or packaging costs. However, the *cost of design* continues to rise. Figure 2.1 shows an estimate of design costs which were estimated to be around US$15M, contained largely through continuing

P. Coussy and A. Morawiec (eds.) *High-Level Synthesis.*

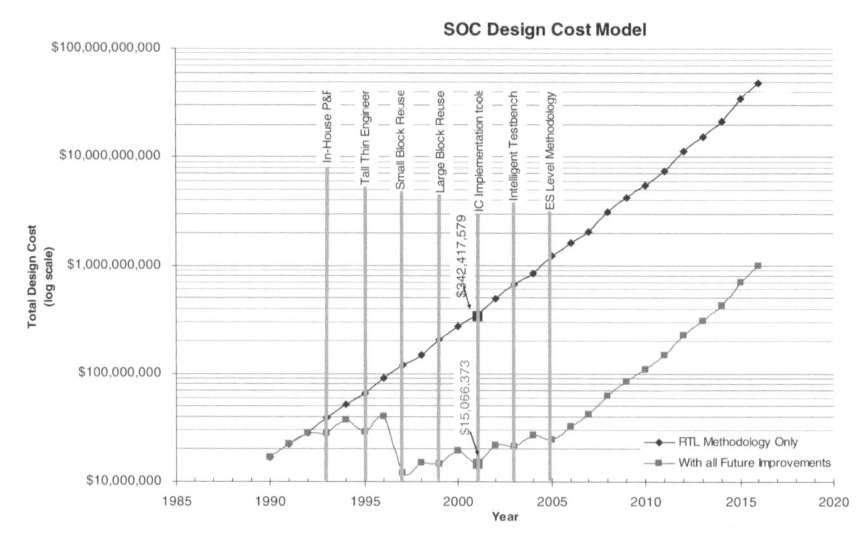

Fig. 2.1 Rising cost of IC design and effect of CAD tools in containing these costs (courtesy: Andrew Kahng, UCSD and SRC)

advances in IC implementation tools. Even more importantly, silicon architectures – that is, the architecture and organization of logic and processing resources on chip – are of critical importance. This is because of a tremendous variation in the *realized* efficiency of silicon as a computational fabric. A large number of studies have shown that energy or area efficiency for a given function realized on a silicon substrate can vary by two to three orders of magnitude. For example, the power efficiency of a microprocessor-based design is typically 100 million operations per watt, where as reprogrammable arrays (such as Field Programmable Gate Arrays or FPGAs) can be 10–20×, and a custom ASIC can give another 10× gain. In a recent study, Kuon and Rose show that ASICs are 35× more area efficient that FPGAs [1]. IC design is probably one of the few engineering endeavors that entail such a tremendous variation in the quality of solutions in relation to the design effort. If done right, there is a space of 10–100× gain in silicon efficiency when realizing complex SOCs. However, realizing the *intrinsic* efficiency of silicon in practice is an expensive proposition and tremendous design effort is expended to reach state power, performance and area goals for typical SOC designs. Such efforts invariably lead to functional, performance, and reliability issues when pushing limits of design optimizations. Consequently, in parallel with the Moore's law, each generation of computer-aided design (CAD) researchers has sought to disrupt conventional design methodologies with the advent of *high-level* design modeling and tools to automate the design process. This pursuit to raise the abstraction level at which designs are modeled, captured, and even implemented has been the goal of several generations of CAD researchers. Unfortunately, thus far, every generation has come away with mixed success leading to the rise of yet another generation that seems to have got it right. Today, such efforts are often lumped under the umbrella term of ESL or Electronic System Level design which in turn means a range of activities from algorithmic design and implementation to virtual system prototyping to function-architecture co-design [43].

2.2 The Vision Behind High-Level Synthesis

Mario Barbacci noted in late 1974 that in theory one could "compile" the instruction set processor specification (then in the ISPS language) into hardware, thus setting up the notion of design synthesis from a high-level language specification. High-level Synthesis in later years will thus come to be known as the process of automatic generation of hardware circuit from "behavioral descriptions" (and as a distinction from "structural descriptions" such as synthesizable Verilog). The target hardware circuit consists of a structural composition of data path, control and memory elements. Accordingly, the process was also variously referred to as a transformation "from behavior to structure." By the early eighties, the fundamental tasks in HLS had been decomposed into hardware modeling, scheduling, resource allocation and binding, and control generation. Briefly, modeling concerned with capturing specifications as program-like descriptions and making these available for downstream synthesis tasks via a partially-ordered description that is designed to expose concurrency available in the description. Task scheduling schedules operations by assigning these to specific clock cycles or by building a function (i.e., a scheduler) that determines execution time of each operation at the runtime. Resource allocation and binding determine the resources and their quantity needed to build the final hardware circuit. Binding refers to specific binding of an operation to a resource (such as a functional unit, a memory, or an access to a shared resource). Sometimes module selection has been used to describe the problem of selecting an appropriate resource type from a library of modules under a given metric such as area or performance. Finally, control generation and optimization sought to synthesize a controller to generate appropriate control signals according to a given schedule and binding of resources. This decomposition of HLS tasks was for problem solving purposes; almost all of these subtasks are interdependent.

Early HLS had two dominant schools of thought regarding scheduling: fixed latency constrained designs (such as early works by Pierre Paulin, Hugo DeMan and their colleagues) and fixed resource constrained designs (such as works by Barry Pangrle, Howard Trickey and Kazutoshi Wakabayashi). In the former case, resources are assigned in a minimal way to meet a clock latency goal, in the latter, minimal time schedules are derived given a set of pre-defined physical resources. The advantage of fixed latency is easy incorporation of the resulting designs into larger timing-constrained constructions. These techniques have met with success in the design of filters and other DSP functions in practical design flows. Fixed resource models allowed a much greater degree of designer intervention in the selection and constraint of underlying components, potentially allowing use of the tools in area or power-constrained situations. They also required more complex scheduling algorithms to accommodate the implied constraints inherent in the chosen hardware models. Improvements in the underlying algorithms later allowed for simultaneous consideration of timing and resource constraints; however, the complexity of such optimization limits their use to relatively small designs or forces the use of rather coarse heuristics as was done in the Behavioral Compiler tool from Synopsys. More recent scheduling algorithms (Wave Scheduling, Symbolic Scheduling, ILP and Interval Scheduling) allow for automated exploration of speculative execution in systematic ways to increase the available parallelism in a design. At the high end of this spectrum, the distinction between static (pre-determined execution patterns) and dynamic (run-time determined execution patterns) are blurred by the inclusion of arbitration and local control mechanisms.

2.3 History

High-level synthesis (HLS) has been a major preoccupation of CAD researchers since the late 1970s. Table 2.1 lists major time points in the history of HLS research through the eighties and the nineties; this list of readings would be typical of a researcher active in the area throughout this period. As with any history, this is by no means a comprehensive listing. We have intentionally skipped some important developments in this decade since these are still evolving and it is too early to look back and declare success or failure.

Early work in HLS examined scheduling heuristics for data-flow designs. The most straightforward approaches include scheduling all operations as soon as possible (ASAP) and scheduling the operations as late as possible (ALAP) [5–8]. These were followed by a number of heuristics that used metrics such as urgency [9] and mobility [10] to schedule operations. The majority of the heuristics were derived from basic list scheduling where operations are scheduled relative to an ordering based on control and data dependencies [11–13]. Other approaches include iteratively rescheduling the designs [14] and scheduling along the critical path through the behavioral description [15]. Research in resource allocation and binding techniques have sought varying goals including reducing registers, reducing functional units, and reducing wire delays and interconnect costs [3–5]. Clique partitioning and clique covering were favorite ingredients to solving module allocation problems [6] and to find the solution of a register-compatibility graph with the lowest combined register and interconnect costs [16]. Network flow formulations were used to bind operations and registers at each time step [18] and to perform module allocation while minimizing interconnect [17].

Given the dependent nature of each task within HLS, researchers have focused on performing these tasks in parallel, namely through approaches using integer linear programming (ILP) [19–22]. In the OSCAR system [21], a 0/1 integer-programming model is proposed for simultaneous scheduling, allocation, and binding. Wilson and co-authors [22] presented a generalized ILP approach to provide an integrated solution to the various HLS tasks. In terms of design performance, pipelining

Table 2.1 Major timepoints in the historical evolution of HLS through the 1980s and 1990s

Year	Authors
1972–75	Barbacci, Knowles: ISPS description
1978	McFarland: ValueTrace (VT) model for behavioral representation
1980	Snow's Thesis that was among the first to show use of CDFG as a synthesis specification
1981	Kuck and co-authors advance compiler optimizations (POPL)
1983	Hitchcock and Thomas on datapath synthesis
1984	Tseng and Siewiorek work on bus-style design generator
1984	Emil Gircyz thesis on using ADA for modeling hardware, precursor to VHDL
1985	Kowalski and Thomas on use of AI techniques for design generation
1985	Pangrle on first look-ahead/clock independent scheduler
1985	Orailoglu and Gajski: DESCART silicon compiler; Nestor and Thomas on synthesis from interfaces
1986	Knapp on AI planning; Brewer on Expert System; Marwedel on MIMOLA; Parker on MAHA pipelined synthesis; Tseng, Siewiorek on behavioral synthesis
1987	Flamel by Tricky; Paulin on force-directed scheduling; Ebcioglu on software pipelining
1988	Nicolau on tree-based scheduling; Brayton and co-authors: Yorktown silicon compiler; Thomas: System architect's workbench (SAW); Ku and DeMicheli on HardwareC; Lam: on software pipelining; Lee on synchronous data flow graphs for DSP modeling and optimization
1989	Wakabayashi on condition vector analysis for scheduling; Goosens and DeMan on loop scheduling
1990	Stanford Olympus synthesis system; McFarland, Parker and Camposano overview; DeMan on Cathedral II
1991	Hilfinger's Silage and its use by DeMan and Rabaey on Lager DSP Synthesis; Camposano: Path based scheduling; Stock, Bergamaschi; Camposano and Wolf book on HLS; Hwang, Lee and Hsu on Scheduling
1992	Gajski HLS book; Wolf on PUBSS
1993	Radevojevic, Brewer on Formal Techniques for Synthesis
1994	DeMicheli book on Synthesis and Optimization covering a good fraction of HLS
1995	Synopsys announces Behavioral Compiler
1996	Knapp book on HLS
....	Another decade of various compiler + synthesis approaches
2005	Synopsys shuts down Behavioral Compiler

was explored extensively for data-flow designs [10, 13, 23–25]. Several systems including HAL [10] and Maha [15] were guided by user-specified constraints such as pipeline boundaries or timing bounds in order to distribute resources uniformly and minimize the critical path delay. Optimization techniques such as algebraic transformations, retiming and code motions across multiplexers showed improved synthesis results [26–28].

Throughout this period, the quality of synthesis results continued to be a major preoccupation for the researchers. Realizing the direct impact of how control structures affected the quality of synthesized circuits, several researchers focused their efforts on augmenting HLS to handle complex control flow. Tree-based scheduling [29] removes all the join nodes from a design so that the control-data flow graph (CDFG) becomes a tree and speculative code motion can be applied. The PUBSS approach [30] extracts scheduling information in a behavioral finite state machine (BFSM) model and generates a schedule using constraint-solving algorithms. NEC created the CVLS approach [31–33] that uses condition vectors to improve resource sharing among mutually exclusive operations. Radivojevic and Brewer [34] provide an exact symbolic formulation that schedules each control path independently and then creates an ensemble schedule of valid control paths. The Waveschedule approach minimizes the expected number of cycles by using speculative execution. Several other approaches [35–38] support generalized code motions during scheduling in synthesis systems where operations can be moved globally irrespective of their position in the input. Prior work examined pre-synthesis transformations to alter the control flow and extract the maximal set of independent operations [39,40]. Li and Gupta [41] restructure control flow to extract common sets of operations with conditionals to improve synthesis results.

Compiler transformations can further improve HLS, although they were originally developed for improving code efficiency for sequential program execution. Prominent among these were variations on common sub-expression elimination (CSE) and copy propagation which are commonly seen in software compilers [1,2]. Although the basic transformations such as dead code elimination and copy propagation can be used in synthesis, other transformations need to be re-instrumented for synthesis by incorporating ideas of mutual exclusivity of operations, resource sharing, and hardware cost models. Later attempts in the early 2000s explored parallelizing transformations to create a new category of HLS synthesis that seeks to fundamentally overcome limitations on concurrency inherent in the input algorithmic descriptions by constructing methods to carry out large-scale code motions across conditionals and loops [42].

2.4 Successes and Failures

While the description above is not intended to be a comprehensive review of all the technical work, it does beg an important question: once the fundamental problems in HLS were identified with cleanly laid out solutions, why didn't the progress in problem understanding naturally lead to tools as had been the case with the standard cell RTL design flows?

There is an old adage in computer science: "Artificial Intelligence can never be termed a 'success' – the techniques that worked such as efficient logic data-structures, data mining and inference based reasoning became valuable on there own – the parts that remain unsolved retain the title 'Artificial Intelligence.'" In many ways, the situation is similar in High Level Synthesis; simple-to-apply techniques were moved out of that context and into general use. For example, the Design Compiler® tool from Synopsys regularly uses allocation and binding optimizations on arithmetic and other replicated units in conventional 'logic optimization' runs. Some of the more clever control synthesis techniques have also been incorporated into that tool's finite state machine synthesis options.

Many of the ideas which did not succeed in the general ASIC context have made a comeback in the somewhat more predictable application of FPGA synthesis with tools such as Mentor's Catapult-C supporting a subset of the C-programming language for direct synthesis into FPGA designs. A number of products mapping designs originally specified in MatLab's M language or in specialized component libraries for LabView have appeared to directly synthesize designs for digital signal processing in FPGA's. Currently, these tools range in complexity from hardware macro-assemblers which do not re-bind operation instances to the fairly complex scheduling supported by Catapult-C. The practicality of these tools is supported by the very large scale of RTL designs that can be mapped into modern large FPGA devices.

On the other hand, the general precepts of High Level Synthesis have not been so well adopted by the design community nor supported by existing synthesis systems. There have been several explanations in the literature: lack of a well-defined or universally accepted intermediate model for high-level capture, poor quality of synthesis results, lack of verification tools, etc. We believe the clearest answer is found in the classical proverb regarding dogs not liking the dogfood. That is, the circuit designers who were the target of such tools and methods did not really care about the major preoccupation of solving the scheduling and allocation problems. For one, this was a major part of the creativity for the RTL implementers who were unlikely to let go of the control of clock cycle boundaries, that is, the explicit specification of which operation happened on which cycle. So, in a way, the targeted users of HLS tools were being told do something differently that they already did very well. By contrast, tools took away the controllability, and due to the semantic gap between the designer intent and the high-level specification, synthesis results often fell short of the quality expectations. A closer examination leads us to point to the following contributing factors:

a. The so-called high-level specifications in reality grew out of the need for simulation and were often little more than an input language to make a discrete event simulator reproduce a specific behavior.
b. The complexity of timing constraint specification and analysis was grossly underestimated, especially when a synthesizer needs to utilize generalized models for timing analysis.
c. Design metrics were fairly naïve: the so-called data-dominated versus control-dominated simplifications of the cost model grossly mis-estimated the true costs and, thus, fell short on their value in driving optimization algorithms. By contrast, in specific application areas such as digital signal processing where the input description and cost models were relatively easier to define, the progress was more tangible.
d. The movement from a structural to a behavioral description – the centerpiece of HLS – presented significant problems in how the design hierarchy was constructed. The parameterization and dynamic elaboration of the major hierarchy components (e.g., number of times a loop body is invoked) requires dramatically different synthesis methods that were just not possible in a description that

essentially looks identical to a synthesis tool. A fundamental understanding of the role of structure was needed before we even began to capture the design in a high-level language.

2.5 Lessons Learnt

The notion of describing a design as a high-level language program and then essentially "compiling" into a set of circuits (instead of assembly code) has been a powerful attractor to multiple generations of researchers into HLS. There are, however, complexities in this form of specification that can ruin an approach to HLS. To understand this, consider the semantic needs when building a hardware description language (HDL) from a high-level programming language. There are four basic needs as shown in Fig. 2.2: (1) a way to specify concurrency in operations, (2) ensure timing determinism to enable a designer build a "predictable" simulation behavior (even as the complete behavior is actually unspecified), (3) ensure effective modeling of the reactive aspects of hardware (non-terminating behavior, event specifications), and (4) capture structural aspects of a design that enables an architect to build larger systems by instantiating and composing from smaller ones.

2.5.1 Concurrency Experiments

Of the four requirements listed in Fig. 2.2, concurrency was perhaps the most dominant preoccupation of HLS researchers since the early years for a good reason: one of the first things that a HLS tool has to do when presented with an

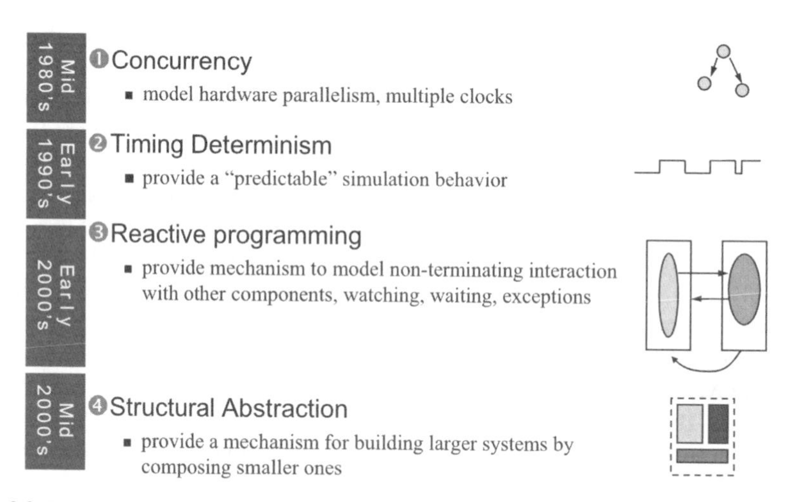

Fig. 2.2 Semantic needs from programming to hardware modeling and time-line over which these aspects were dominant in the research literature

algorithmic description in a programming language is to extract the parallelism inherent in the specification. The most common way was to extract data-flow graphs from the description based on a def-use dependency analysis of operations. Since these graphs tended to be disjoint making it hard for the synthesis algorithms to operate, they were often combined with nodes and edges to represent flow of control. Thus, the combined Control-Data Flow Graphs or CDFG were commonly used. Most of these models did not capture use of any structured memory blocks, which were often treated as separate functional or structural blocks. By and large, CDFGs were used to implement synthesis tasks as graph operations (for example, labeled graphs representing scheduling, and binding results). However, hierarchical modeling was a major issue. Looking back, there were three major lessons that we can point to. First, not all CDFGs were the same. Even if matched structurally, the semantic variations on graphs were tremendous: operational semantics of the nodes, what edges represent, etc. An interesting innovation in this area was the attempt to move all non-determinism (in operations, timing) to the graph model hierarchy in the Stanford Intermediate Format (SIF) graph. In a SIF graph, loops and conditions were represented as separate graph bodies, where a body corresponded to each conditional invocation of a branch. Thus, operationally the uncertainty due to control flow (or synchronization operations) was captured as the uncertainty in calling a graph. It also made SIF graphs DAGs, thus enabling efficient algorithms for HLS scheduling and resource allocation tasks in the Olympus Synthesis System.

The second lesson was also apparent from the Olympus system that employed a version of C, called HardwareC, which enabled specification of concurrent operations at arbitrary levels of granularity: two operations could be scheduled in parallel, sequentially, or in a data-parallel fashion by enclosing them using three different set of parentheses; and then the composition could also be similarly composed in one of three ways, and so on. While it enabled a succinct description of complex dependency relationships (as Series-Parallel graphs), it was counter-intuitive to most designers: a small change on a line could have a significant (and non-obvious) impact on an operation several pages away from the line changed, leading designers to frustrating simulation runs. Experience in this area has finally resulted in most HDLs settling for concurrency specification at an aggregate "process" level, whereas processes themselves are often (though not always, see structural specifications later) sequential.

The third, and perhaps, the most important lesson we learnt when modeling designs was regarding methods used to go from a high-level programming language (HLL) to an HDL. Broadly speaking, there are three ways to do it: (1) as a syntactic add-on to capture "hardware" concepts in the specification. Examples include "process", "channel" in HardwareC, "signals" in VHDL etc. (2) Overload semantics of existing constructs in a HLL. A classic example is that an assignment in VHDL implies placement of an event *in future*. (3) Use existing language level mechanisms to capture hardware-specific concepts using libraries, operator overloading, polymorphic types, etc., as is the case in SystemC. An examination of HDL history would demonstrate the use of these three methods in roughly the same order. While syntactical changes to existing HLL were common-place in the early years of

HDL modeling, later years have seen a greater reliance on library-based HDLs due to a combination of greater understanding of HDL needs combined with advances in HLLs towards sophisticated languages that provide creative ways to exploit type mechanisms, polymorphism and compositional components.

2.5.2 Timing Capture and Analysis for HLS

The early nineties saw an increased focus on the capture of timing behavior in HLS. This was also the time when the term "embedded systems" entered the vocabulary of researchers in this field, and it consequently caused researchers to look at high-level IC design as a *system* design problem. Thus, input descriptions were beginning to look like descriptions of components in temporal interaction with the environment as shown in Fig. 2.3 below. Thus, one could specify and analyze timing requirements separately from the functional behavior of the system design.

Accordingly, the behavioral models evolved: from the early years of functionality and timing models to their convergence into single "operation-event" graphs of Amon and Borriello, we made a full circle to once again separate timing and functional models. Building upon a long line of research on event graphs, Dasdan and Gupta proposed generalized task graph models consisting of tasks as nodes and communications between tasks as edges that can carry multiple tokens. The nodes could be composed according to a classification of tasks: an AND task represents actions that are performed after conjunction of its predecessor tasks have completed, whereas an OR task can initiate once any of its predecessors have completed execution. The tasks could also optionally skip tokens, thereby capturing realistic timing response to events. This structure allowed us to generate discrete event models directly from the task graphs that can be used for "timing simulation" even when the functional behavior of the overall system has not been devised beyond, of course, the general structure of the tasks (Fig. 2.4).

Works such as this enabled researchers to define and make progress on high-level design methodologies that were "timing-driven." While this was a tremendously useful exercise, its applicability was basically limited by the lack of timing detail

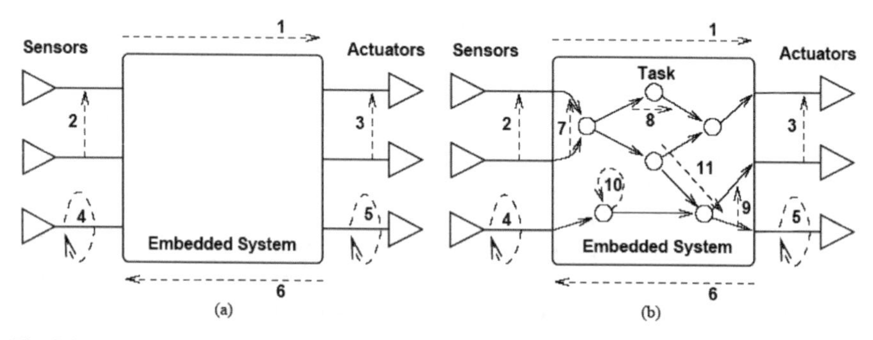

Fig. 2.3 A system design conceptualized as one in temporal interaction with the environment

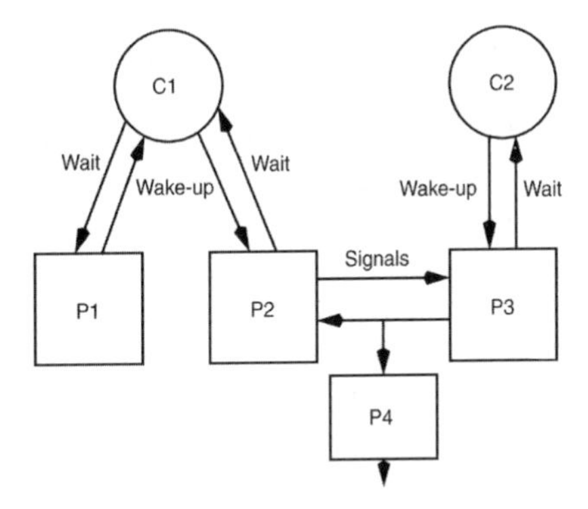

Fig. 2.4 Conceptual model of Scenic consisting of processes, clocks and reactions

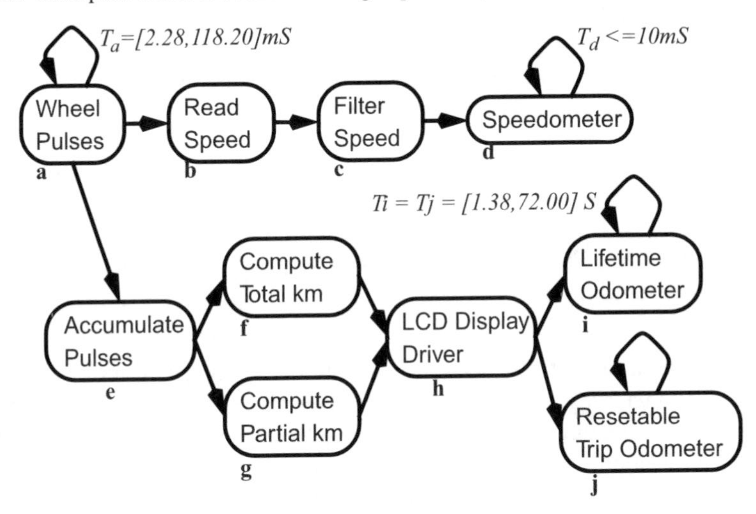

Fig. 2.5 Example of a timing simulation for an automotive information display that uses normally distributed acceleration and deceleration periods (mean: 20 s, deviation: 1 s). The vehicle response is normally distributed as well. The simulation has been created directly from the semantics of the task graph model without detailed functional implementation

available to the system designer at high levels of specification. Consequently, timing analysis needed a lot of detailed specification (related to timing at the interfaces) and solved only a part of the synthesis problem. Conversely, to be useful, one was confronted with the problem of defining time budgets based on sparsely described timing constraints that needed to be decomposed across a number of tasks. Admittedly, this is a harder problem to solve than the original problem of synthesizing a structure of components that could be verified to meet a given timing specification. More importantly, such timing analysis was appearing in the HLS literature around the time when functional verification had taken a dominant role in the broader CAD community of researchers. The separation of function from timing was also problematic for the VLSI system designers that often leverage innovating composition of functionalities to achieve key performance benefits (Fig. 2.5).

Predictably, as it had done in modeling embedded software systems about a decade earlier, the focus on timing behavior gave way to innovations in how *reactive behaviors* were modeled in a programming language. Inspired by the success of synchronous programming languages such as Esterel, Lustre, and Signal in building embedded software and their tools (such as SCADE), the notion of timing abstraction to construct synchronous behaviors in lieu of detailed timing specifications (in the earlier discrete event models) drove new ways to specify HDL models. The new models also crossed paths with the advances in meta-models used in software engineering. Scenic [44] (and its follow on SystemC) represented one such language that provided reactive capture through *watching* and *wait* constructs (built as library extensions). These HDLs which captured the conceptual model of a system were rechristened system-level languages to distinguish these from the more commonly used HDLs such as Verilog and VHDL. While *wait* represented synchronization with a clock, *watching* represented asynchronous conditions. In later years, *watching* was retired in order to simplify the emerging SystemC language that enabled specification of both the hardware and software components of system design.

2.5.3 The Era of Structure: Components, Compositions and Transactions

This brings us to early 2000 and an era of structural compositions characterized by composition/aggregation of models, components and even synthesized elements. UML sought to capture multiple *types* of relationships among components: association, aggregation, composition, inheritance and refinement to describe a system behavior in terms of its compositional elements. Several component composition frameworks appeared in the literature including Polis, Metropolis, Ptolemy, and Balboa. While a description of these is beyond the scope of this work, a common theme among all these frameworks has been attempts to raise the abstraction levels in a way that enables composition of system blocks as robust software components that can be reused across different designs with minimal or no change. Transaction modeling has sought to raise the level of abstraction both in functional behavior of the components as well as their interfaces. Interfaces are constructed to limit the complexity of sub-system design; or rather they are the abstraction enforcers of the design world. Protocols of communication are important to interface abstractions. Early HLS assumed implicit protocols and timing from language level descriptions. Reactive modeling as described in the previous section improved the situation somewhat from the compositionality perspective. More recent effort in Transaction Level Modeling or TLM seeks to orthogonalize the levels of abstractions in computation versus communication in system level models (see Fig. 2.6). This is still an active area of research. It is clear that there needs to be good structural and timing abstractions in order for HLS to succeed.

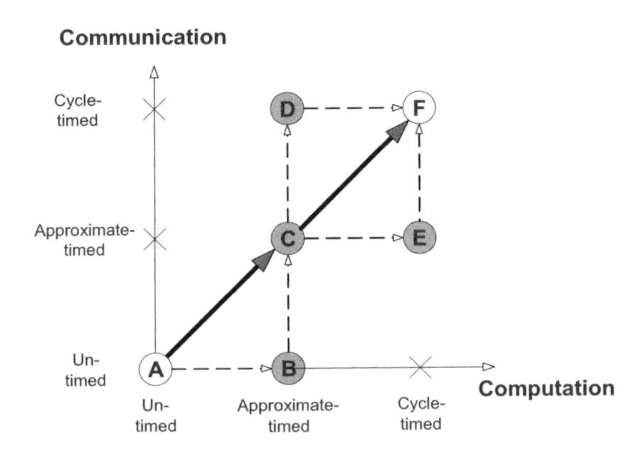

A. **"Specification model"**
"Untimed functioal models"

B. **"Component-assembly model"**
"Architecture model"
"Timed functonal model"

C. **"Bus-arbitration model"**
"Transaction model"

D. **"Bus-functional model"**
"Communicatin model"
"Behavior level model"

E. **"Cycle-accurate computation model"**

F. **"Implementation model"**
"Register transfer model"

Fig. 2.6 A taxonomy of models based on timing abstraction. Models B, C, D and E are often classified as transaction level models (courtesy: Daniel Gajski, UC Irvine)

2.6 Wither HLS?

The goal of hardware compilation of designs from behavioral languages has lead to many valuable contributions in areas beyond the original concept. One example is the class of synchronous languages such as Esterel and Luster which formalize sequential behavior and allow formally verifiable synthesis of both hardware and software (or coupled) systems. While the case for efficient hardware could be disputed, software synthesis from Esterel is an integral part of the control software of many safety critical systems such as the Airbus airliners.

Another interesting related effort is the *BlueSpec* hardware compilation system. Based on an atomic rule-based language scheme, *BlueSpec* allows for an efficient description of cycle-based behaviors which are automatically compiled into efficient hardware architectures that can be reasonably compared to human created designs. Although, in practice, a *BlueSpec* specification is a mixture of behavior and structure, the efficacy of the strategy has been well established in terms of designer efficiency.

On a related tack, SystemC has become the de facto standard for transaction based system modeling which supporting a semi-behavioral hardware compilation scheme. Currently, a hierarchy of transaction specifications cannot be directly synthesized; however, the transaction format does offer several improvements on the procedural languages in early HLS. In particular, they can be annotated with a type hierarchy allowing inference of interfaces and thus timing constraints without losing track of the optimization goals or metrics for the system of transactions. Effectively, alternative interface types offer differing bandwidth and communication latency while requiring accommodation of their timing constraints. It remains to be seen whether these or related ideas can be fleshed out to a practical behavioral synthesis system.

2.7 Conclusions

This brief retrospective is, more than anything else, a personal perspective. This is not just a caveat against inevitable omissions of important work in the area but also an expression of humility for a large number of significant contributions that have continually enabled newer generations of researchers to see farther than their predecessors. Looking back, activity in HLS is marked by an early period of intense activity in synthesis in the eighties, its drop off, divergence from algorithmic optimizations, and a subsequent reemergence as primarily a modeling and architectural specification challenge. Among some of the most exciting developments in the recent years are contributions from computer architecture researchers in defining modeling schemes that rely more on operation-centric behaviors and its early commercialization as *BlueSpec*. While it is too early to tell how HLS will emerge through these efforts, even when it is not called HLS per se, it is clear that the design decisions that affect code transformations, such as transformations of loops and conditionals, and architectural design, such as pipeline structures, are paramount to a successful synthesis solution. In other words, the early attempts at optimization from algorithmic descriptions were somewhat premature and naïve in expectation of a quick success modeled along the lines of logic synthesis. Indeed, a shift in design tools and methods does not happen in isolation from the practitioners who must use these tools. Just as logic synthesis enabled RTL designers to try their hands at what used to be primarily a circuit design activity, the future adoption of HLS will involve *enabling* a new class of practitioners to do things they can not do now. Today, we have broad categories of pain-points in this area: architects have to deal with too many design "knobs" that need to be turned to produce a design that is cost/performance competitive in silicon, whereas ASIC implementers have to understand and carefully apply design optimization effort on things that have a significant impact on the overall system. This is a difficult exercise because the complexity of designs rules out identification of design optimization areas without extensive simulation or emulation of system prototypes. Moving forward, HLS can succeed by enabling a new generation (system architects or ASIC implementers) to do things that they simply cannot be accomplished today. This also entails a tremendous education effort to change the vocabulary of the current generation of system architects and ASIC implementers. Among the number of developments that continue to advance our understanding of the system design process, it is most heartening to see erstwhile computer architects take the lead in defining a meaningful set of problems, models and even solution methods that can lead to design synthesis, design optimization, and design validation for the next generation of tool developers. Such revitalization of the HLS domain holds significant promise for future advancements in how microelectronic systems are architected and implemented on-chip.

Acknowledgement The authors are grateful to Joel Coburn for his constructive suggestions and for his help with research in putting this article together.

References

1. Ian Kuon and Jonathan Rose, Measuring the Gap between ASIC and FPGAs, IEEE Transactions on Computer-Aided Design, February 2007.
2. S. Gupta, R.K. Gupta, N.D. Dutt, and A. Nicolau, SPARK: A parallelizing approach to the high level synthesis of digital circuits, Kluwer, Dordrecht, 2004.
3. G. De Micheli, Synthesis and optimization of digital circuits, McGraw-Hill, New York, 1994.
4. R. Camposano and W. Wolf, High level VLSI synthesis, Kluwer, Dordrecht, 1991.
5. T.J. Kowalski and D.E. Thomas, The VLSI design automation assistant: what's in a knowledge base, Design Automation Conference, 1985.
6. C.J. Tseng and D.P. Siewiorek, Automated synthesis of data paths in digital systems, July 1986.
7. P. Marwedel, A new synthesis for the MIMOLA software system, Design Automation Conference, 1986.
8. H. Trickey, Flamel: a high-level hardware compiler, IEEE Trans. Comput. Aided Des., 6, 259–269, 1987.
9. E. Girczyc, Automatic generation of micro-sequenced data paths to realize ADA circuit descriptions, Ph.D. thesis, Carleton University, 1984.
10. P.G. Paulin and J.P. Knight, Force-directed scheduling for the behavioral synthesis of ASIC's, IEEE Trans. Comput. Aided Des., 8, 661–678, 1989.
11. C.Y. Hitchcock and D.E. Thomas, A method of automatic data path synthesis, Design Automation Conference, 1983.
12. H. De Man, J. Rabaey, P. Six, and L. Claesen, Cathedral-II: A silicon compiler for digital signal processing, IEEE Des. Test Mag., 3, 73—85, 1986.
13. B.M. Pangrle and D.D. Gajski, Slicer: A state synthesizer for intelligent silicon compilation, 1986.
14. I.-C. Park and C.-M. Kyung, Fast and near optimal scheduling in automatic data path synthesis, Design Automation Conference, 1991.
15. A.C. Parker, J.T. Pizarro, M. Mlinar, "MAHA: a program for datapath synthesis", Proc. 23rd IEEE/ACM Design Automation Conference pp. 461–466, Las Vegas NV, June 1986.
16. P.G. Paulin and J.P. Knight, Scheduling and binding algorithms for high-level synthesis, 1989.
17. L. Stok and W.J.M. Philipsen, Module allocation and comparability graphs, IEEE International Sympoisum on Circuits and Systems, 1991.
18. A. Mujumdar, R. Jain, and K. Saluja, Incorporating performance and testability constraints during binding in high-level synthesis, IEEE Trans. Comp. Aided Des., 15, 1212–1225, 1996.
19. C.T. Hwang, T.H. Lee, and Y.C. Hsu, A formal approach to the scheduling problem in high level synthesis, IEEE Trans. Comput. Aided Des., 10, 464–475, 1991.
20. C.H. Gebotys and M.I. Elmasry, Optimal synthesis of high-performance architectures, IEEE J. Solid State Circuits, 1992.
21. B. Landwehr, P. Marwedel, and R. Doemer, Oscar: optimum simultaneous scheduling, allocation and resource binding based on integer programming, European Design Automation Conference, 1994.
22. T.C. Wilson, N. Mukherjee, M.K. Garg, and D. K. Banerji, An ILP solution for optimum scheduling, module and register allocation, and operation binding in datapath synthesis, VLSI Des., 1995.
23. N. Park and A. Parker, Sehwa: A software package for synthesis of pipelines from behavioral specifications, IEEE Trans. Comput. Aided Des., 1988.
24. E. Girczyc, Loop winding – a data flow approach to functional pipelining, International Symposium of Circuits and Systems, 1987.
25. L.-F. Chao, A.S. LaPaugh, and E.H.-M. Sha, Rotation scheduling: A loop pipelining algorithm, Design Automation Conference, 1993.
26. M. Potkonjak and J. Rabaey, Optimizing resource utlization using tranformations, IEEE Trans. Comput. Aided Des., 13, 277–292, 1994.

27. R. Walker and D. Thomas, Behavioral transformation for algorithmic level IC design, IEEE Trans. Comput. Aided Des., 1115–1128, 1989.
28. Z. Iqbal, M. Potkonjak, S. Dey, and A. Parker, Critical path optimization using retiming and algebraic speed-up, Design Automation Conference, 1993.
29. S. Huang et al., A tree-based scheduling algorithm for control dominated circuits, Design Automation Conference, 1993.
30. W. Wolf, A. Takach, C.-Y. Huang, R. Manno, and E. Wu, The Princeton University behavioral synthesis system, Design Automation Conference, 1992.
31. K. Wakabayashi and T. Yoshimura, A resource sharing and control synthesis method for conditional branches, 1989.
32. K. Wakabayashi and H. Tanaka, Global scheduling independent of control dependencies based on condition vectors, Design Automation Conference, 1992.
33. K. Wakabayashi, C-based synthesis experiences with a behavior synthesizer, "Cyber", Design, Automation and Test in Europe, 1999.
34. I. Radivojevic and F. Brewer, A new symbolic technique for control-dependent scheduling, IEEE Trans. Comput. Aided Des., 15, 45–57, 1996.
35. L.C.V. dos Santos and J.A.G. Jess, A reordering technique for efficient code motion, Design Automation Conference, 1999.
36. L.C.V. dos Santos, A method to control compensation code during global scheduling, Workshop on Circuits, Systems and Signal Processing, 1997.
37. L.C.V. dos Santos, Exploiting instruction-level parallelism: A constructive approach, Ph.D. thesis, Eindhoven University of Technology, 1998.
38. M. Rim, Y. Fann, and R. Jain, Global scheduling with code-motions for high-level synthesis applications, IEEE Trans. VLSI Syst., 1995.
39. J. Li and R.K. Gupta, HDL optimizations using timed decision tables, Design Automation Conference, 1996.
40. O. Penalba, J.M. Mendias, and R. Hermida, Maximizing conditional reuse by pre-synthesis transformations, Design, Automation and Test in Europe, 2002.
41. J. Li and R.K. Gupta, Decomposition of timed decision tables and its use in presynthesis optimizations, International Conference on Computer Aided Design, 1997.
42. SPARK parallelizing high-level synthesis framework website, http://mesl.ucsd.edu/spark.
43. B. Baily, G. Martin, A. Piziali, ESL design and verification, Academic Press, New York, 2007.
44. S. Liao, S. Tjiang, R. Gupta, An Efficient Implementation of Reactivity for Modeling Hardware in the Scenic Design Environment, *Design Automation Conference*, 70–75, June 1997.

Chapter 3
Catapult Synthesis: A Practical Introduction to Interactive C Synthesis

Thomas Bollaert

Abstract The design complexity of today's electronic applications has outpaced traditional RTL methods which involve time consuming manual steps such as micro-architecture definition, handwritten RTL, simulation, debug and area/speed optimization through RTL synthesis. The Catapult® Synthesis tool moves hardware designers to a more productive abstraction level, enabling the efficient design of complex ASIC/FPGA hardware needed in modern applications. By synthesizing from specifications in the form of ANSI C++ programs, hardware designers can now leverage a precise and repeatable process to create hardware much faster than with conventional manual methods. The result is an error-free flow that produces accurate RTL descriptions tuned to the target technology.

This paper provides a practical introduction to interactive C synthesis with Catapult® Synthesis. Our introduction gives a historical perspective on high-level synthesis and attempts to demystify the stereotyped views about the scope and applicability of such tools. In this part we will also take a look at what is at stake – beyond technology – for successful industrial deployment of a high-level synthesis methodology. The second part goes over the Catapult workflow and compares the Catapult approach with traditional manual methods. In the third section, we provide a detailed overview on how to code, constrain and optimize a design with the Catapult Synthesis tool. The theoretical concepts revealed in this section will be illustrated and applied in the real-life case study presented in the fourth part, just prior to the concluding section.

Keywords: High-level synthesis, Algorithmic synthesis, Behavioral synthesis, ESL, ASIC, SoC, FPGA, RTL, ANSI C, ANSI C++, VHDL, Verilog, SystemC, Design, Verification, IP, Reuse, Micro-architecture, Design space exploration, Interface synthesis, Hierarchy, Parallelism, Loop unrolling, Loop pipelining, Loop merging, Scheduling, Allocation, Gantt chart, JPEG, DCT, Catapult Synthesis, Mentor Graphics

P. Coussy and A. Morawiec (eds.) *High-Level Synthesis.*

3.1 Introduction

There are a few hard, unavoidable facts about electronic design. One of them is the ever-increasing complexities of applications being designed. With the considerable amount of silicon real-estate made available by recent technologies, comes the need to fill it.

Every new wave of electronic innovation has caused a surge in design complexity, breaking existing flows and commanding change. In the early 1990s, the booming wireless and computer industries drove chip complexity to new heights, forcing the shift to new design methods, pioneering the era of register transfer level (RTL) design.

By fulfilling the natural evolution to raise the design abstraction level every decade or so (transistors in the 1970s, gates in the 1980s and RTL in the 1990s), the move to RTL design also implicitly set an expectation: in its turn, the next abstraction level will rescue stalling productivity.

3.1.1 First-Generation Behavioral Synthesis

If all this sounds familiar, that is because behavioral synthesis – introduced with much fanfare several years ago – promised such productivity gains. Reality proved otherwise, however, as designers discovered that behavioral synthesis tools were significantly limited in what they actually did. Essentially, the tools incorporated a source language that required some timing as well as design hierarchy and interface information. As a result, designers had to be intimately familiar with the capabilities of the synthesis tool to know how much and what kind of information to put into the source language. Too much information limited the synthesis tool and resulted in poor quality designs. Too little information lead to a design that didn't work as expected. Either way, designers did not obtain the desired productivity and flexibility they were hoping to gain.

These first-generation behavioral synthesis tools left the design community with two prejudices: an unfulfilled need for improved productivity and preconceived ideas about the applicability of these tools.

3.1.2 A New Approach to High-Level Synthesis

Acknowledging this unfulfilled need to improve productivity and learning from the shortcomings of initial attempts, Mentor Graphics defined a new approach to high-level synthesis based on pure ANSI C++. Beyond the synthesis technology itself, it was clear that the input language played a pivotal role in the flow and much emphasis was put on this aspect.

The drawbacks of structural languages such as VHDL, (System) Verilog or even SystemC used in first-generation tools are numerous:

- They are foreign to most algorithm developers
- They do not sufficiently raise the abstraction level
- They can turn out to be extremely difficult to write

American National Standards Institute (ANSI) C++ is probably the most widely used design language in the world. It incorporates all the elements to model algorithms concisely, clearly and efficiently. A class library can then be used to model bit-accurate behavior. And C++ has many design and debugging tools that can be re-used for hardware design. With a majority of algorithm developers working in pure C/C++, performing high-level synthesis from these representations allows companies to leverage existing developments and know-how, and to take advantage of abstract system modeling without teaching every designer a new language.

In comparison to first-generation behavioral tools, Catapult proposes an approach where timing and parallelism are removed from the synthesized source language. This is a fundamental difference with tools based on the structural languages mentioned previously which all require some forms of hardware constructs. The Catapult approach allows decoupling implementation information such as complex I/O timing and protocol from the functionality of the source. With this, the functionality and timing of the design can be developed and verified independently.

The flexibility and ease-of-use offered by the synthesis of pure ANSI C++ and Catapult Synthesis' intuitive coding style are a fundamental aspect of this flow.

3.1.3 Datapath Versus Control: Applicability of High-Level Synthesis

If first-generation tools were far from perfect, they nonetheless did reasonably well on pure datapath designs. Reputations – that is the negative ones – can be built in a short lapse of time, and can stick for an inversely long lapse!

Seeing and thinking the world in binary terms is probably too simplistic, if not harmful. It wasn't sufficient for behavioral tools to be good only for datapath designs. They also had to be awful for "control" dominated designs. Insidiously, this polarized the design world into two domains: datapath and control.

Today, many years after the decline of pioneering behavioral synthesis tools, the "datapath versus control" cliché still holds strongly, in ignorance of the advances made by the technology.

But logic designers know that there is more than 1s and 0s to the problem. Tristate, high and low impedance, dreaded X's make timing diagrams look much more... colorful. Similarly, the applicability of high-level synthesis goes much beyond the lazy control/datapath dichotomy.

Algorithms are often assimilated with datapath dominated designs. But many algorithms are purely control oriented, involving mostly decision making as opposed to raw computation. For instance, queuing algorithms such as found in networking devices or rate-matching algorithms in today's modems involve virtually no data processing. They are only about when, where and how to move data; in other words, they are control-oriented. This class of algorithm flows perfectly through modern high-level synthesis tools such as Mentor Graphics' Catapult Synthesis.

It is therefore no surprise that today, industry leaders in electronic design use Catapult Synthesis for all kinds of blocks and systems, ranging from modems such as found in mobile or satellite communications to multimedia encoders/decoders for set-top boxes or smart-phones, and from military devices to security applications.

In Sect. 3.4, we will describe how a complex, hierarchical subsystem consisting of datapath, mixed datapath and control and pure control units can be synthesized with the Catapult Synthesis tool.

3.1.4 Industrial Requirements for Modern High-Level Synthesis Tools

The fact that high-level synthesis tools can provide significant value through faster time-to-RTL and optimized design results is not to be demonstrated anymore. However, there is quite a gap between a working tool and a widely adopted solution which technology alone does not fill.

Saying that a high-level synthesis tool should work doesn't help much when identifying the criteria for successful industrial deployment.

While the high-level synthesis promise is well understood, the impact of such tools on flows and work organizations should not be overlooked. The bottom-line question is the one of risk and reward. High-level synthesis' high reward usually comes through change in existing flows. With millions of dollars at stake on every project, any methodology change is immediately – and understandably – considered a major risk factor by potential users.

Risk minimization, risk minimization and risk minimization are, in that order, the three most important industrial requirements for mainstream adoption of high-level synthesis. Over a decade of experience in this market has taught Mentor Graphics important lessons with this regard.

- Local improvements won't be accepted at the expense of breaking existing methods, imposing new constraints, forcing new languages.
- Intrusive technologies never make it in the mainstream: in their vast majority, designers use pure C/C++; this is how they model and this is what they want to synthesize.
- Non-standard, proprietary language extensions are counter productive and considered an additional risk factor.

- High-level synthesis tools are not used in isolation and should not jeopardize existing flows. They should not only produce great RTL, they should produce RTL that will seamlessly go through the rest of the flow.
- In the semiconductor industry, endorsements and sign-offs are key. Tool and library certification by silicon vendors (ASIC and FPGA) provide user with an important guarantee.
- World class, round the clock, local support is essential to users' security.
- Considering the financial and methodological investment, the reliability and financial stability of the tool supplier matters quite a lot.

If technology matters, the key successful deployment lies beyond raw quality or results. Acknowledging these facts, Mentor Graphics put a lot of emphasis on ease-of-use and user-experience when shaping the Catapult workflow described in the following section.

3.2 The Catapult Synthesis Workflow

The Catapult design methodology is illustrated in Fig. 3.1. The main difference with the traditional design flow is that the manual transformation of the C++ reference into RTL is bridged by an automated synthesis flow where the designer guides synthesis to generate the micro architecture that meets the desired area/performance/power goals. Catapult Synthesis generates the RTL with detailed knowledge of the

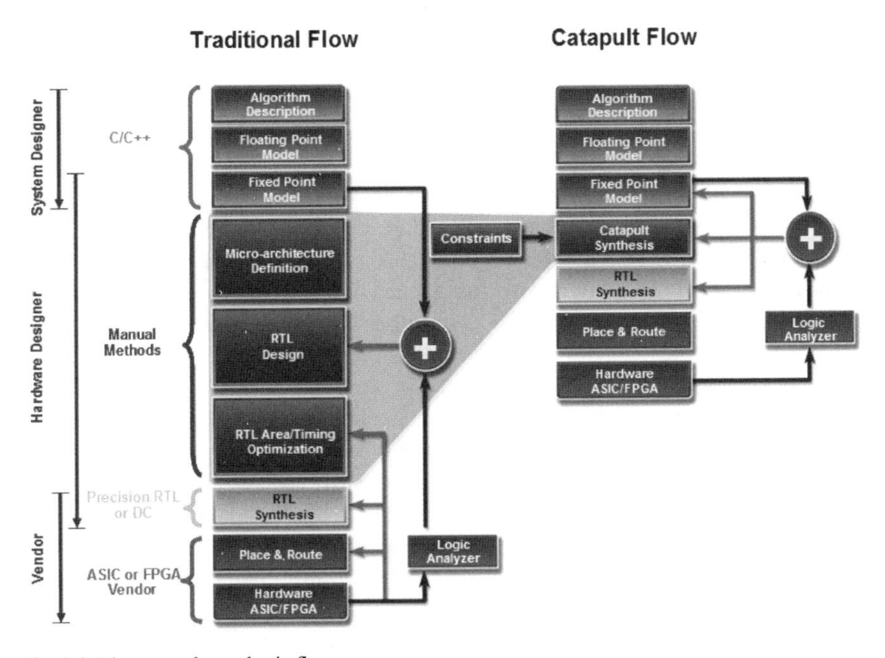

Fig. 3.1 The catapult synthesis flow

delay of each component to eliminate much of the guess work involved in the manual generation of the micro architecture and RTL. The advantages of the Catapult Synthesis flow are reflected both in significantly reduced design times as well as higher quality of designs and the variety of micro architecture that can be rapidly explored.

The flow is decomposed in four major steps. Sections 3.2.1–3.2.4 give an overview of each of these four steps, and Sect. 3.3 walks through a design example, providing more details on the actual synthesis process.

3.2.1 Writing and Testing the C Code

In the Catapult approach, designers start with describing the desired behavior using pure, untimed, ANSI C++. This is a fundamental aspect of the flow. This description is a purely algorithmic specification and requires absolutely no timing or concurrency or target technology information. This makes for far more compact and implementation-independent representations than traditional RTL or "behavioral" specifications written in languages such as VHDL, Verilog or SystemC.

The synthesizable C++ design is modeled with either fixed-point, integer and, in some cases, floating-point arithmetic. Engineers can focus on what matters most: the algorithm, not the low-level implementation details. The execution speed of host-compiled C++ programs allows for thorough analysis and verification of the design, orders of magnitudes more that what can be achieved during RTL simulations.

3.2.2 Setting Synthesis Constraints

Once satisfied with the algorithm, the designer sets synthesis constraints. This entire process only takes a few minutes and can be done over and over for the same design.

The first step is to specify the target technology and desired clock frequency. These details provide Catapult with the needed information to build an optimal design schedule. The designer also specifies other global hardware constraints such as reset, clock enable behavior and process level handshake.

As a next step, individual constraints can be applied to design I/Os, loops, storage and design resources. With this set of constraints the designer can explore the architectural design space. Interface synthesis directives are used to indicate how each group of data is moved in to or out of the hardware design. Loop directives are used to add parallelism to the design, and trade power, performance and area. Memory directives are used to constrain the storage resources and define the memory architecture. Resource constraints are used to control the number of hardware resources that are available to the design.

All these constraints can be set either interactively, through the tool's intuitive graphical user interface as shown in Fig. 3.2, or in batch mode with Tcl scripts.

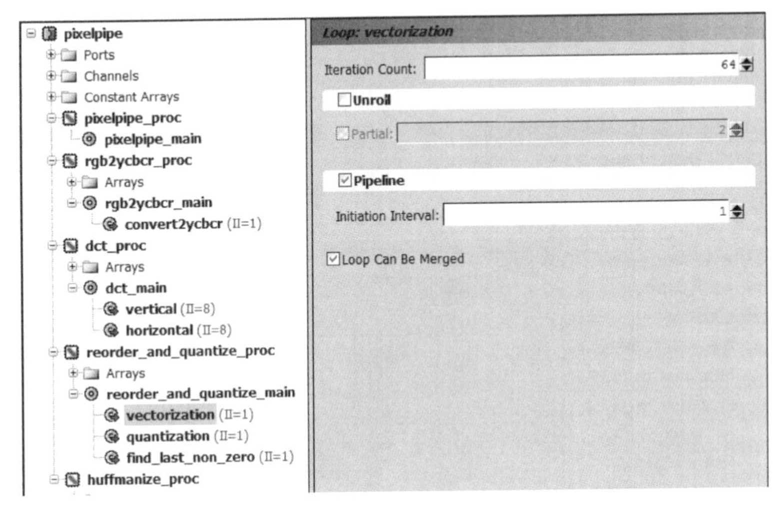

Fig. 3.2 Catapult synthesis – architectural constraints window

3.2.3 Analyzing the Algorithm/Architecture Pair

Catapult Synthesis provides a full set of algorithm and design analysis tools. Amongst them, the Gantt chart (Fig. 3.3) provides full insight on loop profiles, algorithmic dependencies and functional units in the design. In this view, the algorithm is always analyzed with respect to the target hardware and clock speed because these constraints can have major effects on how an algorithm should be structured. Using the Gantt chart designers can easily get information about how the explored algorithm/architecture pair performs with respect to actual goals. This view is also very valuable for tracking design bottlenecks and narrowing on specific areas requiring optimization.

With these analysis tools, designers can always fully understand why and how different synthesis constraints impact the design and what the actual results look like. This "white-box" visibility into the process is an important feature helping with ease-of-use and shortening the learning curve.

Designers are always in control, interacting and iterating, converging towards optimal results.

3.2.4 Generating and Verifying the RTL Design

Once the proper synthesis constraints are set, Catapult generates RTL code suitable for either ASIC or FPGA synthesis tools. In traditional flows, generation of the RTL from the specification is done manually, a process that may require several months

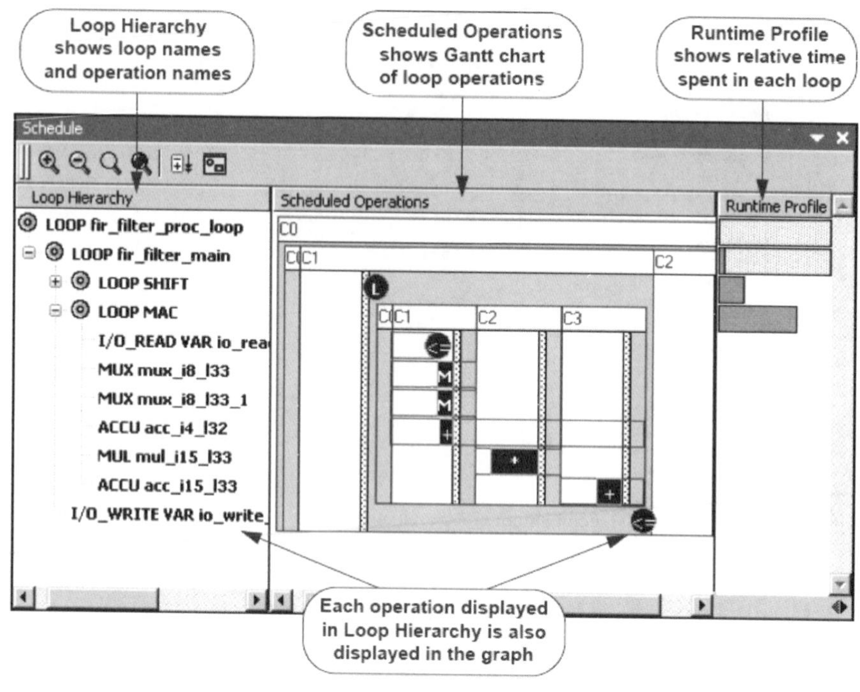

Fig. 3.3 The Gantt chart

to complete. In the Catapult flow, the generation of RTL is accomplished in a matter of minutes.

Catapult generates VHDL, Verilog or SystemC netlists, based on user settings. Various reports are also produced providing both hardware-centric and algorithm-centric information about the design's characteristics.

Finally, Catapult provides an integrated verification flow that automates the process of validating the HDL netlist(s) output from Catapult against the original C/C++ input. This is accomplished by wrapping the netlist output with a SystemC "foreign module" class and instantiating it along with the original C/C++ code and testbench in a SystemC design. The same input stimuli are applied to both the original and the synthesized code and a comparator at each output validates that the output from both are identical (Fig. 3.4). The flow automatically generates all of the SystemC code to provide interconnect and synchronization signals, Makefiles to perform compilation, as well as scripts to drive the simulation.

3.3 Coding and Optimizing a Design with Catapult Synthesis

This section provides an overview of the various controls the user can leverage to efficiently synthesize his design.

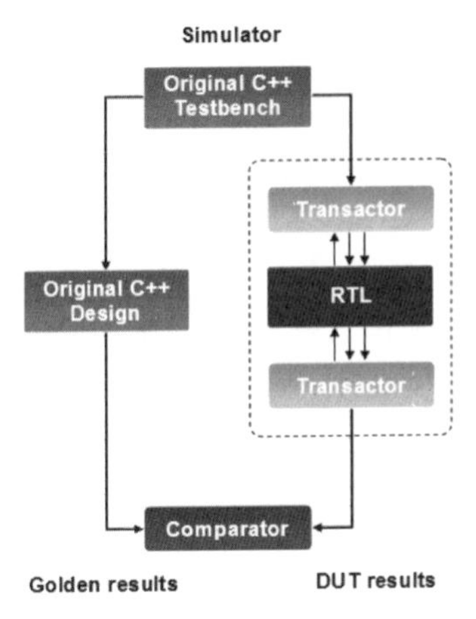

Fig. 3.4 Catapult synthesis' automatic verification flow

3.3.1 Coding C/C++ for Synthesis

The coding style used for functional specification is plain C++ that provides a sequential implementation of the behavior without any notion of timing or concurrency. Both the syntax and the semantics of the C++ language are fully preserved.

3.3.1.1 General Constructs and Statements

Catapult supports a very broad subset of the ANSI C++ language. The C/C++ synthesized top-level function may call other sub-functions, which may be inlined or may be kept as a level of hierarchy. The design may also contain static variables that keep some state between invocations of the function. "if" and "switch" condition statements are supported, as well as "for," "do" and "while" looping statements. "break," "continue" and "return" branching statements are synthesizable as well. The only noticeable restriction is that the code should be statically determinable, meaning that all its properties must defined at compilation time. As such, dynamic memory allocation/deallocation (malloc, free, new, delete) is not supported.

3.3.1.2 Pointers

Pointers are synthesizable if they point to statically allocated objects and therefore can be converted into array indexes. Pointer arithmetic is also supported and a pointer can point to several objects inside of an array.

```
template<class T>
struct rgb_t {
  T r, g, b;
  unsigned char to_gray() {
    float c[3]={0.299, 0.587, 0.114};
    return ((r*c[0]) + (g*c[1]) + (b*c[2]));
  };
};

void convert_to_gray(
       rgb_t<unsigned char> src[IMG_SZ],
       unsigned char        dst[IMG_SZ] )
{
  for (int i=0; i<IMG_SZ; i++) {
    *dst++ = src[i].to_gray();
  }
}
```

Fig. 3.5 Coding style example

3.3.1.3 Classes and Templates

Compound data types such as classes, structs and arrays are fully supported for synthesis. Furthermore, parameterization through C++ templates is also supported. The combination of classes and templates provides a powerful mechanism facilitating design re-use.

The example in Fig. 3.5 gives an overview of some of the coding possibilities allowed by the Catapult synthesizable subset. A struct is defined to model a RGB pixel. The struct is templatized so users can define the actual bitwidth of the R, G and B fields. Additionally, a method is defined which returns a grayscale value from the RGB pixel. The synthesized design is the "convert_to_gray" function. It is implemented as a loop which reads RGB pixels one by one from an input array, calls the "to_gray" method to compute the result and assigns it to the output array using pointer arithmetic.

3.3.1.4 Bit-Accurate Data Types

Hardware designers are accustomed to bit-accurate datatypes in hardware design languages such as VHDL and Verilog. Similarly, bit-accurate data types are needed to synthesize area efficient hardware from C models. The arbitrary-length bit-accurate integer and fixed-point "Algorithmic C" datatypes provide an easy way to model static bit-precision with minimal runtime overhead. Operators and methods on both the integer and fixed-point types are clearly and consistently defined so that they have well defined simulation and synthesis semantics.

The precision of the integer type ac_int<W,S> is determined by template parameters W(integer that gives bit-width) and S (a boolean that determines whether the integer is signed or unsigned).

The fixed-point type ac_fixed<W,I,S,Q,O> has five template parameters which determine its bit-width, the location of the fixed-point, whether it is signed or unsigned and the quantization and overflow modes that are applied when constructing or assigning to object of its type.

The advantages of the Algorithmic C datatypes over the existing integer and fixed-point datatypes are the following:

- Arbitrary-Length: this allows a clean definition of the semantics for all operators that are not tied to an implementation limit. It is also important for writing general IP algorithms that don't have artificial (and often hard to quantify and document) limits for precision.
- Precise Definition of Semantics: special attention has been paid to define and verify the simulation semantics and to make sure that the semantics are appropriate for synthesis. No simulation behavior has been left to compiler dependent behavior. Also, asserts have been introduced to catch invalid code during simulation.
- Simulation Speed: the implementation of ac_int uses sophisticated template specialization techniques so that a regular C++ compiler can generate optimized assembly language that will run much faster than the equivalent SystemC datatypes. For example, ac_int of bit widths in the range 1–32 can run 100$\times$ faster than the corresponding sc_bigint/sc_biguint datatype and 3$\times$ faster than the corresponding sc_int/sc_uint datatype.
- Correctness: the simulation and synthesis semantics have been verified for many size combinations using a combination of simulation and equivalence checking.
- Compilation Speed and Smaller Executable: code written using ac_int datatypes compiles 5$\times$ faster even with the compiler optimizations turned on (required for fast simulation). It also produces smaller binary executables.
- Consistency: consistent semantics of ac_int and ac_fixed.

In addition to the Algorithmic C datatypes, Catapult Synthesis also supports the C++ native types (bool, char, short, int and long) as well as the SystemC sc_int, sc_bigint and sc_fixed types and their unsigned version.

3.3.2 Synthesizing the Design Interface

3.3.2.1 Hardware Interface View of the Algorithm

The design interface is how a hardware design communicates with the rest of the world. In the C/C++ source code, the arguments passed to the top-level function infer the interface ports. Catapult can infer three types of interface ports:

- Input Ports transfer data from the rest of the world to the design. All inputs are either non-pointer arguments passed to the function or pointer arguments that are read only.
- Output Ports transfer data from the design to the rest of the world. Structure or pointer arguments infer output ports if the design reads from them but does not write to them.
- Input ports transfer data both to and from the design. These are pointer arguments that are both written and read.

3.3.2.2 Interface Synthesis

Catapult builds a correspondence between the arguments of the C/C++ function and the I/Os of the hardware design. Once this is established, the designer uses interface synthesis constraints to specify properties of each hardware ports.

With this approach, designers can target and build any kind of hardware interface. Interface synthesis directives give users control other parameters such as bandwidth, timing, handshake and other protocols aspects.

This way the synthesized C/C++ algorithm remains purely functional and doesn't have to embed any kind of interface specific information. The same code can be retargeted based on any interface requirement (bandwidth, protocol, etc...)

Amongst other transformations and constraints, the user can for instance:

- Define full, partial or no handshake on interface signals
- Map arrays to wires, memories, busses or streams
- Control the bandwidth (bitwidth) of the hardware ports
- Add optional start/done flags to the design
- Define custom interface protocols

Hardware specific I/O signals such as clock, reset, enable or and handshaking signals do not need to be modeled either and are added automatically based on user constraints.

3.3.3 Loop Controls

3.3.3.1 Loop Unrolling

Loop unrolling exposes parallelism that exists across different subsequent iterations of a loop by partially or fully unrolling the loop.

The example in Fig. 3.6 features a simple loop summing two vectors of four values. If the loop is kept rolled, then Catapult will generate a serial architecture. As shown on the left, a single adder will be allocated to implement the four additions. The adder is therefore time-shared, and dedicated control logic is built accordingly. Assuming the mux, add and demux logic can fit in the desired clock period, four cycles are needed to compute the results.

On the right-hand side, the same design is synthesized with its loop fully unrolled. Unrolling is applied by setting a synthesis constraint and has the same effect as copying four times the loop body. Catapult can now exploit the operation-level parallelism to build a fully parallel implementation of the same algorithm. The resulting architecture necessitates four adders to implement the four additions and has a latency of one clock cycle.

Partial unrolling may also be used to trade the area, power and performance of the resulting design. In the above example, an unrolling factor of 2 would cause the loop body to be copied twice, and the number of loop iterations halved. The

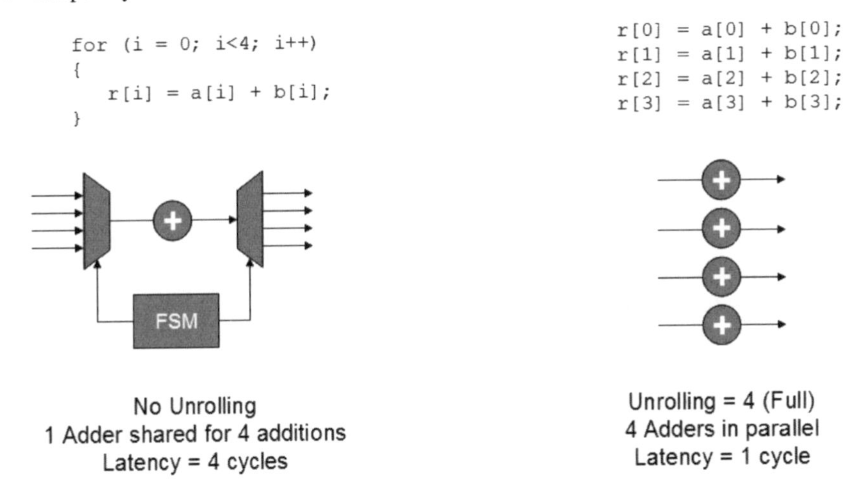

Fig. 3.6 Unrolling defines how many times to copy the body of a loop

synthesized solution would therefore be built with two adders, and have a latency of two cycles.

3.3.3.2 Loop Merging

Loop merging exhibits loops-level parallelism. This technique applies to sequential loops and creates a single loop with the same functionality as the original loops. This transformation is used to reduce latency and area consumption in a design by allowing parallel execution, where possible, of loops that would normally execute in series.

With loop merging, algorithm designers can develop their application in a very natural way, without having to worry about potential parallelism in the hardware implementation.

In Fig. 3.7, the code contains sequential loops. Sequential loops are very convenient to model the various processing stages of an algorithm. By enabling or disabling loop merging, the designer decides if in the generated hardware, the loops should run in parallel (merging enabled) or sequentially (merging disabled). With this technique, the designer maintains the readability and hardware independence of his source code. The transformation and optimization techniques in Catapult can produce a parallelized design which would have required a much more convoluted source description, as shown on the right-hand side.

It should also be noted in this example that Catapult is able to appropriately optimize the intermediate data storage. When sequentially processing the two loops, intermediate storage is needed to store the values of "a." When parallelizing the two loops, values of "a" produced in the first loop can directly be consumed by the second loop, removing the need for storage.

```
for (i = 0; i<32; i++)
{
   a[i] = b[i] * c[i];
}
for (i = 0; i<16; i++)
{
   z[i] = a[i] + x[i];
}
```

```
for (i = 0; i<32; i++)
{
   atmp = b[i] * c[i];

   if (i<16)
     z[i] = atmp + x[i];
}
```

No Merging
Loops execute sequentially
Latency = 48 cycles

Merging Enabled
Loops execute in parallel
Latency = 32 cycles

Fig. 3.7 Merging parallelizes sequential loops

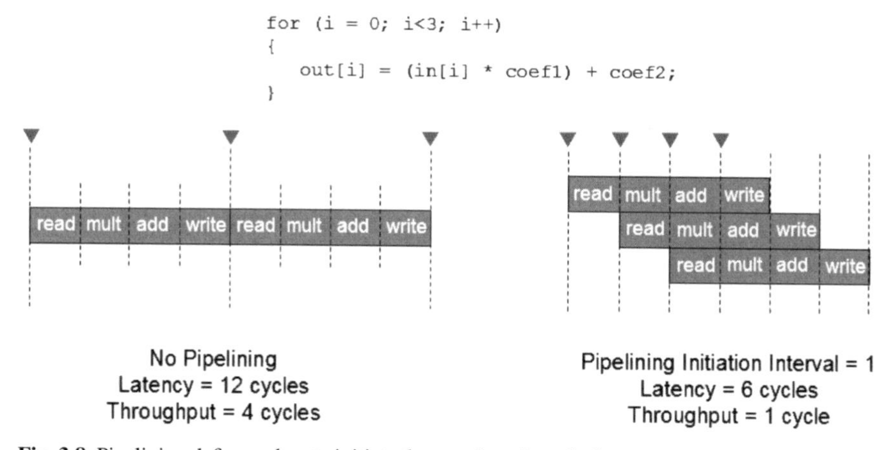

Fig. 3.8 Pipelining defines when to initiate the next iteration of a loop

3.3.3.3 Loop Pipelining

Loop pipelining provides a way to increase the throughput of a loop (or decreasing its overall latency) by initiating the next iteration of the loop before the current iteration has completed. Overlapping the execution of subsequent iterations of a loop exploits parallelism across loop iterations. The number of cycles between iterations of the loop is called the initiation interval. In many cases loop pipelining may improve the resource utilization thus increasing the performance/area metric of the design.

In example Fig. 3.8, a loop iteration consists of four operations: an I/O read to in[i], a multiplication with coef1, an addition with coef2, and finally an I/O write to out[i]. Assuming that each of these operations executes in a clock cycle, and if no loop constraints are applied, the design schedule will look as shown on the left hand side. Each operation happens sequentially, and the start of a loop iteration (shown here with the red triangle) happens after the previous iteration completes. Conceptually, the pipeline initiation interval is equal to the latency of a loop iteration, in this case, four cycles.

By constraining the initiation interval with loop pipelining, designers determine when to start each loop iteration, relative to the previous one. The schedule on the right hand side illustrates the same loop, pipelined with an initiation interval of one cycle: the second loop iteration starts one cycle after the first one.

Pipelining a design directly impacts the data rate of the resulting hardware implementation. The first solution makes 1 I/O access every four cycles, while the second one will make I/O accesses every cycles. Some applications may require a given throughput, therefore commanding the initiation interval constraint. Other designs may tolerate some flexibility, allowing the designers to explore different pipelining scenarios, trading area, bandwidth utilization as well as power consumption.

3.3.4 Hierarchical Synthesis

The proper integration of individual blocks into a sub-system is one of the major challenges in chip design. With its hierarchical synthesis capability Catapult Synthesis can greatly simplify the design and integration tasks, building complex multi-block systems correct-by-construction.

While loop unrolling exploits instruction level parallelism and loop merging exploits loop level parallelism, hierarchy exploits function level (task-level) parallelism. In Catapult, the user can specify which function calls should be synthesized as hierarchical units. The arguments of the hierarchical function define the data flow of the system, and Catapult will build all the inter-block communication and synchronization logic.

Hierarchy generalizes the notion of pipelining, allowing different functions to run in a parallel and pipelined manner. In complex systems consisting of various processing stages, hierarchy is very useful to meet design throughput constraints. When pipelining hierarchical systems, Catapult builds a design were the execution of the various functions overlap in time. As shown in Fig. 3.9, in the sequential source code, the three functions (stage1, stage2 and stage3) execute one after the other. In

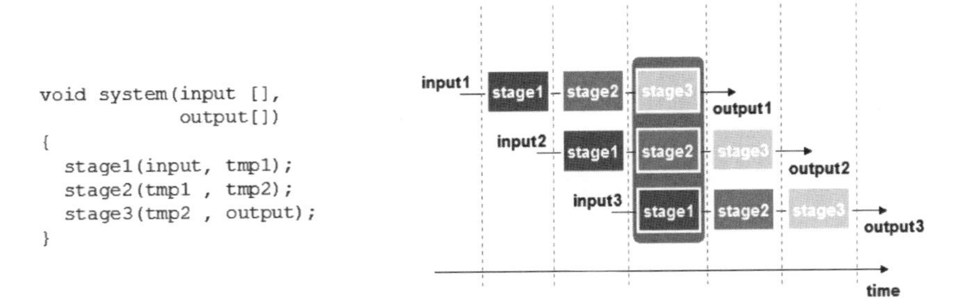

Fig. 3.9 Task-overlapping with hierarchical synthesis

the resulting hierarchical system, the second occurrence of stage1 can start together with the first occurrence of stage2, as soon as first occurrence of stage1 ends.

3.3.5 Technology-Driven Scheduling and Allocation

Scheduling and allocation is the process of building and optimizing the design given all the user constraints, including the specific clock period and target technology. With the clock period defining the maximum register-to-register path, the technology defines the logic delay for each design operation. The design schedule is therefore intimately tied to these clock and technology constraints (Fig. 3.10). This is fundamental to build optimized RTL implementations, allowing efficient retargeting of algorithmic specifications from one ASIC process to another, or even to FPGAs, with always optimal results.

This capability opens new possibilities in the field of IP and reuse. While RTL reuse can provide a quick path to the desired functionality, it often comes at the expense of suboptimal results. RTL IPs maybe reused over many years. Developed on older processes, IPs will certainly work on newer ones, but without taking advantage of higher speeds and better densities, therefore resulting in bigger and slower than needed implementations. In contrast, Catapult can built optimized RTL designs from functional IPs for each process generation, taking reuse to a new level of efficiency.

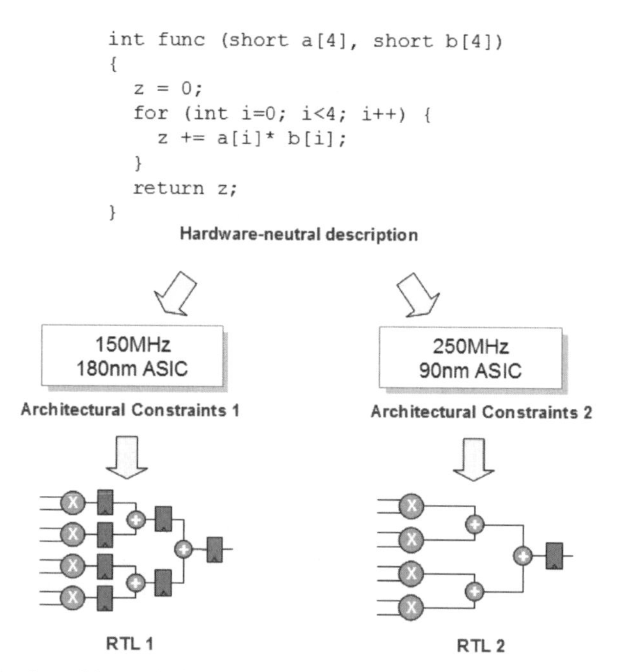

Fig. 3.10 Technology-driven scheduling and allocation

3.4 Case Study: JPEG Encoder

In this section we will show how a sub-system such as a JPEG encoder can be synthesized with Catapult Synthesis.

We chose a JPEG encoder design for this case study, as we felt that the application would be sufficiently familiar to most readers in order to be easily understood without extensive explanations. Moreover, such an encoder features a pedagogical mix of datapath and control blocks, giving a good overview of Catapult Synthesis' capabilities.

3.4.1 JPEG Encoder Overview

The pixel pipe (Fig. 3.11) of the encoder can be broken down in four main stages: first RGB to YCbCr color space conversion block, second DCT (discrete cosine transform), third zigzag reordering combined with quantization and last, the Huffman encoder.

3.4.2 The Top Level Function

The top level function synthesized by Catapult (Fig. 3.12) closely resembles the system block diagram. Four sub-functions implement the four processing stages of the algorithm. The sub-functions simply pass on arrays to each other, mimicking the system data flow.

3.4.3 The Color Space Conversion Block

The color space conversion unit is implemented as a relatively straightforward vector multiplication. Different sets of coefficients are used for Y, Cb and Cr components.

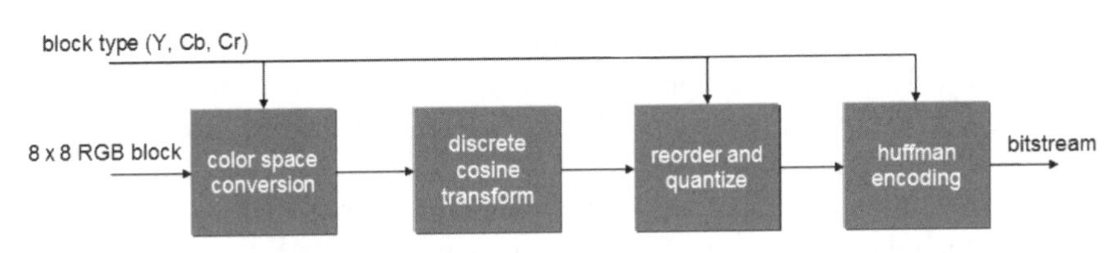

Fig. 3.11 JPEG encoder block diagram

```
void pixelpipe(
      uint2          *type,                  // input  block type, Y, Cr, Cb
      rgb_t           rgb[64],               // input  block (R, G, B)
      codes_t        *codes,                 // output huffman codes
      unsigned char  huffsizes[2][2][256],   // input  huffman tables [DC,AC][LUMA,CHROMA]
      unsigned int   huffcodes[2][2][256] )  // input  huffman tables [DC,AC][LUMA,CHROMA]
{
  char  ycbcrOut[64];
  short dctOut[64];
  short vectorOut[64];
  int   quantizeOut[64];
  uint6 last_non_zero_index;
  uint2 ycbcrType, quantizeType;

  // convert 8x8 RGB block to 8x8 YCbCr
  rgb2ycbcr(rgb, type, ycbcrOut, &ycbcrType);

  // run 2D 8x8 DCT on the block
  dct(ycbcrOut, dctOut);

  // zig-zag and quantize the results
  reorder_and_quantize(&ycbcrType, dctOut, quantizeOut, &last_non_zero_index, &quantizeType);

  // run-length and Huffman encode
  huffmanize(&quantizeType, quantizeOut, &last_non_zero_index, codes, huffsizes, huffcodes);
}
```

Fig. 3.12 C++ source code for the synthesized top level

```
void rgb2ycbcr( rgb_t    rgb[64],
                uint2   *typeIn,
                char     ycbcr[64],
                uint2   *typeOut )
{
  ac_fixed<16,1,true> coeffs[3][3] = {
   { 0.299    ,  0.587    ,  0.114     },
   {-0.168736, -0.331264,  0.5       },
   { 0.5      , -0.418688, -0.081312 } };

  ac_fixed<8,8,false> tmp;
  uint2 k = *typeIn;
  // normalize values by subtracting 128 and convert to YCbCr
  convert2ycbcr:
  for (unsigned int i=0; i<64; i++) {
    tmp = coeffs[k][0]*rgb[i].r + coeffs[k][1]*rgb[i].g + coeffs[k][2]*rgb[i].b + ((k)?128:0);
    tmp -= 128;
    ycbcr[i] = tmp.to_int();
  }
  *typeOut=k;
}
```

Fig. 3.13 C++ source code for the rgb2ycbcr function

In the C source (Fig. 3.13), the RBG input is modeled as an array of structs. The rgb_t struct contains three fields: r, g and b. By default, Catapult assumes the R, G and B components are mapped to three different interface resources. Using interface synthesis constraints, it is possible to merge them all side-by-side on the same resource and map this resource to a memory.

This way, the color space conversion block will get all its input from a single memory, with every read returning all three R, G and B color components over a $3 \times 8 = 24$ bit data bus (Fig. 3.14).

The function itself is pipelined with an initiation interval of 1, to create a continuously running design with a throughput of 1 memory access per cycle. By the same effect, outputs will be produced at a constant rate of one sample per cycle.

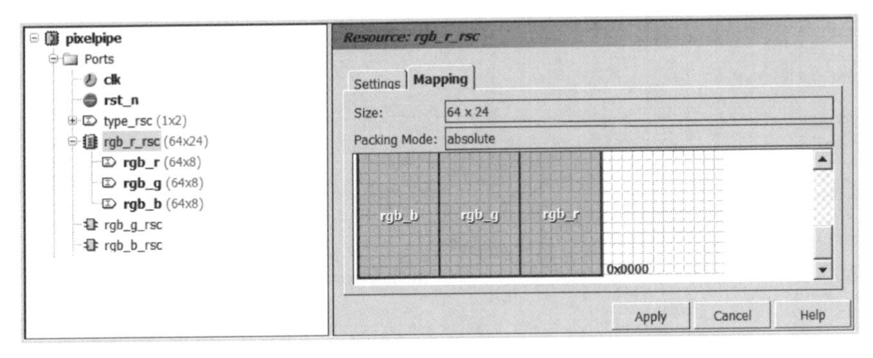

Fig. 3.14 Mapping side-by-side R, G and B pixels in a memory

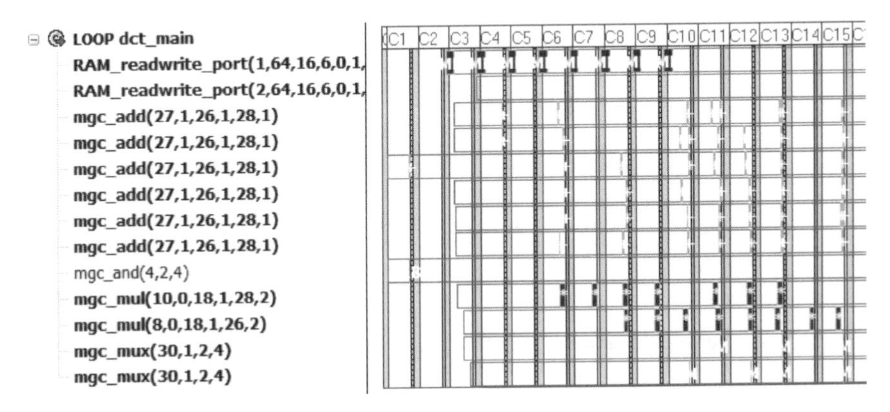

Fig. 3.15 Gantt chart of the horizontal DCT – throughput 1 sample per cycle

3.4.4 The DCT Block

The DCT is based on a standard 2D 8×8 Chen implementation. It decomposes in a vertical and a horizontal pass, with a transpose buffer in between. In this datapath dominated design, it easily possible to explore different micro-architectures to trade-off performance (latency and throughput) versus the number of computational resources such as adders or multipliers.

The smallest implementation allowing a constant throughput of one sample per cycle can be scheduled with only 2 multipliers and 6 adders, and has an overall latency of 82 cycles to process a full 8×8 block. Figure 3.15 shows a partial view of the corresponding Gantt chart. The left column lists the resources used to build the micro-architecture. The right part shows how and when these operators are used to cover specific operations from the reference algorithm. The Gantt chart shows that the two multipliers are time-shared to implement 16 different multiplications. Similarly, the six adders implement 48 different additions.

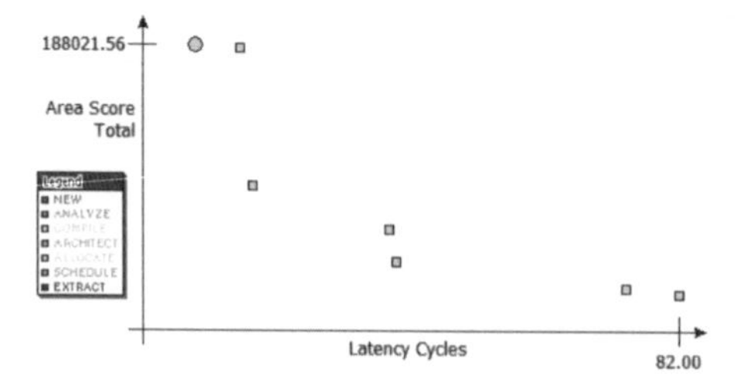

Throughput	Latency	Relative Area
1 s/cycle	82	1.00 x
1 s/cycle	74	1.18 x
2 s/cycle	39	1.90 x
2 s/cycle	38	2.80 x
4 s/cycle	17	4.02 x
4 s/cycle	15	7.86 x
8 s/cycle	8	7.95 x

Fig. 3.16 Catapult XY plot and table of results of the horizontal DCT

```
const unsigned zigzagpath_vector[64] = {
   0,  1,  8, 16,  9,  2,  3, 10,
  17, 24, 32, 25, 18, 11,  4,  5,
  12, 19, 26, 33, 40, 48, 41, 34,
  27, 20, 13,  6,  7, 14, 21, 28,
  35, 42, 49, 56, 57, 50, 43, 36,
  29, 22, 15, 23, 30, 37, 44, 51,
  58, 59, 52, 45, 38, 31, 39, 46,
  53, 60, 61, 54, 47, 55, 62, 63
};

vectorization:
for (unsigned short i = 0; i < 64; i++) {
  datat[i] = datai[zigzagpath_vector[i]];
}

uint1 LC = (*block_type==LUMA)?LUMA:CHROMA;
quantization:
for (unsigned short i = 0; i < 64; i++) {
  datao[i] = datat[i] / qvector[LC][i];
}
```

Fig. 3.17 C++ source code for the reorder and quantize block

After this first implementation is obtained, the user can easily trade area and latency through simple scheduling options. With the same throughput requirements, a design with only 74 cycles of latency can be built with eight adders instead of six.

By increasing or decreasing the throughput constraints, it is possible to further explore the design space. Figure 3.16 shows the full table of results obtained, as well as a screenshot of the Catapult built-in XY plot tool used to compare and contrast the various solutions. The last solution, featuring a throughput of eight samples per cycles is effectively processing entire rows of the 8×8 data set.

3.4.5 The Reorder and Quantize Block

The zigzag reordering and quantization steps are fairly simple. The first step reorders the DCT results according to a predefined "zigzag" sequence and the second one quantizes those results based on luminance and chrominance quantization tables.

As shown in Fig. 3.17, these two steps are naturally coded as two sequential loops, one for each step. Without loop merging, the two loops run sequentially, 135 cycles are required to process a full 8×8 block and the throughput is not constant.

With loop merging, Catapult is able to fold the two sequential loops into a single one effectively exploiting loop level parallelism. Once properly pipelined, the result is a continuously running design which simultaneously reorders and quantizes data at a constant rate of one sample per cycle and with a latency of only 67 cycles for a full block.

3.4.6 The Huffman Encoder Block

Compared to the other blocks in the JPEG pixel pipe, the Huffman encoder is much more of a control-oriented, decision making algorithm. The run-length part of the encoder scans values as they arrive, counting the number of consecutive zeros. When a non-zero value is found, it is paired with the number of preceding zeros. This pair of symbols is then Huffman encoded, forming a bitstream of codewords (Fig. 3.18). In the C program, the function returns the bitstream as an array of struct. Catapult interface synthesis directives are used to build a streaming interface with

```
for (unsigned short i=0; i<64; i++) {
  int value = block[i];
  if (i==0) {
    // ------------------------------------------
    // Huffman encode the DC Coefficient
    int dcval           = value;
    int diff            = dcval - dc[*block_type];
    dc[*block_type]     = dcval;

    huf_code = huffencode(huffcodes[LC][DC],diff);
    huf_size = huffencode(huffsizes[LC][DC],diff);
  } else {
    // ------------------------------------------
    // Huffman encode the AC Coefficient
    if (value) {
      // if data is not 0...
      // ...insert the (num_zeros,block[i]) code
      new_code    = true;
    } else {
      // if data is 0...
      if (num_zeros==15 && i<huf_lnzi) {
        // if we found 16 consecutive zeros
        // and if we are not yet in the last block of zero's
        // ...insert the special (15,0) code
        new_code    = true;
      } else if (i==63) {
        // if we reached the end of the last block of zero's
        // ...insert the special (0,0) code
        new_code    = true;
      } else {
        // else keep counting zero's !
        new_code    = false;
      }
    }
    huf_code   = (new_code) ? huffencode(huffcodes[LC][AC],num_zeros,value) : 0;
    huf_size   = (new_code) ? huffencode(huffsizes[LC][AC],num_zeros,value) : 0;
    num_zeros  = (new_code || (i>=huf_lnzi)) ? 0 : (int)num_zeros+1;
  }
  codes->code[i] = huf_code; // the code        (the bits)
  codes->size[i] = huf_size; // the code length (the number of bits)
}
```

Fig. 3.18 C++ source code for the run-length encoder

handshake. Every cycle the encoder outputs a codeword with an additional flag, indicating whether the current output data is valid or not.

3.4.7 Integrating the Hierarchical System

When performing top-down hierarchical synthesis, Catapult starts by independently synthesizing each of the four sub-functions. Then Catapult integrates all the sub-blocks, building the appropriate inter-block communication and creating the needed synchronization logic. Top-level control structures are synthesized to guarantee safe and efficient data exchange between blocks.

When two blocks exchange data through an array, Catapult distinguishes two cases, depending if the producer and consumer access the array in the order or not. If they do, then a streaming communication can be synthesized. If the two blocks access the array in different order, then intermediate storage is required to allow the two blocks to run in parallel. Catapult can automatically build ping-pong memories, round robin memories and other kinds of interleaved structures.

In our JPEG encoder, the array written by the quantization block and read by the Huffman encoder is accessed in the same order by blocks, from index 0 up to 63, with constant increments. Catapult will therefore build a streaming connection between both blocks.

However, while the DCT outputs results from index 0 up to 63, the reordering block reads those values in a zigzag order. In this case intermediate storage will be required, for instance in the form of a ping-pong buffer and its associated control and synchronization logic (Fig. 3.19).

3.4.8 Generating the Design

Catapult outputs VHDL, Verilog and SystemC netlists, both RTL and behavioral, as well as various scripts and makefile needed to use the design in various simulation and synthesis tools.

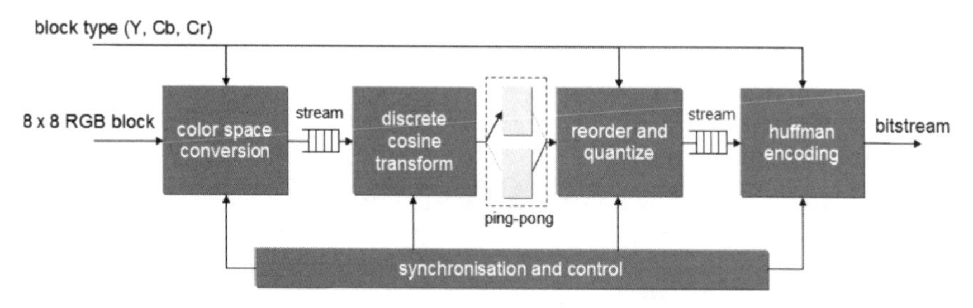

Fig. 3.19 Hardware integration and communication architecture of the JPEG encoder

```
#ifdef CCS_SCVERIFY
  testbench::exec_pixelpipe(&block_type, rgb, codes, huffsizes, huffcodes);
#else
  pixelpipe(&block_type, rgb, codes, huffsizes, huffcodes);
#endif
```

Fig. 3.20 Instrumented testbench for automatic verification

In this example, once all the constraints are set, it takes a little over 3 min of synthesis runtime, on an average workstation, to produce the desired design implementation, turning 469 lines of C++ code modeling the JPEG encoder into 11,200 lines of RTL VHDL.

3.4.9 Verifying the RTL

Once the RTL is generated, Catapult provides a push-button verification flow allowing simulation of the generated design against the original design and testbench.

For this matter the testbench calling the synthesized C function should be instrumented to call the verification infrastructure instead of just the reference algorithm when running the automatic verification flow (Fig. 3.20).

Besides this simple change, the rest of the flow is fully automated and the user simply needs to run the Catapult generated makefile which will take core of compiling and linking the proper C, SystemC and HDL design files within the specified simulation environment.

The difference in simulation performance between the C design and the equivalent RTL gives another good idea of the benefits of designing in C instead of HDL. In this example, a trivial testcase which runs in a 1/10th of a second, runs in about 2:30 min on an average workstation, showing a 1,500× difference. Not only edits are more quickly done in C than in HDL, they can also be much more rapidly and thoroughly verified.

3.5 Conclusion

In this paper, we gave in depth overview of Catapult Synthesis, an interactive C synthesis tool which generates production quality results up to 20× fasters than with manual approaches.

While much debate has occurred about the applicability and the maturity of behavioral synthesis tools, the success of Catapult in the market place and its endorsement by leading semiconductor vendors demonstrate the viability of this design methodology which is now clearly used beyond the traditional circle of visionaries and early adopters.

This success was built on state-of-the-art technology, resulting from many man/years of internal research and development. But synthesizing RTL from abstract specifications is not an end in itself. There far more other real-life constraints which technology-alone doesn't address. Mentor Graphics and the Catapult Synthesis team have always recognized the importance of complying with industrial requirements, such as integration in flows, vendor sign-off, risk-management, knowledge transfer, reliable support and, last but not least, clear ROI.

Acknowledgments The author would like to acknowledge the Catapult Synthesis team, and most specifically, Bryan Bowyer, Andres Takach and Shawn McCloud for their direct or indirect contributions to this work.

Chapter 4
Algorithmic Synthesis Using PICO

An Integrated Framework for Application Engine Synthesis and Verification from High Level C Algorithms

Shail Aditya and Vinod Kathail

Abstract The increasing SoC complexity and a relentless pressure to reduce time-to-market have left the hardware and system designers with an enormous design challenge. The bulk of the effort in designing an SoC is focused on the design of product-defining application engines such as video codecs and wireless modems. Automatic synthesis of such application engines from a high level algorithmic description can significantly reduce both design time and design cost. This chapter reviews high level requirements for such a system and then describes the PICO (Program-In, Chip-Out) system, which provides an integrated framework for the synthesis and verification of application engines from high level C algorithms. PICO's novel approach relies on aggressive compiler technology, a parallel execution model based on Kahn process networks, and a carefully designed hardware architecture template that is cost-efficient, provides high performance, and is sensitive to circuit level and system level design constraints. PICO addresses the complete hardware design flow including architecture exploration, RTL design, RTL verification, system validation and system integration. For a large class of modern embedded applications, PICO's approach has been shown to yield extremely competitive designs at a fraction of the resources used traditionally thereby closing the proverbial design productivity gap.

Keywords: SoC design, ASIC design, ESL synthesis, Algorithmic synthesis, High level synthesis, Application engine synthesis, C-to-RTL, PICO, Architecture exploration, Soft IP, Kahn process networks, System integration, Software drivers, System modeling, System validation, Transaction level models, Task level parallelism, Instruction level parallelism, Pipeline of processing arrays, Data streams, RTL verification, Co-simulation, Reusable hardware interfaces

P. Coussy and A. Morawiec (eds.) *High-Level Synthesis.*

4.1 Introduction

The recent explosion in consumer appliances, their design complexity, and time-to-market pressures have left the system designers facing an enormous design productivity gap. System and register-transfer level (RTL) design and verification are increasingly the bottleneck in the overall product cycle. The EDA community has been trying to get around this bottleneck for over a decade, first with behavioral synthesis [1], and then with intellectual property (IP) reuse [2]. However, both those approaches have their limitations. In general, behavioral synthesis is a very difficult problem and has yielded poor cost and performance results compared to hand designs. IP reuse, on the other hand, has worked to a limited extent in System-on-Chip (SoC) designs, where standard IP blocks on a common embedded platform may be shared across various generations of a product or even across families of products.

A typical platform SoC comprises four different types of IP as shown in Fig. 4.1. These are:

1. *Star IP such as CPUs and DSPs:* Star IP needs significant investment in terms of building the hardware, the software tool chain as well as the creation, debugging and compatibility of operating system and application software. This type of IP is usually designed manually, doesn't often change, and is very hard to alter when it does. Therefore, this IP is typically reused across several generations of a product.
2. *Complex application engines such as video codecs and wireless modems:* These IP blocks are critical for differentiating the end product and change rapidly with each revision in functionality, target technology, or both. Additionally, signifi-

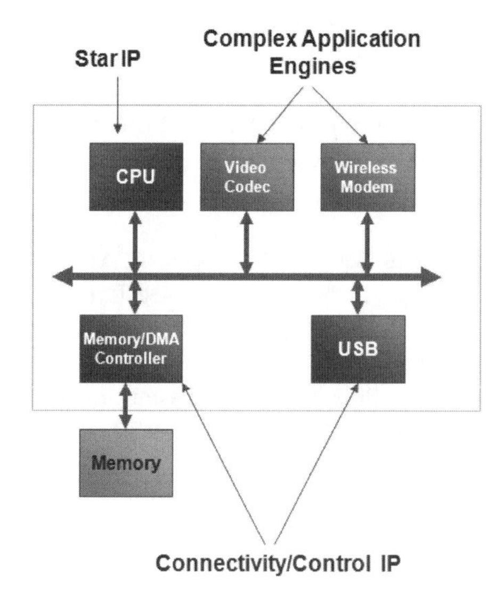

Fig. 4.1 An SoC embedded platform with application engines

cant investment is continually being made to improve their power, performance and area across product generations. Therefore, direct reuse of this IP is quite limited.

3. *Connectivity and control IP such as USB port and DMA:* This is system level glue that never defines the functionality nor differentiates the end product. This IP, therefore, is typically reused to reduce cost and provide standardization. It does sometimes need a limited amount of tailoring.
4. *Memory:* Memory takes up the largest amount of silicon area, but also neither defines the function nor differentiates the end product. Memories are almost always compiled and built bottom-up. Their models are generated from the transistor level behavior.

Each of these different types of IP needs to be integrated into an SoC. The availability of standard interfaces (memory, streaming, bus) based on industry standard protocols, such as OCP [3], make this integration more straightforward.

Unlike other IP elements of the platform SoC, IP reuse of product-defining *application engines* is hard because every new product context requires some specialization and adaptation to meet the new design objectives. For this reason, and because they critically define the SoC functionality, the bulk of the SoC design effort is focused on the design and verification of application engines.

4.1.1 Application Engine Design Challenges

Complex application engines such as multi-standard codec and 3 G wireless modems used in the next generation consumer devices place extreme requirements on their designs – they require very high performance at very low power and low area. For example, software defined radio for 4 G wireless modem requires 10–100 GOPs (giga operations per second) at a budget of 100–500 mW of power [4] – that is, about 100 MOPs mW^{-1}. Off-the-shelf solutions such as general-purpose processors or DSPs cannot satisfy such extreme requirements. Embedded DSPs are unable to provide the high performance. On the other hand, high end DSPs such as IBM Cell processor can provide the high performance but their power consumption is very high (in the 10 MOPs mW^{-1} range).

The solution is to build application-specific or custom processors, or dedicated hardware systems to meet the extreme performance-power-area goals. Typically, direct hardware implementations can achieve 100–1,000 MOPs mW^{-1} and provide 2–3 orders of magnitude better area and power compared to embedded processors or DSPs.

Customization, however, has its cost. Manual design of application engines using current design methodologies is very expensive in terms of both design time and non-recurring engineering (NRE) cost leading to SoCs that take millions of dollars and years to design. This is not sustainable for two reasons. First, SoCs are growing in complexity because of the insatiable demand for more and more features and the high degree of chip integration made possible by Moore's law. For example, a cell-phone chip now contains multiple modems, imaging pipeline for a camera, video codecs, music players, etc. A video codec used to be a whole chip a few years back, and now it is a small part of the chip. Second, there is relentless pressure to reduce time-to-market and lower prices.

It is clear that automation is the key to success. Automatic application engine synthesis (AES) from a high level algorithmic description significantly reduces both design time and design cost. There is a growing consensus in the design community that hardware/software co-design, high level synthesis, and high level IP reuse are together necessary to close the design productivity gap.

4.1.2 Application Engine Design Space

Application engines like multi-standard video codecs are large, complex systems containing a significant number of processing blocks with complex dataflow and control flow among them. Externally, these engines interact with system CPU, system bus and other application engines. The ability to synthesize complex application engines from C algorithms automatically requires a careful examination of the type of architectures that lend themselves well to such automation techniques.

Broadly speaking, there are three main approaches for designing application engines [4] (see Fig. 4.2).

1. *Dedicated hardware accelerators:* They provide the highest performance and the lowest power. Typically, they are 2–3 orders of magnitude better in power and performance than a general purpose processor. They are non-programmable but can provide limited amount of multi-modal execution based on configuration parameters. There are two approaches for automatic synthesis of dedicated hardware blocks:

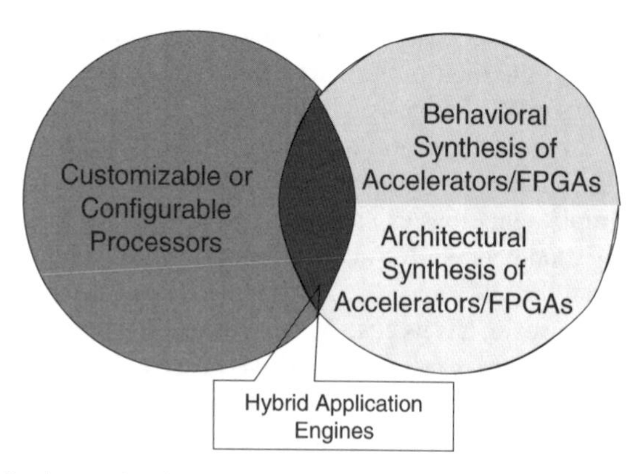

Fig. 4.2 The application engine design space

   (a) *Behavioral synthesis*: This is a bottom-up approach in which individual blocks are designed separately. C statements and basic blocks are mapped to datapath leading to potentially irregular datapath and interconnect. The datapath is controlled by a monolithic state machine which reflects the control flow between the basic blocks and can be fairly complex.
   (b) *Architectural synthesis*: This is a top-down approach with two distinguishing characteristics. First, it takes a global view of the whole application and can optimize across blocks in order to provide high performance. Second, it uses an efficient, high performance architecture template to design datapath and control leading to more predictable results. PICO's approach for designing dedicated hardware accelerators falls in this category.

2. *Customizable or configurable processors:* Custom or application-specific processors can give an order of magnitude better performance and power than general-purpose processor while still maintaining a level of programmability. This approach is well-suited for the following two cases

   (a) The performance requirements are not very high and power requirements are not very stringent.
   (b) Standards or algorithms are still in flux, and flexibility to make algorithmic changes after fabrication is needed.

3. *Hybrid approach:* In our view, this is the right approach for synthesizing complex application engines. An efficient architecture for these engines is a combination of

   (a) Programmable processor(s), typically custom embedded processor, for parts of the application that don't require high performance
   (b) Dedicated hardware blocks to get high performance at low power and low area
   (c) Local buffers and memories for high bandwidth

This approach allows a full spectrum of designs to be explored that trade-off among multiple dimensions of cost, performance, power and programmability.

4.1.3 Requirements of a Production AES System

In addition to generating competitive hardware, a high level synthesis system needs to fit in a SoC design flow for it to be practically useful and of significant benefit to designers. We can identify a number of steps in the SoC design process. These steps, along with the capabilities that the synthesis system must provide for each step, are described below.

1. *Architecture exploration for application engines:* Architecture and micro-architecture choices have a great impact on the power, performance and area

of a design, but there is no way to reliably predict this impact without actually doing the design. A high level synthesis system makes it possible to do design space exploration to find an optimal design. However, the system must be structured to make it easy to explore multiple designs from the same C source code. For example, a system that requires users to control the scheduling of individual operations in order to get good results is not very useful for architectural exploration because of the amount of time it takes to do one design. Therefore, design automation is the key to effective exploration.

2. *High level, multi-block IP design and implementation*: This is, of course, the main purpose of a high level synthesis system. It must be able to generate designs that are competitive with manual designs for it to be widely acceptable in production environments.
3. *RTL verification*: It is unrealistic to expect that designers would write testbenches for the RTL generated by a synthesis system. They should verify their design at the C level using a C test bench. The synthesis system should then automatically generate either an RTL test bench including test vectors or a C-RTL co-simulation test bench. In addition, the synthesis system should provide a mechanism to test corner cases in the RTL that cannot be exercised using the C test bench.
4. *System modeling and validation (virtual platform) support:* Currently, designers have to manually write transaction level models (TLM) for IP they are designing in order to incorporate them in system level platforms. This is in addition to implementing designs in RTL. Generating transaction level models directly from a C algorithm will significantly reduce the development time for building these models.
5. *SoC integration*: To simplify the hardware integration of the generated IP into an SoC, the system should support a set of standard interfaces that remain invariant over designs. In addition, the synthesis system should provide software device drivers for easy integration into a CPU based system.
6. *RTL to GDSII design flow integration:* The generated RTL should seamlessly go through the existing RTL flows and methodologies. In addition, the RTL should close timing in the first pass and shouldn't present any layout problems because it is unrealistic to expect that designers will be able to debug these problems for RTL they didn't write.
7. *Soft IP reuse and design derivatives:* One of the promised benefits of high level synthesis system is the ability to reuse the same C source for different designs. Examples include designs at different performance points (low-end vs. high-end) across a product family or design migration from one process node to another process node. As an example of the requirement placed on the tool, support for process migration requires that there is a methodology to *characterize* the process and then feed the relevant information to the tool so that it is retargeted to that process.

4.2 Overview of AES Methodology

Figure 4.3 shows the high level flow for synthesis of application engines following the hybrid approach outlined in Sect. 4.1.2. Typically, the first step in the application engine design process is high level partitioning of the desired functionality into hardware and software components. Depending on the application, an engine may consist of a control processor (custom or off-the-shelf) and one or more custom accelerator blocks that help to meet one or more design objectives such as cost, performance, and power. Traditionally, the accelerator IP is designed block by block either by reusing blocks designed previously or by designing new hardware blocks by hand keeping in view the budgets for area, cycle-time and power. Then the engine is assembled together, verified, and integrated with the rest of the SoC platform, which usually takes up a significant fraction of the overall product cycle. The bottlenecks and the risks in this process clearly are in doing the design, verification and integration of the various accelerator blocks in order to meet the overall functionality specification and the design objectives. In the rest of the paper, we will focus our attention on these issues.

In traditional hardware design flows, substantial initial investment is made to define a detailed architectural specification of various accelerator blocks and their interactions within the application engine. These specifications help to drive the manual design and implementation of new RTL blocks and their verification test benches. In addition, a functional executable model of the entire design may be used to test algorithmic coverage and serve as an independent reference for RTL verification.

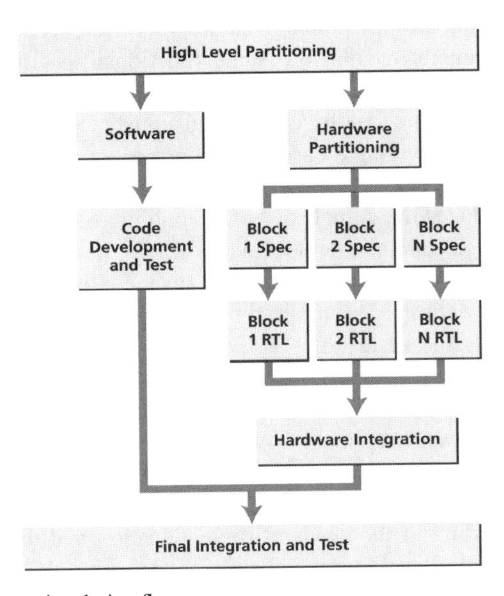

Fig. 4.3 Application engine design flow

In design flows based on high level synthesis, on the other hand, an automatic path to RTL implementation and verification is possible starting from a high level, synthesizable specification of functionality together with architectural information that helps in meeting the desired area, performance and power metrics. The additional architectural information may be provided to a HLS tool in various ways. One possible approach is to combine the hardware and implementation specific information together with the input specification. Some tools based on SystemC [5] require the user to model the desired hardware partitioning and interfaces directly in the input specification. Other tools require the user to specify detailed architectural information about various components of the hardware being designed using a GUI or a separate design file. This has the advantage of giving the user full control of their hardware design but it increases the burden of input specification and makes the specification less general and portable across various implementation targets. It also leaves the tool with very little freedom to make changes and optimizations in the design in order to meet the overall design goals. Often, multi-block hardware integration and verification becomes solely the responsibility of the user because the tool has little or no control over the interfaces being designed and their connectivity.

4.2.1 The PICO Approach

PICO [6] provides a fully automated, performance-driven, application engine synthesis methodology that enables true algorithmic level input specification and yet is sensitive to physical design constraints. PICO not only produces a cost-effective C-to-RTL mapping but also guarantees its performance in terms of throughput and cycle-time. In addition, multiple implementations at different cost and performance tradeoffs may be generated from the same functional specification, effectively reusing the input description as flexible algorithmic IP. This methodology also reduces design verification time by creating customized verification test benches automatically and by providing a correct-by-construction guarantee for both RTL functionality and timing closure. Lastly, this methodology generates standard set of interfaces which reduces the complexity of assembling blocks into an application engine and final integration into the SoC platform.

The key to PICO's approach is to use an advanced parallelizing compiler in conjunction with an optimized, compile-time configurable architecture template to generate hardware as shown in Fig. 4.4. The behavioral specification is provided using a subset of ANSI C, along with additional design constraints, such as throughput and clock frequency. The RTL design creation can then be viewed as a two step process. In the first step, a retargetable, optimizing *compiler* analyzes the high level algorithmic input, exposing and exploiting enough parallelism to meet the required throughput. In the second step, an *architectural synthesizer* configures the architectural template according to the needs of the application and the desired physical design objectives such as cycle-time, routability and cost.

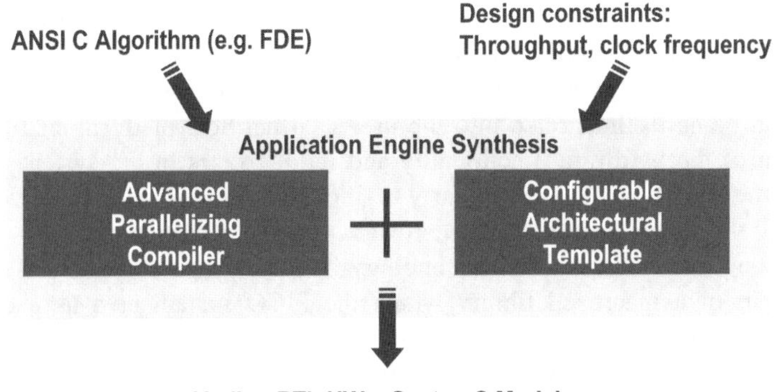

Fig. 4.4 PICO's approach to high level synthesis

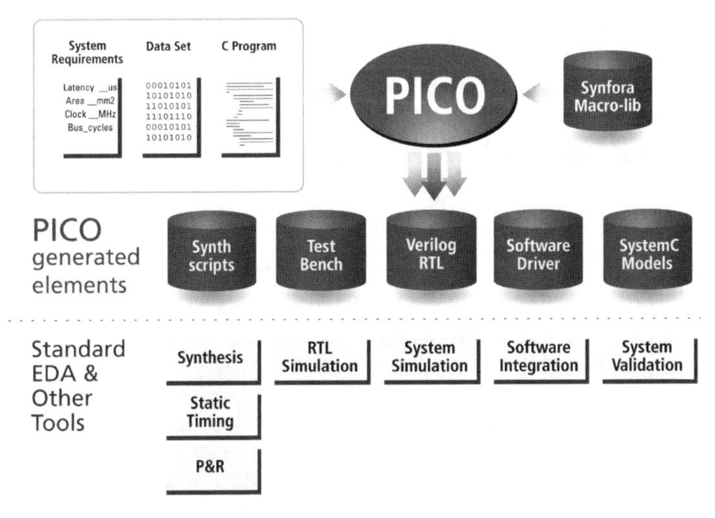

Fig. 4.5 System level design flow using PICO

4.2.2 PICO's Integrated AES Flow

Figure 4.5 shows the overall design flow for creating RTL blocks using PICO. The user provides a C description of their algorithm along with performance requirements and functional test inputs. The PICO system automatically generates the synthesizable RTL, customized test benches, synthesis and simulation scripts, as well as software integration drivers to run on the host processor. The RTL implementation is cost-efficient and is guaranteed to be functionally equivalent to the algorithmic C input description by construction. The generated RTL can then be taken through standard simulation, synthesis, place and route tools and integrated into the SoC through automatically configured scripts.

Along with the hardware RTL and its related software, PICO also produces SystemC-based TLM models of the hardware at various levels of abstraction – untimed programmer's view (PV), and timed programmer's view (PV+T). The PV model can be easily integrated into the user's virtual SoC platform enabling fast validation of the hardware functionality and its interfaces in the system context, whereas the PV+T model enables early verification of the performance, the parallelism and the resources used by the hardware in the system context.

The knowledge of the target technology and its design trade-offs is embedded as part of a macrocell library which the PICO system uses as a database of hardware building blocks. This library consist of pre-verified, parameterized, synthesizable RTL components such as registers, adders, multipliers, and interconnect elements that are carefully hand-crafted to provide the best cost-performance tradeoff. These macrocells are then independently characterized for various target technology libraries to obtain a family of cost-performance tradeoff curves for various parametric settings. PICO uses this characterization data for its internal delay and area estimation.

4.2.3 PICO Results and Experience

The PICO Express™ tool incorporating our approach has been used extensively in production environments. Table 4.1 shows a representative set of designs done

Table 4.1 Some example designs created using PICO Express™

Product	Design	Area	Performance	Time vs. hand design
DVD	Horizontal–vertical filter	60–49 K gates, 40% smaller than target	Met cycle budget and frequency target	v1: 1 month v2: 3 days vs. 2–3 months
Digital Camera	Pixel scaler	Met the target	Multiple versions designed at different targets	2–3 weeks Multiple revisions within hours
Set-top box	HD video codec	200 K gates, 5% smaller than hand design	Same as hand design	<2 months to design and verify
Camcorder	High-perf. video compression	1 M gates, met the target	Same as hand design	Same design time with significantly less resources
Video Processing	Multi-standard deblocking, deringing and chroma conversion	Same as hand design	30% higher than hand design	3–4× productivity improvement
Multi-media cell phone	High bandwidth 3 G wireless baseband	400 K gates, same as hand design	Same as hand design	2 months vs. >9 months
Wireless LAN	LDPC encoder for 802.11n	60 K gates, 6% over hand design	Same as hand design, low power	<1 month to design and verify

using PICO Express. These designs range from relatively small horizontal-vertical filter for a DVD player with ∼49 K gates to large designs with more than 1 M gates for high performance video compression. In all cases, designs generated using PICO Express met the desired performance targets with an area within 5–10% of the hand-design except in one case where the PICO design had significantly less area. In all cases, PICO Express provided significant productivity improvements ranging from 3–5× for the initial design and more than 20× for derivative designs. As far as we know, no other HLS tool can handle many of these designs because of their complexity and the amount of parallelism needed to meet performance requirements. Users' experience with PICO Express is described in these papers [7, 8].

4.3 The PICO Technology

In this section, we will describe the key ingredients of the PICO technology that help to meet the application engine design challenges and the requirements of a high level synthesis tool as outlined in Sect. 4.1.

4.3.1 The Programming Model

The foremost goal of PICO has been to make the programming model for designing hardware to be as simple as possible for a large class of designs. PICO has chosen C/C++ languages as the preferred mode of input specification at the algorithmic level. The goal is not to replace Verilog or VHDL as hardware specification languages necessarily, but to raise the level of specification to a point where the essential algorithmic content can be easily manipulated and explored without worrying about the details of hardware allocation, mapping, and scheduling decisions.

Another important goal for PICO's programming model is to allow the user to specify the hardware functionality as a sequential program. PICO automatically extracts parallelism from the input specification to meet the desired performance based on its analysis of program dependences and external resource constraints. However, the functional semantics of the hardware generated still corresponds to the input sequential program. On one hand, this has obvious advantages for understandability and ease of design and debugging, while on the other hand, this allows the tool to explore and throttle the parallelism as desired since the input specification becomes largely independent of performance requirements. This approach also helps in verifying the final design against the input functional specification in an automated way.

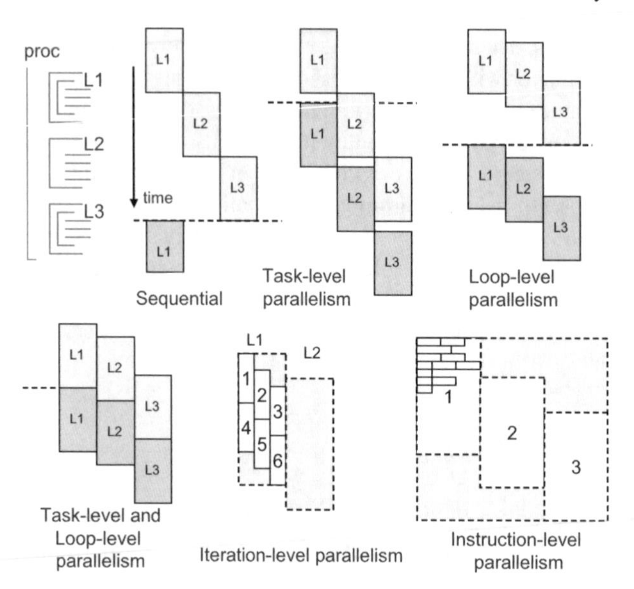

Fig. 4.6 Multiple levels of parallelism exploited by PICO

4.3.1.1 Sources of Parallelism

A sequential programming model may appear to place a severe restriction to the class of hardware one can generate or the kind of parallelism one can exploit in those hardware blocks. However, this is not actually so. A very large class of consumer data-processing applications such as those in the fields of audio, video, imaging, security, wireless, networking, etc. can be expressed as a sequential C program that process and transform arrays or streams of data. There is tremendous amount of parallelism in these applications at various levels of granularity and PICO is able to exploit them all using various techniques.

As shown in Fig. 4.6, a lot of these applications consist of a sequence of transformations expressed as multiple loop-nests encapsulated in a C procedure that is designated to become hardware. One invocation of this top level C procedure is called one *task* which processes one block of data by executing each loop-nest once. This would be followed by the next invocation of the code processing another block of data. PICO, however, converts the C procedure code to a hardware pipeline where each loop-nest executes on a different hardware block. This enables procedure level *task parallelism* to be exploited by pipelining a sequence of tasks through this system, increasing overall throughput considerably.

At the loop-nest level, PICO provides a mechanism to express streaming data that is synchronized with two-way handshake and flow control in the hardware. In the C program, this manifests itself simply as an intrinsic function call that writes data to a stream and another intrinsic function call that reads data from that stream. Streams may be used to communicate data between any pair of loop-nests as long as temporal causality between the production and the consumption of data is maintained during sequential execution. The advantage of the fully synchronized communication in hardware is that the loop-nests can be executed in parallel with local transaction level flow control which exploits *producer-consumer parallelism* at the loop level.

Within a single hardware block implementing a loop-nest, PICO exploits *iteration level parallelism*, by doing detailed dependence analysis of the iteration space and transforming the loop-nest to run multiple iterations in parallel even in the presence of tight recurrences. Subsequently, the transformed loop iteration code is scheduled using software-pipelining techniques that exploit *instruction level parallelism* in providing the shortest possible schedule while meeting the desired throughput.

4.3.1.2 The Execution Model

Given the parallelism available in consumer application at various levels, the PICO compiler attempts to exploit this parallelism without violating the sequential semantics of the application. This is accomplished by following the well-defined, parallel execution model of Kahn process networks [9], where a set of sequential processes communicate via streams with block-on-read semantics and unbounded buffering. Kahn process networks have the advantage that they provide *deterministic parallelism*, i.e., the computation done by the process network is unchanged under different scheduling of the processes. This property enables PICO to parallelize a sequential program with multiple loop-nests to a Kahn process network implemented in hardware where each loop-nest computation is performed by a corresponding hardware block that communicates with other such blocks via streams. Since the process network is derived from a sequential program, it still retains the original sequential semantics even under different parallel executions of its hardware blocks. Each hardware block, in turn, runs a statically parallelized implementation of the corresponding loop-nest that is consistent with its sequential semantics using software-pipelining techniques. In this manner, iteration level and instruction level parallelism are exploited at compile-time within each hardware block, and producer–consumer and task level parallelism are exploited dynamically across blocks without violating the original sequential semantics.

The original formulation of Kahn process networks captured infinite computation using unbounded FIFOs on each of the stream links. However, PICO is able to restrict the size of computation and buffering provided on each link by imposing additional constraints on the execution model. These constraints are described below:

- *Single-task execution*: Each process in a PICO generated process network is able to execute one complete invocation to completion without restarting. This corresponds to the single task invocation of the top level C procedure in the input specification, where each loop-nest in that procedure executes once and the procedure terminates. In actual hardware execution, multiple tasks may be overlapped in a pipelined manner depending on resource availability, but this constraint ensures that the initiation and termination states of each task are well-defined, just as they are in C.

- *Sequential semantics*: By construction, all process networks generated by PICO have sequential semantics as specified by the input C program, i.e., it is possible to execute one task by running the various processes to completion in the original C program order without deadlock, albeit using large stream buffers.
- *Self-cleaning*: It is required for the PICO input programs to execute such that no excess tokens are accumulated in any of the inter-process stream buffers between two tasks. The tool verifies this property during simulation. This property can also be stated as a constraint where the total number of tokens written to a stream during a task execution must equal the total number of tokens read out from that stream during that task. However, any residual state left within the processes from executing one task can still be carried to the next task via its internal registers and shared memories. The purpose of this constraint is to ensure that FIFO queues do not grow unbounded as more and more tasks are run through the system. However, it does not constrain the instantaneous access pattern of a stream during the course of a single task. That constraint is enforced by the following property.
- *Rate-matching*: PICO ensures that, on an average, the *rate* of production and consumption of tokens in each stream is matched at a granularity that is much smaller that the entire task. This is achieved by adjusting the rate of computation of each process by varying the throughput of the pipelined schedule running within each process. Of course, this is only a heuristic because the actual token production or consumption may be data dependent. But this analysis ensures that even during the course of a task the FIFOs can be bounded.

Given the above constraints, it is guaranteed that the size of the buffer needed on each stream link is bounded for a deadlock-free execution. Furthermore, rate matching ensures that the required buffer size is proportional to the computation of a single iteration rather than the entire loop-nest. PICO produced process networks can then be analyzed (either statically or via simulation) to determine the maximum size of a FIFO buffer needed on each stream link in order to execute in a deadlock-free manner. During execution with bounded buffers, the processes may block on write when the output link's FIFO buffer is full, but they are guaranteed to make progress eventually. PICO can also generate models that predict the overall performance of the application taking both read and write stalls into account and verify it against the desired performance specification.

4.3.2 The Architecture Template

The general structure of the hardware generated by PICO from a top level C procedure is shown in Fig. 4.7. This architectural template is called a *pipeline of processing arrays* (PPA). Each of the top level loop-nests in the C procedure is

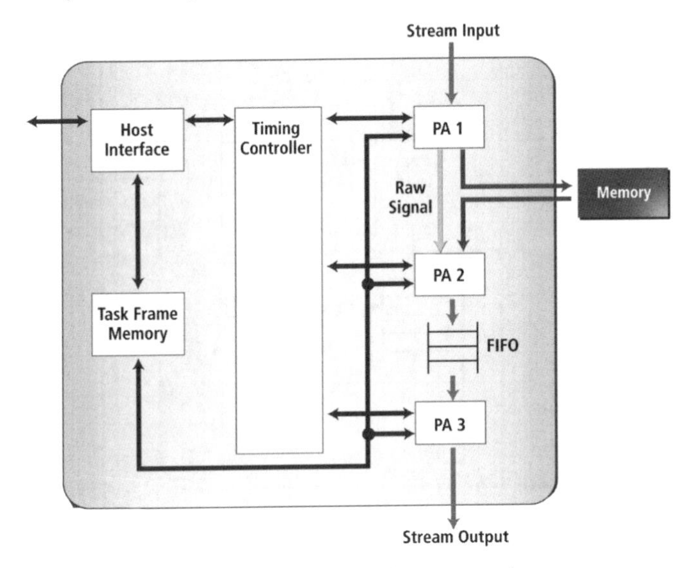

Fig. 4.7 The pipeline of processing arrays (PPA) architecture template

mapped to a hardware block or a *processing array* (PA) that communicates with other PAs via one or more streams, memories or raw signals. The communication is not restricted to a sequential pipeline – in general, it is a directed graph. However, the sequential semantics of the original C procedure is preserved by means of a *timing controller* that enforces essential control dependencies between the PAs. When the communication between two PAs is via memories or raw signals, the timing controller sequentializes the two blocks through handshaking control signals. This ensures consistent sequential semantics for shared arrays and scalar values that are produced by the first PA and consumed by the second. However, PAs that communicate via data streams are allowed to run in parallel because they can synchronize at the level of individual data elements.

The host interface and the task frame memory shown in Fig. 4.7 serve to provide smooth integration of the PPA hardware into a SoC using memory mapped IO. The host interface consists of a slave memory port that presents the local address space of the PPA at a configurable global memory address. The local address space of the PPA includes its local data memories and the task frame memory containing externally visible configuration registers. This IO mapping enables the external host processor to control and manage the PPA accelerator via an API library that provides methods to start/stop the PPA and exchange data with it via load/store transactions to the appropriate addresses.

Figure 4.8 shows the architecture template of an individual processing array. Each array is structured like a wide Very Long Instruction Word (VLIW) processor customized to execute only one program – a loop iteration. The control of each array is optimized into a simple finite-state machine generated according to the operation schedule, while its datapath is configured to implement the dataflow of the schedule using RTL components drawn from PICO's pre-characterized macrocell library.

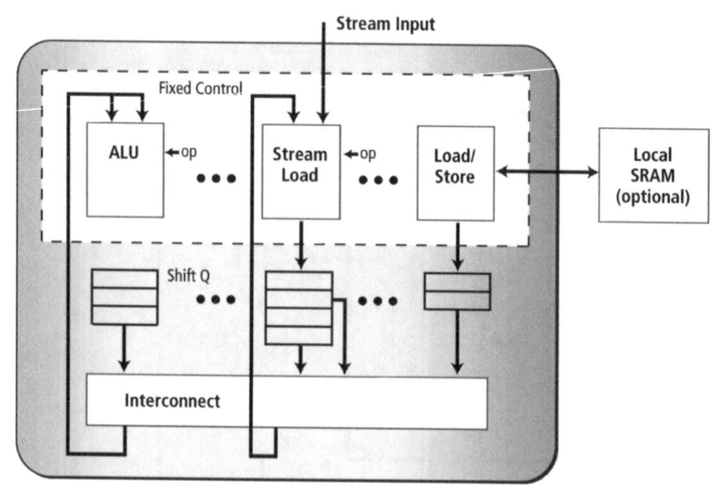

Fig. 4.8 The processing array (PA) architecture template

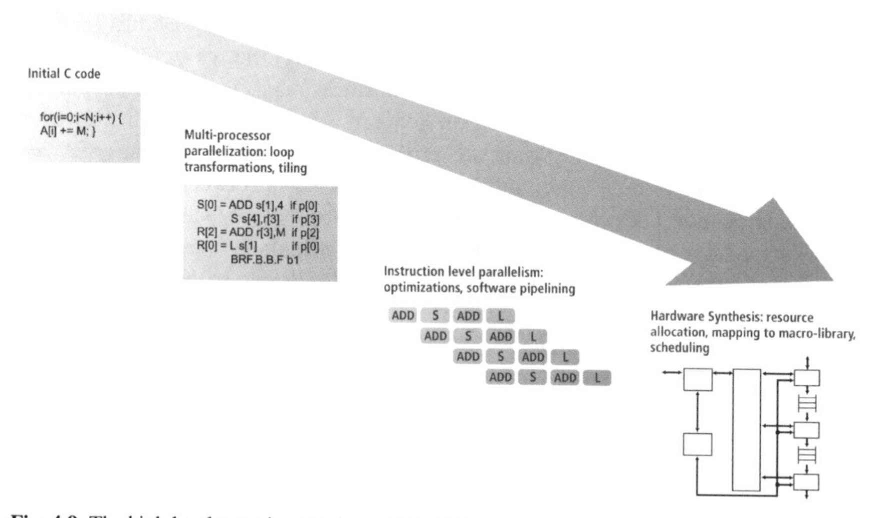

Fig. 4.9 The high level steps in mapping of C to RTL

4.3.3 C-to-RTL Mapping Process

The process of converting a top level C procedure into a PPA is a complex one with various analysis and transformation steps. The details of this process have been described in another paper [10], but we will outline the major steps here (see Fig. 4.9).

1. *Preprocessing*: The first step in the C-to-RTL conversion process is to analyze the input program for data dependences and identify which loop-nests can be parallelized and which loop-nests have sequential dependences between them. Within each loop-nest, extensive loop dependence analysis is done using the Omega library [11] framework to identify critical recurrences and loop transformations that help to alleviate those recurrences.
2. *Loop perfectization and rate matching*: The original C program may consist of multi-dimensional, non-perfect loop-nests with out-of-loop code and straight-line code between loops. The PICO compiler analyzes several candidate choices for creating perfect loop-nests out of this code that are each matched for production and consumption rates of streams and budgeted for achieving the desired overall performance based on data dependences. Out of the succeeding candidates, the one with the least estimated area is chosen for further transformations.
3. *Iteration scheduling*: The appropriate loop transformations prescribed by the previous step are carried out. The result is a single-dimension, perfect loop corresponding to each process node that can be scheduled at a steady throughput rate.
4. *Operation level optimizations*: Several classical and instruction level optimizations are applied to each loop code using PICO's extensive instruction level parallelization technology. This includes transformations to eliminate branches as well as anti- and output-dependences.
5. *Function unit allocation*: PICO allocates an optimal set of function units for the datapath of each PA based on the exact mix of the operations used in the application, the desired throughput for each type of operation and the available function units in the macrocell library that can execute those operations. In addition, several heuristics are used to further share the function units based on predicate analysis.
6. *Operation scheduling*: Given the set of data function units, the PICO compiler schedules and maps the operations of each loop into a modulo-schedule [12] with the desired throughput computed earlier. The scheduler attempts to chain operations and insert pipeline stages intelligently [13] based on actual delay characteristics of function unit macrocells for the given target technology library.
7. *Register allocation*: Following the modulo-schedule, the pipeline structure of the PA is determined and register allocation of program variables can be undertaken. PICO uses a specialized, distributed register file structure called ShiftQs [14] that share registers to store multiple variables, have custom read/write port connectivity and are optimized for register cost.
8. *Netlist generation:* The final step in the process is to generate the interconnect between the function units and the ShiftQs based on the dataflow of the operation schedule and output the resulting netlist as structural RTL.

4.3.4 Design and Verification Flow

The overall design flow for the PICO tool is shown in Fig. 4.10. The user writes application code as a top level C procedure that can be functionally verified using a

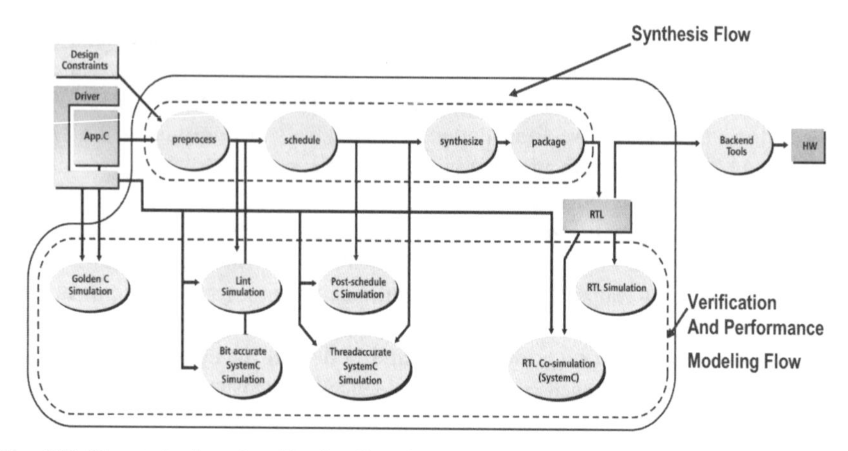

Fig. 4.10 The synthesis and verification flows in PICO

driver written in C/C++. The user also sets design constraints such as the desired throughput, target clock frequency, and the target technology library. These constraints can be supplied through a GUI or a scripted interface based on TCL. The tool then sequences through a series of design and verification phases that transform the input C specification into a RTL implementation.

The synthesis flow performs the C-to-RTL mapping steps described in Sect. 4.3.3. The *preprocess* phase performs preprocessing, the *schedule* phase performs loop perfectization, rate matching and iteration scheduling, and the *synthesis* phase performs the remaining optimization and transformation steps.

Figure 4.10 also shows that several simulation-based verification phases may be interleaved with the synthesis phases. Multiple verification phases help to catch bugs early in the design flow and build confidence in the translation process. The various simulation phases fall into two main categories – C-based simulations, and SystemC-based simulations. These are described below in the order of their occurrence in the verification flow:

1. *Golden C simulation*: This is a direct simulation of the input C code along with its driver using a standard C compiler such as gcc. The results of this phase serve as a reference for subsequent phases.
2. *Lint simulation*: This is a C-based simulation after the preprocessing phase whose purpose is to check for static and dynamic errors in the user's specification. This phase catches errors such as reading uninitialized variables, or out-of-bounds array access. Certain restrictions on the input C syntax imposed by PICO are also checked and appropriate warnings or errors are issued. Finally, this phase also checks for overflow or underflow of variable values from the bitwidth precision implied by their C data type or specified by the user using pragmas. The usual C semantics is to follow modulo arithmetic and ignore such precision violations,

but in our experience, these violations are typically a source of bugs in the RTL implementation. Therefore, it is very useful to flag them early.

3. *Bit-accurate SystemC simulation*: In the preprocessing phase, PICO also creates a fast, bit-accurate, single-threaded SystemC model of the hardware block for the purpose of validating its external interfaces in the system context. This is also known as the programmer's view (PV). As the name suggests, the bit-accurate model models the protocol and bitwidth precision of all external interfaces precisely at the transaction level. Interfaces such as external memories, streams and raw signals are modeled as individual channels. This model can be easily integrated into user's system validation platform for firmware testing long before the hardware is available.
4. *Post-schedule C simulation*: As the name suggests, this C-based simulation is carried out immediately after the schedule phase. Aside from serving to verify the validity of the transformations performed during the scheduling phase, this phase also generates input/output functional test-vectors for each PPA task invocation that are used later for RTL verification.
5. *Thread-accurate SystemC simulation*: After the schedule phase, PICO generates a multi-threaded SystemC model that models the process network view of the PPA hardware accurately. This is also known as the timed programmer's view (PV+T). Each loop-nest representing the computation of a PA is encapsulated in its own SystemC thread and the communication between them is setup using stream, memory, or register channels. Each such thread executes as a Kahn process at an average computation rate prescribed by the rate matching phase. The thread-accurate model can be used as a fast resource and performance prediction model for architectural exploration because it models all hardware resources including computation process nodes, memories and stream buffers precisely. Although the model is not meant to be cycle-accurate so that it can execute quickly (usually 10–100$\times$ over RTL), each thread is synchronized to execute in line with its prescribed rate of computation. As such, this model provides an excellent insight into the achievable task and loop level parallelism and the stream and memory resource contention in the final hardware before the actual hardware is available.
6. *RTL simulation*: PICO automatically generates a RTL test-bench along with the hardware which can run the functional test-vectors collected during the Post-scheduling C simulation and check them for correct execution. Environmental variations such as stream handshake delays and abort conditions can also be added with appropriate randomization in order to test the robustness of the design.
7. *RTL co-simulation*: The final test of the generated RTL is to test it within the context of the original driver code using SystemC. PICO automatically generates a set of transactors that translate the transaction level interactions to and from the driver code to signal level interface protocol accepted by the PPA RTL. This co-simulation tests the PPA interfaces for correctness and provides a view of the overall performance of the PPA design in the context of its deployment.

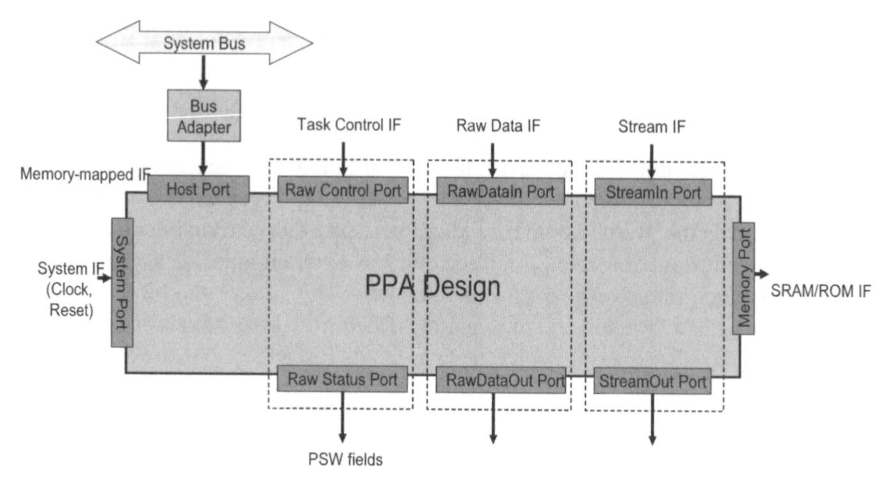

Fig. 4.11 Standard interfaces for PICO designed IP

4.3.5 Reusable Interfaces

A key benefit of using a configurable architectural template for RTL design is that PICO can restrict the generated designs to conform to a small set of interfaces that are typical of those used in today's SoC designs. As shown in Fig. 4.11, these interfaces are currently provided for a memory mapped host processor access channel (the PPA is a slave on this bus), streaming data input and output, and local SRAM/ROM. Additional raw control and data interfaces may also be configured when direct hardware control is desired.

The compiler understands the semantics of these interfaces at the transaction level and the translation from the C operations to transactions on the appropriate interfaces. This enables the designer to program at the algorithmic level, that is, the input specification remains independent of the exact RTL implementation and signal/cycle level behavior so that it may be reused and retargeted easily to different cost-performance points.

The interface RTL is designed by configuring a parameterized interface template which is pre-verified across the range of parameterization. The combination of these ensures that there will be no protocol errors in the RTL operation. The protocols used should be familiar to SoC designers. They are easy to adapt to other flavors of these interfaces or more complex buses. There are many advantages to using these common interfaces:

- Designers need not spend time redesigning interface protocols, or verifying them; any adapter to PICO interfaces only needs to be designed and verified once.
- Opportunities for miscommunication in interfacing between blocks are substantially reduced.

- As each interface provides a simple abstraction to the transaction level, PICO can deliver an automated verification flow where transactions in the C code are mapped to signal level transfers in the RTL test bench.
- PICO can easily be extended to adapt its native interfaces to any interface protocol by creating a protocol adapter as a piece of IP, without the designer needed to specify the protocol in C.
- Each block designed by PICO will have a common "look and feel" even though it is highly customized and very efficiently designed; this will reduce effort and time in design, verification, and integration.

By standardizing and pre-verifying interface designs that historically result in many defects, the PICO strategy of reusable interfaces supports quality and reusability best practices for SoC design.

4.4 Conclusion

PICO takes a novel approach to hardware design from algorithmic specifications, dramatically improving the cost of design, verification and integration of RTL blocks into SoC platforms. The approach relies on aggressive compiler technology, a parallel execution model based on Kahn process networks, and a carefully designed architecture template that is cost-efficient, provides high performance, and is sensitive to circuit level and system level design constraints.

From a user's perspective, PICO translates a familiar, easy-to-use, general-purpose programming model to a robust and aggressively parallel hardware execution model that can be easily explored at the architectural level to generate a range of cost, performance and power tradeoff points from the same software IP. The user provides design constraints and architectural guidance to the compiler at a high level, while the compiler exploits parallelism in the application automatically to meet those constraints and hides the details of hardware mapping, scheduling and synchronization.

Finally, the PICO tool addresses the complete hardware design flow including architecture exploration, RTL design, RTL verification, system validation and system integration. We believe that this degree of flow integration is essential for the success of a high level synthesis tool. For a large class of modern embedded applications designed using PICO, this approach has been shown to yield extremely competitive designs at a fraction of the resources used traditionally thereby closing the proverbial design productivity gap.

Acknowledgments PICO started as a research project at Hewlett-Packard Laboratories in the mid-1990s under the leadership of (late) Dr. Bob Rau who has done pioneering work in the field of VLIW computing. A lot of the ideas and the vision of applying that body of work to embedded system design can be traced back to him. We are also indebted to the founding team of Synfora including Craig Gleason, Darren Cronquist, Mukund Sivaraman, Dave Goldberg, and Jeff Benis as well as former colleagues at HP Labs including Mike Schlansker and Rob Schreiber for contributing towards the PICO technology in very significant ways. Finally, we would like to thank the entire Synfora team for making the PICO vision a reality and the PICO tool successful in the EDA marketplace.

References

1. R. Camposano, From behavior to structure: high-level synthesis, *IEEE Design & Test*, **7**(5): 8–19, 1990.
2. W. Savage, J. Chilton, R. Camposano, IP reuse in system on a chip era, in: *Proc. Intl. Symp. System Synthesis*, 2000, pp. 2–7.
3. Open Core Protocol Specification, Release 2.2, OCP International Partnership Association, Inc., http://www.ocpip.org.
4. V. Kathail, C. Gleason, S. Mahlke, M. Kudlur, K. Fan, Automated architecture synthesis from C algorithms, Tutorial in: *Intl. Conf. on Compilers, Architectures and Synthesis for Embedded Systems (CASES)*, Seoul, S. Korea, October 2006.
5. SystemC Language Reference Manual, IEEE Std. 1666^{TM}-2005. http://www.systemc.org.
6. V. Kathail, S. Aditya, R. Schreiber, B. R. Rau, D. C. Cronquist, M. Sivaraman, PICO: Automatically Designing Custom Computers, *IEEE Computer*, **35**(9):39–47, 2002.
7. N. Chawla, R. Guizzetti, Y. Meroth, A. Deleule, V. Gupta, V. Kathail, P. Urard, Multimedia application specific engine design using high level synthesis, *Proc. DesignCon 2008*, Santa Clara, California, February 2008.
8. M. Fillinger, P. Thiruchelvum, Using PICO Express to Reduce Design Time for Complex Application Engines, in: *10th Sophia Antipolis MicroElectronics Forum (SAME)*, Sophia-Antipolis, France, October 2007.
9. G. Kahn, The semantics of a simple language for parallel programming, in: *Proc. IFIP Congress 74, Information Processing*, Stockholm, Sweden, 1974, pp. 471–475.
10. R. Schreiber, S. Aditya, S. Mahlke, V. Kathail, B. R. Rau, D. Cronquist, M. Sivaraman, PICO-NPA: High-level synthesis of non-programmable hardware accelerators, *Journal of VLSI Signal Processing*, **31**:127–142, 2002.
11. The Omega Project. University of Maryland, http://www.cs.umd.edu/projects/omega.
12. B. R. Rau, Iterative modulo scheduling, *International Journal of Parallel Processing*, **24**:3–64, 1996.
13. M. Sivaraman, S. Aditya, Cycle-time aware architecture synthesis of custom hardware accelerators, in: *Proc. Intl. Conf. on Compilers, Architectures and Synthesis for Embedded Systems (CASES)*, 2002. pp. 35–42.
14. S. Aditya, M. S. Schlansker, ShiftQ: A buffered interconnect for custom loop accelerators, in: *Proc. Intl. Conf. on Compilers, Architectures and Synthesis for Embedded Systems (CASES)*, 2001, pp. 158–167.

Chapter 5
High-Level SystemC Synthesis with Forte's Cynthesizer

Michael Meredith

Abstract This chapter will describe the SystemC-based design style and verification approach used with the Cynthesizer high-level synthesis product from Forte Design Systems. It will outline the SystemC and C++ constructs that are supported for synthesis and cover C++ techniques used for creating modular, reusable interfaces. Techniques for specifying constraints and controlling the high-level synthesis results will be described.

Keywords: SystemC, Synthesizable constructs, Modular interfaces, Sockets, Scheduling, Loop transformations, Verification

5.1 Introduction

Forte uses standard SystemC semantics along with a set of tool-control directives, a target clock period and a process technology library as input to its Cynthesizer high-level synthesis tool. Using SystemC allows the synthesis of a broad range of functionality at multiple levels of abstraction due to the range of hardware functionality that can be expressed in SystemC.

By using the high-level language constructs of C++ along with the hardware constructs of SystemC, hardware designers using Cynthesizer are able to design at a level of abstraction higher than RTL, and maximize their ability to reuse the source code that is the product of their engineering effort.

The SystemC and C++ constructs that are used to make this possible are described below, and a number of constructs that cannot be used for synthesis are identified.

P. Coussy and A. Morawiec (eds.) *High-Level Synthesis.*

5.2 C++ Support

Because SystemC is a class library implemented in C++, the advantages of high-level C++ constructs are available to hardware designers working in SystemC. Cynthesizer supports a large number of these constructs but, just as there are SystemVerilog constructs that are only intended for verification, there are C++ constructs that are only appropriate for modeling and testbench construction, not for synthesis.

5.2.1 Synthesizable High-Level C++ Constructs

The C++ constructs that are within the synthesizable subset can be used in ways that give SystemC synthesis advantages unattainable in any other hardware design language.

- *Encapsulation*: C++ classes can be used in SystemC synthesis to manage the complexity inherent in hardware design.

  Algorithmic functionality can be captured in a class for reuse. Functions providing a public API for use of the algorithm can be made externally available using the C++ "public" access control. Internal computation functions and storage of internal state needed by the algorithm can be made private.

  Interface functionality can be encapsulated as discussed earlier creating a modular, reusable interface. Modular interfaces expose a transaction-level function call interface to the designer which allows them to be used without requiring the designer to be expert in the details of the interface protocol.
- *Construction of custom data types*: Operator overloading is a C++ technique whereby a class can provide a custom implementation for such built-in operators as "*" (multiply) and "+" (add). This allows the construction of user defined datatypes such as for complex arithmetic, and matrix arithmetic. Arithmetic operations can be performed on these datatypes using conventional C++ syntax, e.g., a = b + c; which promotes ease-of-use and improves a reader's ability to understand the code.
- *Development of configurable IP*: C++ provides template specialization as a way to write a single body of code which can represent a wide range of behaviors depending on the specific template parameters selected. As a simple example, template specialization can be used to build a filter class that can operate on any given datatype, including user-defined custom datatypes.

  A more sophisticated example is the cynw_float parameterized floating-point class that Forte has developed. It allows the user to specify template parameters to choose the exponent and mantissa widths as well and configure options such as normalization and rounding behaviors.

Supported C++ Constructs	
Arithmetic operators	Integer data types
Logical operators	Structures
References	Classes
Statically-determinable pointers	Inheritance
if-else statements	Operator overloading
switch-case statements	Inferring memories from arrays
do, while, and for loops	Inferring registers from arrays
break and continue statements	Template classes and functions
	Template specialization

5.2.2 Non-Synthesizable C++ Constructs

One characteristic of the synthesis process is that it uses the source code of the high-level design without access to information that can only be determined at simulation time. In other words, the synthesis process can only take advantage of language features that can be resolved statically and information that can be determined by inspection of the source code. This is the source of most of the restrictions on C++ constructs that can be used:

- *Pointer arithmetic*: In the processor-based execution environment in which the C and C++ languages were originally envisioned, all variables, structures, arrays and other storage elements are defined to exist within a single uniform address space.

  A hardware implementation may include multiple separate memories of different kinds as well as storage elements implemented directly with flip-flops. Clearly, in this environment making decisions based on the value of the address of a variable is meaningless. Consequently, pointer arithmetic is not supported for SystemC synthesis.
- *Pointer dereferencing*: Similarly, accessing a specific storage element by its address assumes a processor-based execution environment. Therefore, in general, passing pointers and dereferencing them is not supported for SystemC synthesis. Nevertheless, under some circumstances the target of the pointer can be unambiguously determined by a static analysis of the source code. For instance, if the address of an array is passed directly to a subroutine it is usually possible to statically determine the identity of the array in question. In such cases the use of the pointer will be supported by synthesis.

- *Dynamic memory allocation*: For reasons similar to those limiting the use of pointers, allocation of storage elements using malloc(), free(), new, and delete is not supported for SystemC synthesis. One notable exception is that allocation of sub-modules using new is supported.
- *Virtual functions*: Virtual functions select the behavior of a particular object based upon run-time determination of its class identity. Since this cannot, in general, be determined statically, use of virtual functions is not supported for SystemC synthesis.

5.3 Synthesizable Module Structure

Synthesizable SC_MODULES can include multiple SC_CTHREAD processes, and multiple SC_METHOD processes. In addition they can include submodules along with signals and channels to provide internal interconnect. Because SC_MODULES are C++ classes, they can also include data members of any synthesizable data type to provided internal state, and member functions that can be used by the processes to implement the required behavior.

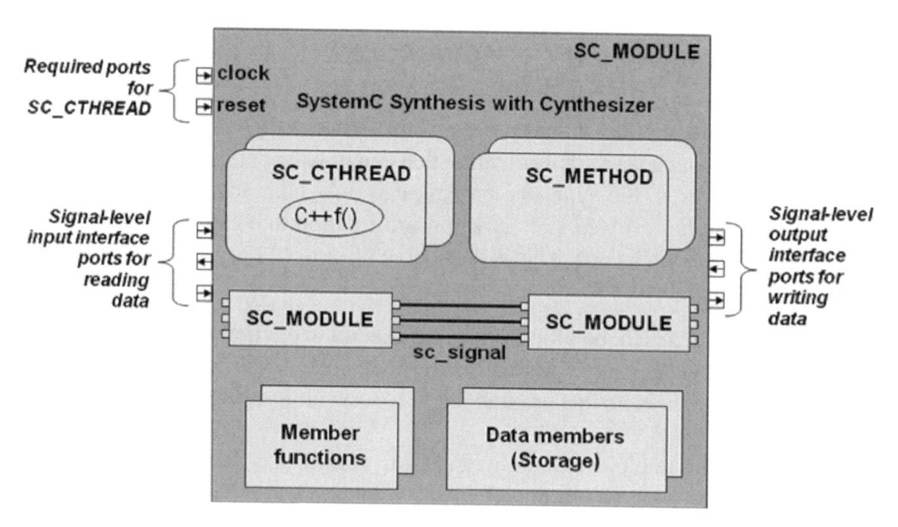

5.4 Concurrent Processes

Among the required hardware semantics provided by SystemC are process constructs that allow a designer to directly express concurrent behaviors. Two of these process constructs, SC_CTHREAD, and SC_METHOD, are appropriate for synthesis to hardware and are supported by Cynthesizer. A module may contain any combination of these. Use of multiple SC_CTHREADs and/or SC_METHODs in a single module is fully supported.

This allows SystemC synthesis using Cynthesizer to encompass the areas traditionally considered separately as the behavioral level and the register-transfer level. Using Cynthesizer, an engineer can combine high-level behavioral design with low-level register-transfer level design.

Ordinarily, an engineer who wants to do a pure RTL design will choose a conventional HDL such as SystemVerilog or VHDL. SystemC is more often used when high-level synthesis is needed for a substantial part of the design. Typically a complex algorithm or a complex control structure is defined using an SC_CTHREAD, or multiple concurrently executing SC_CTHREADs. These are combined with SC_METHODS for implementation of small parts of the design that can be better specified at a low level. Examples of these low-level parts of the design might include the clock boundary crossing logic or an asynchronous bypass path.

5.4.1 SC_CTHREAD Processes

The SC_CTHREAD construct implements a clocked process. The declaration of the process includes specification of a signal that will be used as the clock for the process. The semantics of SC_CTHREAD guarantee that the behavior of the process will be synchronized to this clock.

In addition the reset_signal_is() function specifies a signal that will be used to reset the process. Whenever the reset signal is asserted, the process is restarted in its initial state. This allows explicit initialization behaviors to be written that determine the state of the flip-flops of the design when it comes out of reset. During simulation, within the body of the subroutine that is the behavior of the SC_CTHREAD process, execution proceeds sequentially until the process hits a wait() statement upon which the process is suspended until the next clock cycle.

These characteristics make SC_CTHREAD ideal for high-level synthesis of abstract, untimed behaviors combined with detailed cycle-accurate, pin-level interfaces.

Synthesizer interprets all behavior in an SC_CTHREAD process that occurs before the first wait() statement as reset conditions. Synthesis requires that this reset code be schedulable in a single clock cycle.

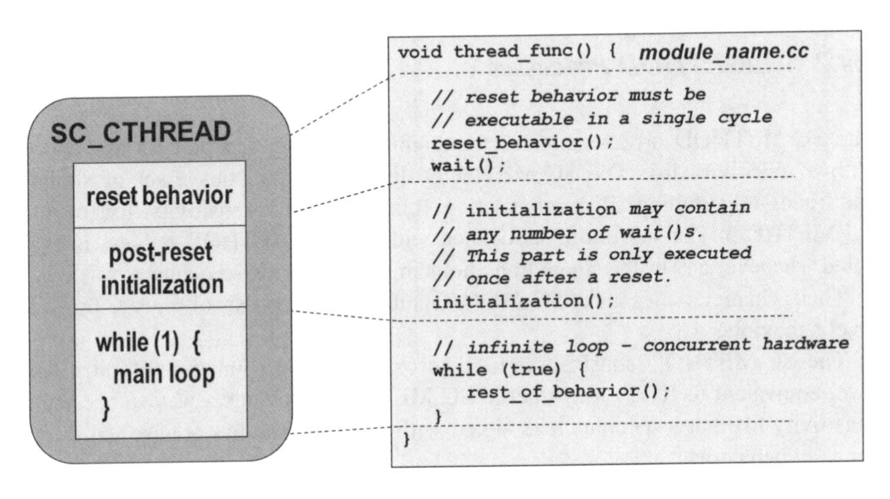

```
SC_MODULE(sub)
{
  // ports
  sc_in_clk clk;
  sc_in<bool> rst;
. . .
  SC_CTOR(sub)
  {
    SC_CTHREAD( thread_func, clk.pos() );
    reset_signal_is( rst, 1 );
  }
  void thread_func()
  {
    // reset behavior
. . .
    wait();

    while(1)
    {
. . .
    }
  }
};
```

The "SC_CTHREAD" statement associates the thread_func() function with the positive edge of the signal clk. Cynthesizer implements such a thread as a circuit synchronous to that clock edge.

The "reset_signal_is" statement makes the "1" level of the rst signal reset the thread.

5.4.2 SC_METHOD Processes

The SC_METHOD process construct implements a triggered process associated with a sensitivity list. The SC_METHOD declaration includes a set of signals and rising-edge/falling-edge information that define the sensitivity list of the SC_METHOD. The subroutine associated with the SC_METHOD process is executed whenever any of the signal transitions in its sensitivity list occurs.

These characteristics make SC_METHOD ideal for synthesis of register-transfer level behaviors.

The SC_METHOD construct is used to express design functionality at a low level equivalent to RTL for synthesis. SC_METHOD provides a way to specify a sensitivity list that a specific clock signal with a thread, and has precise semantics for reset behavior.

Cynthesizer can be used to synthesize synchronous SC_METHODS using static sensitivity to a clock as follows.

```
SC_CTOR(sync)
{
  CYN_DEFAULT_INPUT_DELAY(.1,"delay");
  SC_METHOD( sync_method );
  sensitive_pos( clk );
  dont_initialize();
}
```

Asynchronous SC_METHODS using static sensitivity to a number of inputs can also be synthesized as follows.

```
SC_CTOR(async)
{
  CYN_DEFAULT_INPUT_DELAY(.1,"delay");
  SC_METHOD( async_method );
  sensitive << input_1 << input_2;
  dont_initialize();
}
```

SystemC semantics for SC_METHOD also provide for dynamic sensitivity using the next_trigger() function. Dynamic sensitivity is not supported for synthesis.

SystemC semantics for SC_METHOD also provide that each SC_METHOD will be executed once at the beginning of simulation. This is meaningless in the context of synthesis, so disabling this behavior using the dont_initialize() function is recommended.

5.5 Modular Interfaces

In addition to simple signals carrying single-bit or scalar values, designers using Cynthesizer can implement high-level channels for communication. By encapsulating the low-level signals, ports, and protocol functions in modular interface socket classes, the designer is relieved of the tedious connection of individual signals, and can connect an entire interface, such as a connection to a memory, with a single binding function. In addition, the modular interface code can be thoroughly tested once, and then reused many times without modification, avoiding numerous common errors and reducing debug time.

These modular interfaces consist of C++ classes containing synthesizable SystemC code using constructs such as signals, ports, and synthesizable protocol code. Common interfaces are provided by Forte. Interfaces conforming to specific corporate standards can be written in SystemC by a corporate CAD department, and project-specific interfaces can be written by any engineer.

The abstraction and modularity capabilities of C++ and SystemC offer a unique advantage for high-level hardware design when they are used in this way to encapsulate interfaces for ease-of-use and for reuse.

The key technique is to use the C++ class mechanism to encapsulate the signal-level connections (i.e., ports) along with the code that implements the signal-level protocol.

In general, there are two complementary interfaces (e.g., input and output) that are implemented as two modular interface "socket" classes. These are connected by binding calls to a modular interface "channel" class. The processes in the modules containing the sockets call transaction-level interface functions defined in the socket classes to execute their interface behaviors.

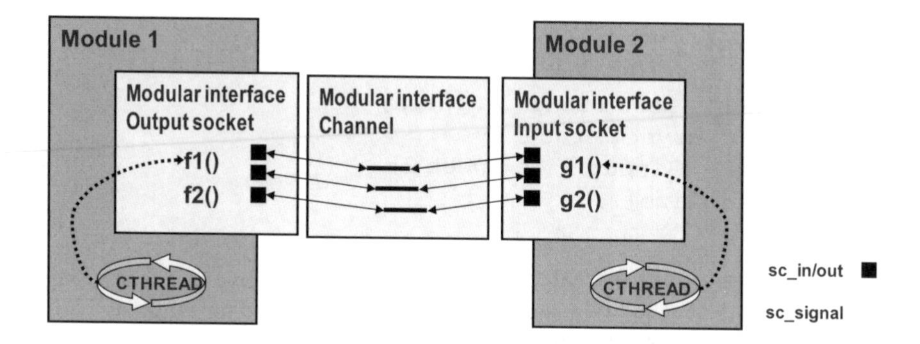

5.5.1 Modular Output Socket

In its simplest form, an output socket for a ready/valid handshake interface might look like the following.

```
// Output socket.
template <class T>
class RV_out
{
public:
sc_in<bool> rdy;
  sc_out<bool> vld;
  sc_out<T> data;

  RV_out( const char* name=0 )
  {}

  // reset function called from SC_CTHREAD
  // establishes initial conditions
  void reset()
```

```
  {
    vld = 0;
    data = 0;
  }

  // put function called from SC_CTHREAD
  // executes output protocol
  void put( const T& val )
  {
    do { wait(); } while ( !rdy.read() );
    data.write( val );
    vld = 1;
    wait();
    vld = 0;
  }
};
```

Note that the sc_in/sc_out ports are incorporated into the modular interface socket as data members. The two transactions that the port implements, reset() and put(), are also implemented as member functions.

5.5.2 Modular Input Socket

The corresponding input socket implements the reciprocal protocol. Note that the direction of the ports is reversed from that of the output socket.

```
// Output socket
template <class T>
class RV_in
{
  public:
    RV_in( const char* name=0 )
    {}

    sc_out<bool> rdy;
    sc_in<bool> vld;
    sc_in<T> data;

    //
    // Protocol transaction functions
    //
    void reset()
    {
        rdy = 0;
    }
```

```
    T get()
    {
        wait();
        rdy = 1;
        do { wait(); } while ( !vld.read() );
        rdy = 0;
        return data.read();
    }
};
```

5.5.3 Use of Modular Interfaces

The modular interface socket can be used in a design in a way that is similar to how a simple sc_in or sc_out port would be used. The instantiation and binding of the socket look just like an sc_in or sc_out port. To execute the protocol, the SC_CTHREAD calls the transaction functions of the modular interface socket as follows.

```
SC_MODULE(sub)
{
  sc_in_clk clk;
  sc_in<bool> rst;
  RV_in< sc_uint<8> > din;
  RV_out< sc_uint<8> > dout;

  SC_CTOR(sub)
  {
    SC_CTHREAD( thread_func, clk.pos() );
    reset_signal_is( rst, 1 );
  }
  void thread_func()
  {
    // reset behavior
    din.reset();
    dout.reset();
    wait();

    while (1)
    {
      sc_uint<8> d = din.get();
      dout.put( d+1 );
    }
  }
};
```

5.5.4 Channel

The signals that are needed to provide connectivity for this interface can also be encapsulated in a channel class as follows.

```
// Channel class
template <class T>
class RV
{
  public:
    sc_signal<bool> rdy;
    sc_signal<bool> vld;
    sc_signal<T> data;
};
```

5.5.5 Binding

The addition of a couple of binding functions to the modular interface socket allows the entire interface to be bound using a single function call. This reduces the number of lines of code needed to use an interface, allows interchange of different interfaces with minimal code modification, and prevents trivial errors due to misspelling and misconnecting individual signals. For our example the binding functions in the output port are as follows.

```
// Output socket.
template <class T>
class RV_out
{
. . .

  //
  // Binding functions.
  //
  template <class C>
  void bind( C& c )
  {
    rdy(c.rdy);
    vld(c.vld);
    data(c.data);
  }

  template <class C>
  void operator() ( C& c )
```

```
    {
      bind(c);
    }
};
```

Note that the addition of these functions allows the binding to be done using the conventional SystemC port binding syntax:

```
socket.bind(channel);
```

or

```
socket(channel);
```

Also note that the binding functions are defined as templates. This lets the same ports and binding functions to be used for port-to-port binding in a hierarchical design.

5.6 Structural Hierarchy

In addition to the process control constructs, SystemC synthesis supports the SystemC constructs for construction of structural hierarchies. An engineering team can attack a large design problem using structural decomposition, breaking the problem down into multiple smaller modules that communicate through user-defined interfaces. Individual sub-modules can be assigned to different team members if desired supporting a conventional team structure and concurrent design approach. Each module can contain any number of cooperating SC_CTHREADs, SC_METHODs, and sub-modules. Communication between modules is achieved using a port-to-signal binding mechanism of a kind that is familiar to RTL designers, or even designers using schematics.

Here is an example of a hierarchical design using modular interfaces as described previously.

```
SC_MODULE(parent)
{
  // ports
  sc_in_clk clk;
  sc_in<bool> rst;
  RV_in< sc_uint<8> > din;
  RV_out< sc_uint<8> > dout;

  // submodules
  sub_module m_sub1;
  sub_module m_sub2;
```

```
  // signals and channels
  RV< sc_uint<8> > chan;

  SC_CTOR(parent)
    : m_sub1("sub1"),
      m_sub2("sub2"),
      chan("chan")
  {
    // bind first module using bind() function
    m_sub1.clk.bind(clk);
    m_sub1.rst.bind(rst);
    m_sub1.din.bind(din);    // socket-to-socket
    m_sub1.dout.bind(chan); // socket-to-channel

    // bind second module using socket() syntax
    m_sub2.clk(clk);
    m_sub2.rst(rst);
    m_sub2.din(chan);
    m_sub2.dout(dout);
  }
};
```

This use of SystemC constructs rather than tool constructs for implementation of hierarchy and communication improves the overall verification process dramatically. The complete structural hierarchy can be simulated at a behavioral level, accurately representing the concurrency of all the modules and threads, and accurately verifying the pin-level communication protocols between them. This allows the functional verification to be performed using high-speed behavioral simulation, and eliminates the need for many slow RTL simulations.

5.7 Creating RTL with Predictable Timing Closure

One of the challenges in RTL design is to ensure that the RTL you have written will have successful timing closure through logic synthesis at the specified clock rate when implemented in the chosen process technology. High-level synthesis has to meet the same challenge to be practical for wide deployment.

Cynthesizer achieves this by combining a number of steps. First, the timing information about the cells in the target process technology library are used as an input to the high-level synthesis process. This information is read in a Liberty format .lib file provided by the chosen foundry.

Second, Cynthesizer has advanced datapath optimization technology that it uses to build a library of gate-level functional units such as adders, multipliers, multiplexors, etc based on the cells available in the target technology .lib file. These functional units are optimized for a specific clock frequency, and may be implemented in a pipelined manner, where each pipeline stage is designed to fit within the designated clock period.

Functional unit library compilation is performed in advance of high-level synthesis once per process technology and clock period to speed the synthesis process. All the tools needed for library compilation to be performed by the user are included with Cynthesizer. No additional tool needs to be purchased.

Cynthesizer also creates custom functional units as needed during high-level synthesis. These include non-square parts (i.e., a 12-bit by 3-bit adder) as well as parts to implement more complex expressions. Cynthesizer automatically identifies useful expressions in the algorithm of the design (such as "a + (b * c) – 3)" and builds gate-level parts on the fly that implement them.

Third, Cynthesizer uses this detailed timing information when it schedules the operations of the algorithm to ensure that no combinatorial path in use exceeds the clock period. Additional user controls are available to allow the user to adjust the "aggressiveness" with which Cynthesizer fills each clock period with logic. These controls can be used to make downstream timing closure even easier, thereby reducing processing time in downstream tools such as logic synthesis.

Cynthesizer produces RTL produced that has a structural character. Adders, multipliers, multiplexors, etc are instantiated with a finite state machine determining what values are presented to each instantiated part in each clock cycle. This ensures that the timing assumptions made during high-level synthesis are maintained during logic synthesis.

5.8 Scheduling

It has been noted that a primary benefit of using behavioral synthesis is the ability to write clocked processes whose functionality takes more than one clock cycle. This gives the user the ability to control the latency and throughput of the resulting circuit without performing detailed resource assignment and scheduling by hand.

At the same time, I/O activity at the ports of the module being synthesized must conform to a specified protocol in order to have the synthesized block interoperate with other blocks. The protocol mandates that certain relationships between I/O events must be held constant. For instance, the data value must be presented on the data bus in the same cycle as the data_valid line is driven to true.

5.8.1 Mixed-Mode Scheduling

Cynthesizer addresses these requirements by providing a number of directives that give the user high-level control of its scheduling. The Cynthesizer scheduler allows different code blocks in a single SC_CTHREAD to be scheduled differently according the user requirements. A "code block" is defined as any section of C++ code delimited by "{" and "}." Thus it can be a clause of an if-else statement, the body of a loop, or any other set of statements that the user chooses to group together.

Note that while the protocol can be written in-line as it is shown here, protocols are typically encapsulated into modular interface classes for ease-of-use and for ease-of-reuse.

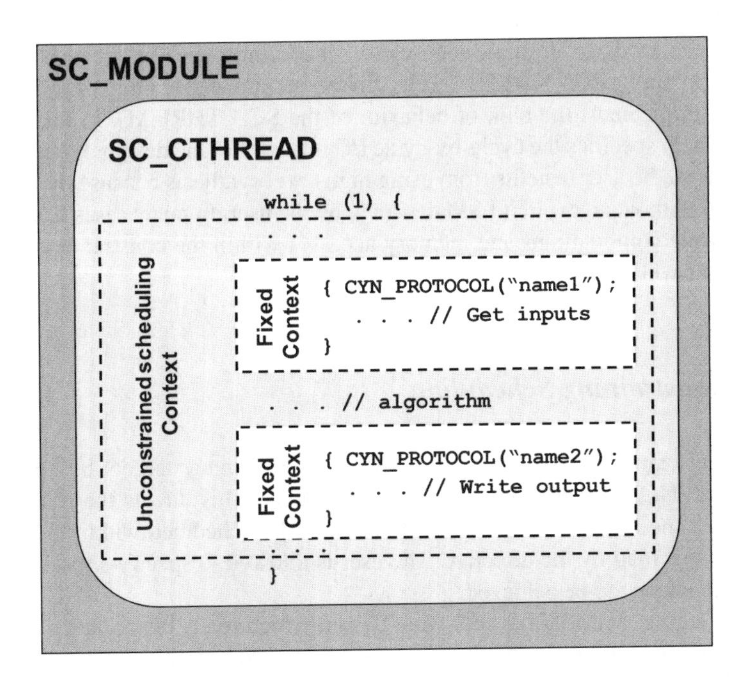

5.8.2 *Unconstrained Scheduling*

To begin with, it is assumed that all the code in the design, unless otherwise identified, is completely untimed, and that the scheduler of the high-level synthesis process has complete freedom to implement the functionality in as many or as few clock cycles as it chooses. No guarantees of any cycle-by-cycle timing are made in this unconstrained code, although the order of operations determined by the dependency relationships within the code is maintained.

By default, without any scheduling constraints, Cynthesizer will optimize for area, taking as many cycles as necessary to complete the computation with a minimal set of functional units.

5.8.3 Scheduling for Protocol Accuracy

In order to give the user maximum control of cycle-by-cycle timing for implementing protocols, Cynthesizer allows the specification of cycle-accurate blocks of code by the use of the CYN_PROTOCOL directive. This directive, associated with a particular code block directs Cynthesizer not to insert any clock cycles within that code block except for those specified by the user with wait() statements. Within these protocol blocks, scheduling ensures that the ordering of port and signal I/O and the given wait()s is held constant.

For some kinds of designs, such close scheduling control is needed that it is desirable to apply a CYN_PROTOCOL directive to the entire body of the while(1) loop that implements the bulk of behavior of the SC_CTHREAD. In this case the user precisely specifies the cycle-by-cycle I/O behavior of the design. Even with this tight control, the user benefits from using high-level synthesis because the design is expressed without an explicit FSM designed by the user. In many cases Cynthesizer can schedule computations and memory accesses within the constraints of the I/O schedule as well.

5.8.4 Constraining Scheduling

Scheduling can be constrained to achieve specific latency targets by applying a CYN_LATENCY directive to a specific code block. This directs the scheduler to ensure that the behavior of the given block is to be scheduled within the number of cycles specified by the directive. The user is allowed to specify a minimum and maximum latency to be achieved.

For example, consider the following design which reads in six data values and outputs a computed result. The data is expressed as a structure:

```
struct data_struct;
{
  sc_uint<8> A;
  sc_uint<8> B;
  sc_uint<8> C;
  sc_uint<8> D;
  sc_uint<8> E;
  sc_uint<8> F;
  sc_uint<8> G;
}
```

The module has a modular interface input port and a modular output port:

```
RV_IN<data_struct> in_port;
RV_OUT< sc_uint<28> > out_port;
```

The main while loop of the SC_CTHREAD is:

```
while( true )
{
  sc_uint<28> X;

  // read the data from the input port
  struct data_struct data = in_port.get();

  {
  // do the computation in 4 cycles
  CYN_LATENCY( 4, 4, "algorithm_latency" );
  X = ( A + B + C ) * ( D + E + F ) * G;
  }

  // write the result to the output port
  out_port.put(X);
}
```

This can be implemented by Cynthesizer using two adders and one multiplier to perform this computation in the specified four cycles using the following schedule. This produces an overall throughput of one value per six cycles.

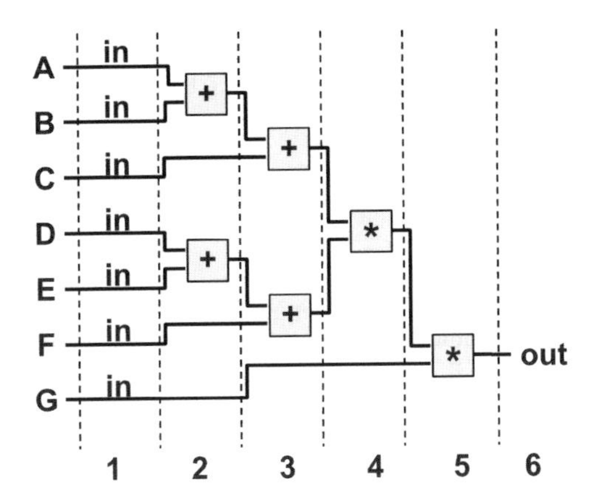

If, on the other hand a slower circuit were acceptable, a 6-cycle latency for the computation (resulting in an overall throughput of one value per eight cycles) could be achieved by specifying:

```
CYN_LATENCY( 6, 6, "algorithm_latency" );
```

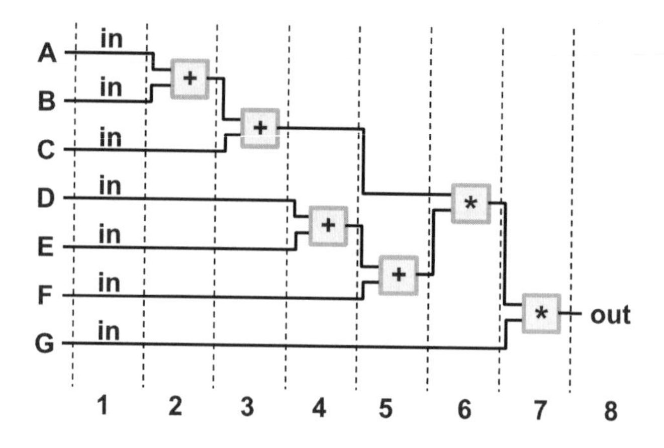

Cynthesizer could achieve this with the following schedule.

Note that Cynthesizer would automatically produce a new FSM and datapath to meet the desired latency without the user rewriting the algorithm.

Also note that this example is extremely simplified. In reality, more than one operation will often be chained within a single clock cycle depending on the relationships between the required latency, the clock period, the propagation delay through the adders and multipliers and their relative sizes. For instance, if the clock cycle were long enough, and the target process technology were fast enough the design could be scheduled in a single cycle using four adders and two multipliers.

```
CYN_LATENCY( 1, 1, "algorithm_latency" );
```

5.9 Loops

Unlike RTL, where loops are seldom used, looping constructs are common in high-level design. These include loops with non-constant bounds, where the loop termination condition depends on the state of the design and the input data, as well as simple for-loops with constant bounds.

5.9.1 Supported Loop and Loop Termination Statements

Cynthesizer supports loops of all forms in the SystemC input code. All the C++ loop statements may be used:

- "for" loops
- "while" loops
- "do/while" loops

The "continue" and "break" statements may be freely used for loop termination if desired.

5.9.2 Directives for Loop Control

Loops can be handled in three ways depending on the parallelism desired by the user.

5.9.3 Default Loop Implementation

The default behavior is for Cynthesizer to implement a loop as a looping structure in the finite-state machine that is built in the synthesized RTL. In this case there will be at least one cycle per iteration of the loop. This will introduce the minimum parallelism with the one instance of the needed hardware being used over and over for each iteration of the loop.

5.9.4 Unrolling

Unrolling a loop creates additional copies of the hardware that implements the loop body. These copies can operate in parallel, performing the computation of several iterations of the loop at the same time.

Loop unrolling is controlled using the CYN_UNROLL directive. The simplest form of the directive

```
CYN_UNROLL(ON,"tag");
```

specifies that the loop be completely unrolled. As a convenience, ALL can be specified to completely unroll an entire loop nest.

```
CYN_UNROLL( ALL, "tag" )
```

For example the following would result in four multipliers being used.

```
for ( int i=0; i<4; i++ )
{
  CYN_UNROLL( ON, "example_loop" );
  array[i] = array[i] * 10;
}
```

As if it had been written as follows:

```
array[0] = array[0] * 10;
array[1] = array[1] * 10;
array[2] = array[2] * 10;
array[3] = array[3] * 10;
```

Loops can also be partially unrolled, creating parallel hardware for fewer than the total number of iterations of the loop using the directive of the form: CYN_UNROLL (CONSERVATIVE, N, "tag");

So, the following loop

```
for ( int i=0; i<4; i++ )
{
  CYN_UNROLL( CONSERVATIVE, 2, "example_loop" );
  array[i] = array[i] * 10;
}
```

Would be implemented as if it had been written as follows:

```
for ( int i=0; i<2; i=i+2 )
{
  array[i] = array[i] * 10;
  array[i+1] = array[i+1] * 10;
}
```

5.9.5 Pipelining

Cynthesizer can automatically perform loop pipelining. This can be applied to any loop within the design. Pipelining the implementation of an entire thread can be accomplished by applying the pipelining directive to the while(1) loop that constitutes the bulk of the thread behavior. Consider our earlier example scheduled with a computational latency of 4. Recall that this consumed two adders and one multiplier to produce a throughput of one value each six cycles.

We could pipeline this earlier example as follows.

```
while(true)
{
  CYN_INITIATE( CONSERVATIVE, 2, "main_loop" );
  struct data_struct data = in_port.get();
  sc_uint<28> X = (A + B + C) * (D + E + F) * G;
  out_port.put(X);
}
```

This constrains the synthesis schedule to initiate a new iteration of the loop every two cycles. This would result in the following schedule.

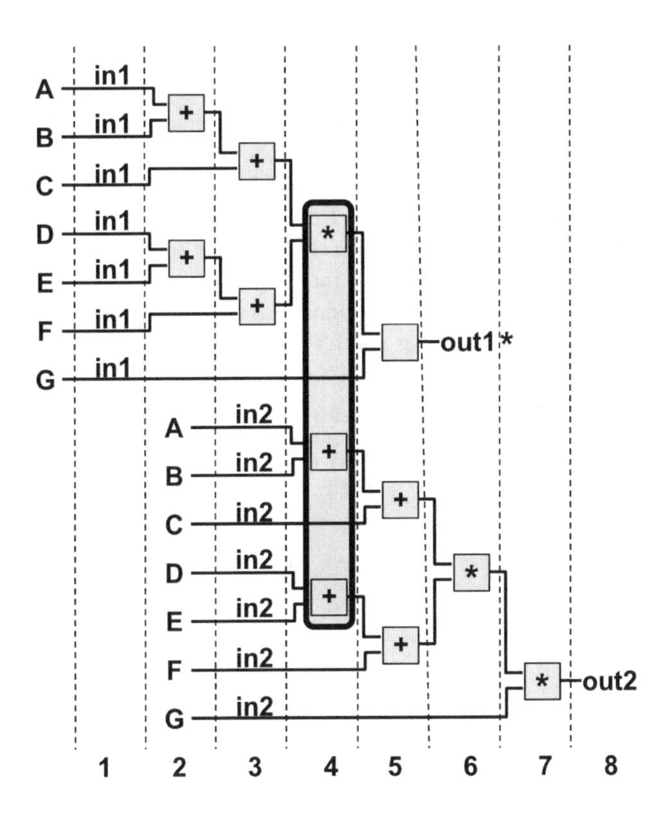

Note that the maximum resource utilization occurs beginning in cycle 4 where two adders and one multiplier are used. By pipelining the design, we are able to achieve a throughput of two values every eight cycles without using any additional multipliers or adders. This is a 50% increase in throughput with no increase in computing resources. Note again, this is done without any need to recode the algorithm.

5.10 Verification

The key verification advantage of SystemC high-level synthesis using Cynthesizer is that the designer is able to:

- Design at a high level
- Verify the algorithm and the interface protocols using high-speed behavioral simulation

- Synthesize RTL that implements the SystemC semantics that were simulated
- Use the same testbench for high-level simulation and RTL simulation

The design can comprise a single module or multiple cooperating modules. In the case of multiple modules, the high-level SystemC simulation ensures that the modules are operating correctly individually and working together properly. This simulation validates the algorithms, the protocol implementations at the interfaces, and the interactions of the modules operating concurrently.

The modules can then be synthesized, and the resulting RTL can be verified using the same testbench that was used at the high level. This is made possible by the mixed-mode scheduling described earlier in which the algorithm is written as untimed SystemC while the interfaces are specified as cycle-accurate SystemC. Multiple testbench configurations may be constructed to verify various combinations of high-level modules and RTL modules.

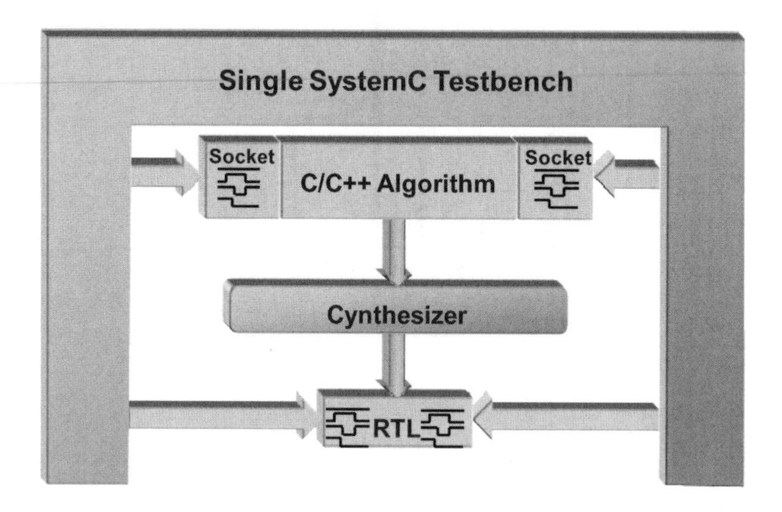

Cynthesizer incorporates a complete dependency management and process automation system that automatically generates needed cosimulation wrappers and testbench infrastructure to automate verification of multiple configurations of high-level and RTL modules without any need to customize the testbench source code itself.

5.11 Conclusion

This chapter has outlined the synthesizable constructs of C++ and SystemC supported by the Forte Design Systems in its Cynthesizer product. It has described specific techniques that can be used to encapsulate synthesizable communication protocols in C++ classes for maximum reuse and techniques used to automatically produce well-structured RTL for predictable timing closure. Finally, some of the user-visible mechanisms for controlling scheduling and the architecture of loop implementation have been discussed along with a brief discussion of verification issues automation incorporated in the Cynthesizer product.

Hopefully, this has enabled the reader to understand how SystemC synthesis with Cynthesizer can be used to implement a broad range of functionality at multiple abstraction levels and how the use of high-level C++ and SystemC constructs raises the level of abstraction in hardware design.

Chapter 6
AutoPilot: A Platform-Based ESL Synthesis System

Zhiru Zhang, Yiping Fan, Wei Jiang, Guoling Han, Changqi Yang, and Jason Cong

Abstract The rapid increase of complexity in System-on-a-Chip design urges the design community to raise the level of abstraction beyond RTL. Automated behavior-level and system-level synthesis are naturally identified as next steps to replace RTL synthesis and will greatly boost the adoption of electronic system-level (ESL) design. High-level executable specifications, such as C, C++, or SystemC, are also preferred for system-level verification and hardware/software co-design.

In this chapter we present a commercial platform-based ESL synthesis system, named AutoPilot^{TM} offered by AutoESL Design Technologies, Inc. AutoPilot is based on the xPilot system originally developed at UCLA. It automatically generates efficient RTL code from C, C++ or SystemC descriptions for a given system platform and simultaneously optimize logic, interconnects, performance, and power. Preliminary experiments demonstrate very promising results for a wide range of applications, including hardware synthesis, system-level design exploration, and reconfigurable accelerated computing.

Keywords: ESL, Behavioral synthesis, Scheduling, Resource binding, Interface synthesis

6.1 Introduction

The rapid increase of complexity in System-on-a-Chip (SoC) design urges the design community to raise the level of abstraction beyond RTL. Electronic system-level (ESL) design automation has been widely identified as the next productivity boost for the semiconductor industry. However, the transition to ESL design will not be as well accepted as the transition to RTL in the early 1990s without robust synthesis technologies that automatically compile high-level functional descriptions into optimized hardware architectures and implementations.

P. Coussy and A. Morawiec (eds.) *High-Level Synthesis.*

Despite the past failure of the first-generation behavioral synthesis technology during the mid-1990s, we believe that behavior-level and system-level synthesis and optimizations are now becoming imperative steps in EDA design flows for the following reasons:

- *Embedded processors are in almost every SoC*: With the coexistence of microprocessors, DSPs, memories and custom logic on a single chip, more software elements are involved in the process of designing a modern embedded system. It is natural to use C-based languages to program software for embedded processors. Moreover, the automated C-based synthesis allows the designer to quickly experiment different hardware/software boundaries and explore various area/power/performance tradeoffs using a single functional specification.
- *Huge silicon capacity requires higher level of abstraction*: Design abstraction is one of the most effective methods for controlling rising complexity and improving design productivity. For example, the study from NEC [10] shows that a 1M-gate design typically requires about 300K lines of RTL code, clearly beyond what can be handled by a human designer. However, the code density can be improved by more than 7X when moved to the behavior level. This results in a human-manageable 40K lines of behavioral description.
- *Verification drives the acceptance of SystemC*: Transaction-level modeling (TLM) with SystemC [2] has become a very popular approach to system-level verification [8]. Designers commonly use SystemC TLMs to describe virtual software/hardware platforms, which serve three important purposes: early embedded software development, architectural modeling and functional verification.

  The wide availability of SystemC functional models directly drives the needs for SystemC-based synthesis solutions, which automatically generate RTL code through a series of formal constructive transformations. This avoids the slow and error-prone manual process and simplifies the design verification and debugging effort.
- *Accelerated computing or reconfigurable computing needs C/C++ based compilation/synthesis to FPGAs*: Recent advances in FPGAs have made reconfigurable computing platforms feasible to accelerate many high-performance computing (HPC) applications, such as image and video processing, financial analytics, bioinformatics, and scientific computing applications.

  Since HDLs are exotic to most application software developers, it is essential to provide a highly automated compilation/synthesis flow from C/C++ language to FPGAs.

In this chapter we present a platform-based ESL synthesis system named *AutoPilot™*, offered by AutoESL Design Technologies, Inc. AutoPilot is capable of automatically generating efficient RTL code from an untimed or partially timed C, C++ and SystemC description for the target hardware platform. It performs platform-based behavioral and system synthesis, tightly integrates with a modern leading-edge C/C++ compiler, and embodies a class of novel, near-optimal, and highly-scalable synthesis algorithms.

The synthesis technology was originally developed in the UCLA xPilot system [5], and has been licensed by AutoESL for the commercialization. In its current stage, AutoPilot exhibits the following key features and advantages:

- *Unified C/C++/SystemC design flow*: AutoPilot accepts three kinds of standard C-based design entries: C, C++ and SystemC. It also supports a variety of abstraction models including pure untimed functional model, partially timed transactional model, and fully timed behavioral or structural model. The broad coverage of languages and abstraction models allows AutoPilot to target a wide range of applications, including hardware synthesis, system-level design exploration and high-performance reconfigurable computing.
- *Utilization of state-of-the-art compiler technologies*: AutoPilot incorporates a leading-edge commercial-strength C/C++ compiler in the synthesis loop. Many state-of-the-art compiler techniques (intra-procedural and inter-procedural) are utilized to analyze, transform and aggressively optimize the input behaviors.
- *Platform-based and implementation-aware synthesis*: AutoPilot takes advantage of the target platform information to carry out more informed synthesis and optimization. The timing, area and power for the available computation resources and communication interfaces are all characterized.

  In addition, AutoPilot has tight integration with several downstream RTL synthesis and physical synthesis tools to assure better quality-of-result and higher degree of automation.
- *Interconnect-centric and power-aware optimization*: AutoPilot is able to generate an optimized microarchitecture with consideration of the on-chip interconnects at the high level and maximize both data locality and communication locality to achieve faster timing and power closure. Furthermore, it can carry out aggressive power optimization using fine-grain clock gating and power gating.

The reminder of this paper is organized as follows: Sect. 6.2 presents an overview of the AutoPilot design flow. Sections 6.3 and 6.4 briefly discuss the system front-end and highlight the synthesis engine, respectively. The preliminary experimental results are reported in Sect. 6.5.

6.2 Overall Design Flow

The overall design flow of the AutoPilot synthesis system is shown in Fig. 6.1. AutoPilot accepts synthesizable C, C++, and/or SystemC as input and performs four major steps to generate the cycle-accurate RTLs, which includes compilation and elaboration, advanced code transformation, core behavioral and communication synthesis, and microarchitecture generation.

In the first step the behavioral description is parsed by a GCC-compatible front-end compiler, with the extensions to handle the bit-accurate integer data types. For SystemC designs, elaboration will be invoked to extract processes, ports, channels, and interconnection topologies and construct a detail-rich system-level synthesis data model.

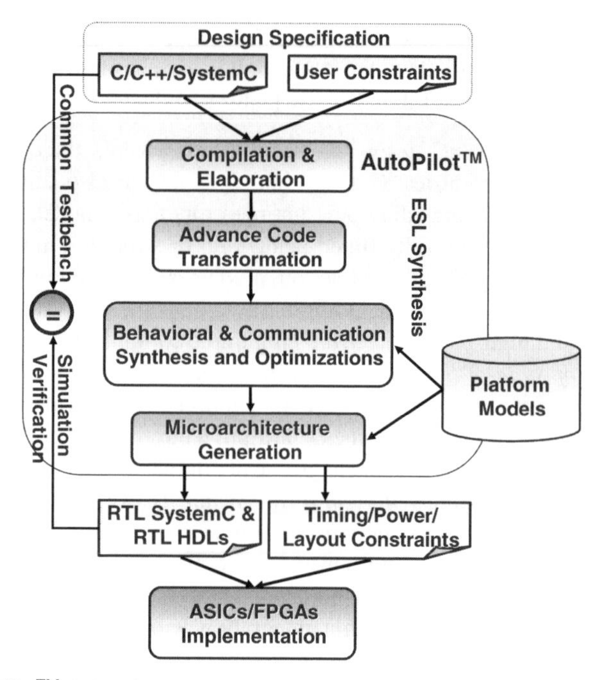

Fig. 6.1 AutoPilot^{TM} design flow

On top of the synthesis data model, AutoPilot applies a set of advanced code transformations and analyses to optimize the input behavior, including traditional compilation techniques such as constant propagation and dead code elimination, and hardware-specific passes such as bitwidth analysis and optimization. The AutoPilot front-end will be discussed in Sect. 6.3.

The code transformation phase is followed by the core hardware synthesis phase. AutoPilot performs platform-based synthesis and interconnect-centric optimizations during scheduling and resource binding; these take into account the user-specified frequency/latency/throughput/resource constraints and generate optimized microarchitectures. We shall discuss more details of the synthesis engine in Sect. 6.4.

At the back-end, AutoPilot outputs RTL VHDL/Verilog code together with constraint files (e.g., multicycle path constraints, physical location constraints, etc.) to leverage the existing logic synthesis and physical design toolset for final implementation on either ASICs or FPGAs. It is worth noting that RTL SystemC code is also generated, which can be directly compiled and simulated with the original C/SystemC test bench to verify the correctness of the synthesized RTLs.

6.3 AutoPilot Front-End

In this section we discuss three major aspects of the AutoPilot front end, i.e., the language support, compiler optimizations, and the platform modeling.

6.3.1 Language Coverage

6.3.1.1 C/C++ Support

AutoPilot has a broad coverage of the C and C++ language features. It provides comprehensive support for most of the commonly-used data types, operators, struct/class constructs, and control flow constructs. Due to the fundamental difference between the memory models of software and hardware, AutoPilot currently disallows the usage of dynamic pointers, dynamic memory allocations, and function recursions.

Designers can fully control the data precisions of a C/C++ specification. AutoPilot directly supports single and double precision floating-point types. In addition, it adds the capabilities (compared to xPilot) in compiling and synthesizing bit-accurate fixed-point data types, for which standard C and C++ language lack native support.

- Arbitrary-precision integer (APInt) data types: The user can specify that an integer type's precision (bit width) is any number of bits up to eight million. For example, *int*24 declares an 24-bit signed integer value. Constant values will be zero or sign extended to the indicated bit width if necessary.
- Arbitrary-precision fixed point (APFixed) data types: AutoPilot provides a synthesizable templatized C++ library, named *APFixed*, for the designer to describe fixed-point math. APFixed library implements the common arithmetic routines via operator overloading and supports the standard quantization and saturation modes.
- IEEE-754 standard single and double precision floating point data types are fully supported in AutoPilot for FPGA platforms. Common floating-point math routines (e.g., square root, exponentiation, logarithm, etc.) can be also synthesized.

6.3.1.2 SystemC Support

AutoPilot fully supports the OCSI synthesizable subset [1] for the SystemC synthesis.

Designers can make use of SystemC bit-accurate data types (i.e., *sc_int/sc_uint*, *sc_bigint/sc_biguint*, and *sc_fixed/sc_ufixed*) to define the data precisions. Multi-module hierarchical designs can be specified and synthesized with the *SC_MODULE* constructs. Within each module, multiple concurrent processes can be declared with the *SC_METHOD* and *SC_CTHREAD* constructs.

6.3.2 Advanced Code Transformations

A variety of compiler optimization techniques are applied to the behavioral description code with the objective to reduce the code complexity, maximize the data locality, and expose more parallelism. The following transformations and analyses are particularly instrumental for AutoPilot hardware synthesis.

- Traditional optimizations such as constant propagation, dead code elimination, and common subexpression elimination that avoid functional redundancy.
- Strength reductions that replace expensive operations (e.g., multiplications and divisions) with simpler low-cost operations (e.g., shifts, additions and subtractions).
- Transformations such as if-conversion and tree height reduction that explicitly expose fine-grain operator-level parallelism.
- Coarse-grain code restructuring by loop transformations such as loop unrolling, loop flattening, loop fusion, etc.
- Analyses such as bitwidth analysis, alias analysis, and dependence analysis that help to reduce the data widths and analyze the data and control dependences.

These transformation are either performed locally within the function bodies, or applied intraprocedurally across the function call hierarchy.

6.3.3 Platform Modeling

AutoPilot takes full advantage of the target platform information to carry out more informed synthesis and optimization. The platform specification describes the availabilities and characteristics of the important system building blocks, including the on-chip computation resources and the selected communication interfaces.

Component pre-characterization is involved in the modeling process. Specifically, it characterizes the delay, area, and power for each type of hardware resource, such as arithmetic units (e.g., adders and multipliers), memories (e.g., RAMs, ROMs and register files), steering logic (multiplexors), and interface logics (e.g., FIFOs, and bus interface adapters). The delay/area/power characteristic functions are derived by varying the bit widths, number of input and output ports, pipeline intervals and latencies, etc. To facilitate our interconnect-centric synthesis. The heterogeneous resources distribution map and the distance-based wire delay lookup tables are also constructed.

AutoPilot greatly extends the platform modeling capabilities in xPilot. It can support advanced ASIC process (e.g., TSMC 90 and 65 nm technologies), a wide range of FPGA device families (e.g., Xilinx Virtex-4/Virtex-5, Altera Stratix II/Stratix III) and various accelerated computing platforms (e.g., Nallatech [4] and XDI [3] acceleration boards).

6.4 AutoPilot Hardware Synthesis Engine

This section highlights several important features of the AutoPilot synthesis engine, including scheduling, resource binding, pipelining, and interface synthesis.

6.4.1 Scheduling

An efficient and versatile scheduler is implemented in the AutoPilot system to exploit parallelism in the behavior-level design and determine the time at which different computations and communications are performed. The core scheduling algorithm is based on a mathematical programming formulation. It has significant advantages over the prior approaches in two major aspects:

- *Versatility*: Our scheduler is able to model a rich set of scheduling constraints (including cycle time constraint, latency constraints, throughput constraint, I/O timing constraints, and resource constraints) in the constraint system, and express different performance metrics (such as worst-case and average-case latency) in the objective function. Moreover, several important synthesis optimizations such as operation chaining, structural pipelining, behavioral template, slack distribution, etc., are all naturally encoded in a single unified mathematical framework.
- *Efficiency and scalability*: Our scheduler is highly efficient and scalable when compared to the other constraint-driven approaches. For instance, typical ILP formulations uses discrete 0–1 variables to model the assignment relationships between operations and time steps, this requires lots of variables and complex equations to express one scheduling constraint since all feasible time steps should be considered. In our formulation, variables directly represent operation execution time and are independent of the final schedule latency. This leads to much more compact constraint system, and the mathematical programming model can be efficiently solved in a few seconds for very complex designs, as evidenced by the Xilinx MPEG-4 design (to be discussed in Sect. 6.5).

The first generation of our scheduler was based on the SDC-based scheduling algorithm and the technical details are available in [7].

6.4.2 Resource Binding

Resource binding determines the numbers of functional units and registers, and the sharing among compatible operations and data transfers. It has a dramatic impact on the final design quality as they determine the interconnection network with wires and steering logic.

AutoPilot is capable of providing optimized binding for various functional units and memory blocks, such as integer and floating-point arithmetic units, transcendental functions, black-box IP blocks, registers, register files, RAMs/ROMs, etc. AutoPilot's binding algorithm can also generate different microarchitectures. For example, it has an option to generate a distributed register-file microarchitecture (DRFM) to optimize both data and communication localities.

DRFM has a semi-regular structure which consists of one or multiple islands. As illustrated in Fig. 6.2, each DRFM island contains a local register file (LRF),

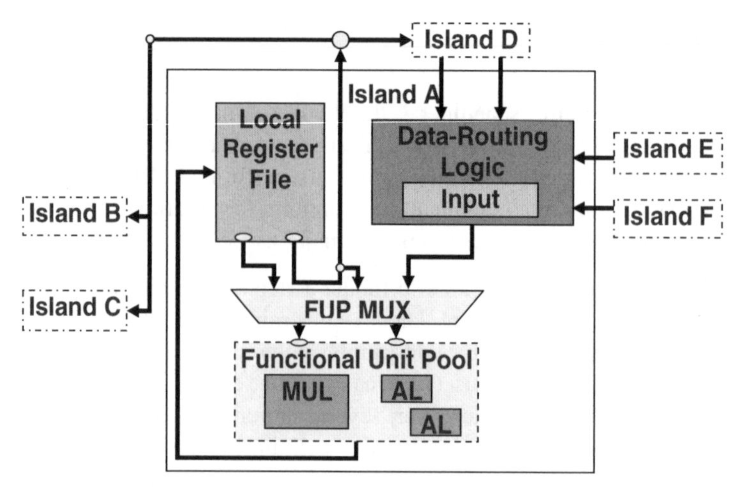

Fig. 6.2 Distributed register-file microarchitecture

a functional unit pool (FUP), and data-routing logic. The LRF serves as the local storage in an island. Each register file allows a variable number of read ports but only a fixed number (typically one) of write ports. The LRF stores the results produced from the local computation units in FUP and provides data to both local FUP and the external islands. By clustering LRF and FUP into a single island, we are able to maximize both data/computation locality and communication locality. This also helps us avoid, to a large extent, the centralized memory structures and global communications which often become the bottlenecks limiting system efficiency in performance, area, and power. To handle the necessary inter-island communications, we use the data-routing logic to route data from the external islands.

DRFM is a semi-regular microarchitecture. The configurations of the LRF, FUP and the data-routing logic are application-specific. One important objective that DRFM-based resource binding tries to minimize is the inter-island connections. This will simplify the data-routing logic in each island and reduce the overall complexity of the resulting datapath.

The technical details of the DRFM-based resource binding algorithm are available in [6].

6.4.3 Pipelining

AutoPilot's synthesis engine (during scheduling, resource binding, and microarchitecture generation) supports several forms of pipelining to improve the system performance.

```
void block_idct(short input[8][8], short output[8][8]) {
    short buffer[8][8];
    idct_row(input, buffer);
    idct_col(buffer, output);
}
```

Fig. 6.3 Pseudo-code for an IDCT block

- *Loop pipelining* allows multiple successive iterations of a loop to operate in parallel by executing one iteration before the previous iteration has completed. As a result, the loop throughput as well as the loop latency can be both improved.
- *Hierarchical functional pipelining* pipelines a function so that the same functional body can start processing new input data before its completion on the current data set. Given a target throughput constraint (in terms of the number of cycles after which new data can be introduced), the pipelining can be applied hierarchically to the callee functions.
- *Multi-function pipelining* executes two or more communicating functions concurrently in a streamed manner. For example, Fig. 6.3 illustrates an 8×8 inverse discrete cosine transform (IDCT) algorithm. Multi-function pipelining will pipeline the execution of row-based transform (*idct_row*) and column-based transform (*idct_col*) and automatically insert the ping-pong memory buffer to hold the intermediate data produced and consumed by these two functions. With this pipeline, the overall throughput of the entire *block_idct* function can be significantly increased.

6.4.4 Interface Synthesis

With AutoPilot's platform-based synthesis methodology, designers are not required to hard code any target-specific interface timing behaviors into the source code.

Designers can simply use the standard function parameters to expose the desired inputs and outputs to the external circuits. AutoPilot interface synthesis is responsible for converting the parameter reads and writes into the actual interface accesses. For example, based on the specified communication interfaces in the platform library, a store operation on a scalar pointer (e.g., $*p = x$) can be turned into a direct wire connection, or a FIFO write, or even a bus transfer (pipelined transfer and burst-mode transfer are both supported).

This capability is particularly convenient for the C and C++ design entries. SystemC-based designs can benefit from this feature as well, although it provides users an array of language constructs to specify the cycle-true and pin-accurate interface connections.

6.5 Experimental Results

We have used AutoPilot to synthesize several real-world complex designs for both FPGAs and ASICs for a wide range of applications, including multimedia image/video processing, digital signal processing, machine learning, financial engineering, and VLSI CAD algorithms.

In this section we report preliminary synthesis results on FPGAs to demonstrate the usage of AutoPilot for three important usage models – hardware synthesis, system-level design exploration, and reconfigurable accelerated computing.

6.5.1 Hardware Synthesis

6.5.1.1 MPEG-4 Simple Profile Decoder

We used AutoPilot to synthesize a real industrial design, the MPEG-4 simple profile decoder from Xilinx [9]. As shown in Fig. 6.4 (from [9]), the entire design contains several pipelined modules, which are interconnected by FIFOs or object FIFOs to form a block-level pipeline.

In our experiments, the same system-level architecture is used, while each submodule is synthesized by AutoPilot system from a C language specification. Manual changes are needed only in a few places to convert the dynamic pointers to synthesizable static pointers.

The synthesis results are reported in Table 6.1. AutoPilot automatically generates more than 10X lines of VHDL code over the original C specification. Targeting a Xilinx Virtex II-pro FPGA (v2p30), the total resource usage is around 7K slices. It is worth mentioning that final area can be significantly reduced with further

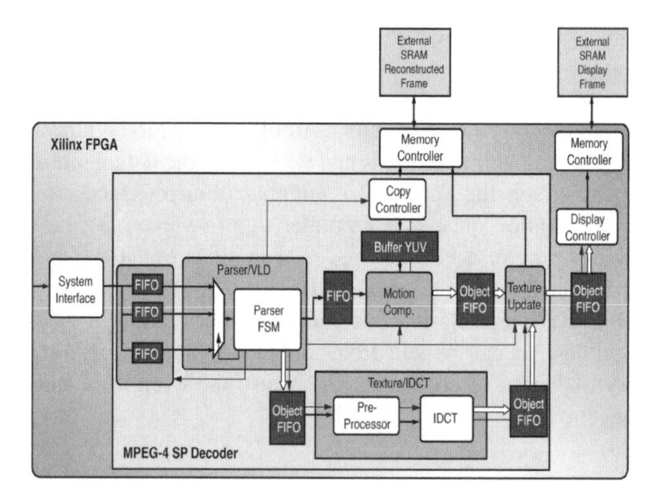

Fig. 6.4 Xilinx MPEG-4 simple profile decoder top-level block diagram

Table 6.1 MPEG-4 simple profile decoder synthesis results

Module	C source file	C line#	VHDL line#	Slices
Motion Comp.	motion_comp.c	312	4,681	899
Parser/VLD	bitstream.c	439	6,093	
	motion_decode.c	492	10,934	2,693
	parser.c	1,095	12,036	
	texture_vld.c	504	6,089	
Texture/IDCT	texture_idct.c	1,819	11,537	2,032
Copy control/	copy_control.c	287	2,815	
texture update	texture_up.c	220	2,736	1,407
Total		5,168	56,921	7,031

Table 6.2 Alternate HW/SW implementations for MPEG-4 decoder

	Seven MicroBlazes	Single PowerPC	PowerPC + HW MotionComp
Throughput	1.18	3.06	3.53
Speedup	–	+68.4%	+15.3%

code refinement such as bitwidth annotations on the function parameters. The main purpose of this experiment is to demonstrate that AutoPilot can quickly synthesize complex vanilla C code into hardware and meet the performance target. We set the final frequency target as 8 ns, and the Xilinx ISE v8.1 static timing analyzer reports positive slacks for all the final modules. The final performance can be estimated for each module using the reported frequency and latency results. Overall, the throughput requirement of 30 frames per second will be easily achieved for a 352×288 frame size (CIF format).

6.5.2 System-Level Design Exploration

AutoPilot can also facilitate the quick system-level exploration for embedded designs. To demonstrate this advantage, we have explored three alternative implementations of the MPEG-4 simple profile decoder on a Xilinx Virtex II-pro development board. The first design comprises seven MicroBlaze soft-core processors, and each processor implements a sub-module of the MPEG-4 decoder. The second design uses a single PowerPC core on Xilinx FPGAs to execute the entire MPEG-4 C program. The third implementation is a hybrid hardware/software design which offloads the motion compensation block onto the FPGA fabrics using the AutoPilot synthesis.

As shown in Table 6.2, the PowerPC version is about 2.6X faster than the soft-core processor network. The speedup is primarily due to the higher clock frequency (up to 450 MHz) of the hard-core PowerPC. Also, the computation workloads on the seven MicroBlazes are not evenly distributed and thus degrades the performance of the processor pipeline.

According to profiling results, the motion compensation module contributes to approximately 16% of the total software decoding time. After we synthesize this block on FPGA for the third design, a 15% throughput increase can be observed, which implies that the latency of the time-consuming motion compensation process has been effectively hidden by the automatic synthesis. Interestingly, the size of the resulting hardware block (around 900 slices) is smaller than a MicroBlaze processor. The performance/area tradeoff of this kind can be easily achieved with the aid of the AutoPilot synthesis.

6.5.3 FPGA-Based Accelerated Computing

One innovation forefront in the High-Performance Computing (HPC) field is to harness FPGA to accelerate domain-specific applications by one or multiple orders of magnitude over the general-purpose microprocessors.

The automatic synthesis support of high-level programming languages (such as C, C++, and FORTRAN) is paramount important to allow the software designs to develop algorithms and implement on FPGAs.

6.5.3.1 Lithographic Aerial Image Simulation

In this case study we use AutoPilot to accelerate a lithographic aerial image simulation application, which is an essential component in most DFM (Design for Manufacturability) flows. The lithography simulation itself is a very computationally demanding process and often requires clusters with hundreds CPUs to achieve acceptable turn-around time.

The kernel of the simulation engine is a nested loop illustrated in Fig. 6.5. Abundant data-level parallelism can be exposed by careful loop unrolling and

```
for (x = 0; x < pixel_max; ++x) {
  for (y = 0; y < pixel_max; ++y) {
    // Initialize pixel intensities.
    I[x][y] = 0;
    for (k = 0; k < K; ++k) {
     // Initialize partial sum.
     I_k[x][y] = 0;
     // Core computation.
     for (n = 0; n < 4*N; ++n) {
       addr_x = 5 * x - rect_x[n] + c;
       addr_y = 5 * x - rect_y[n] + c;
       I_k[x][y] += (-1)^n * kernel[k][addr_x][addr_x];
     }
     I[x][y] += I_k[x][y] * I_k[x][y];
     }
  }
}
```

Fig. 6.5 Pseudo-code for the simulation kernel

array/memory partitioning. Loop pipelining and multi-function pipelining are also applied to further increase the performance.

The whole algorithm is written in 2,226 lines of C code and synthesized by AutoPilot, which generates about 24K lines of VHDL code. The accelerator has been implemented on XtremeData XD1000™ development system [3]. The development system uses a dual Opteron™ motherboard and one of the Opteron processors is replaced by an XD1000 co-processor module. The XD1000 co-processor is built around an Altera Stratix II EP2S180, and is compatible with Opteron Socket 940. The FPGA co-processor communicates with the host Opteron CPU via the HyperTransport™ links.

We use Altera Quartus II v6.0 to implement the generated RTLs on the Stratix II FPGA. Table 6.3 shows the resource usage of the synthesized accelerator, which consumes around 30% of the device resources in ALUT logic and memory bits. The final clock frequency is above 100 MHz.

To measure the performance speedup, we conduct experiments on a 200×200 um chip layout specified in GDSII format. We divide the image into $1{,}000 \times 1{,}000$ nm regions and simulate each region with a kernel look-up table sized 2,000 nm by 2,000 nm. We also generate a number of layouts with different densities (N). The software implementation runs on the AMD Opteron 248 processor at 2.2 GHz with a 4 GB DDR memory. The program is compiled through GCC-O3.

Table 6.3 Resource usage of the synthesized accelerator with 5×5 partitioning

	ALUTs	Memory bits	Fmax (MHz)
Accelerator	23,641	2,883,296	117.01

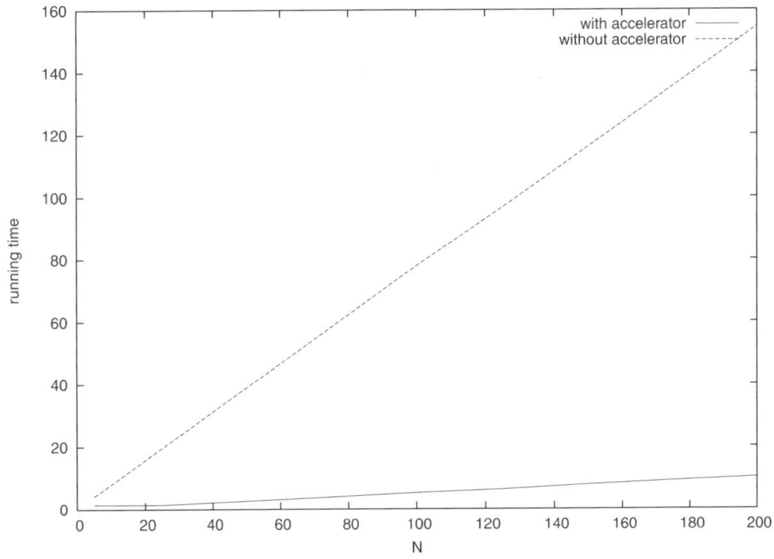

Fig. 6.6 Execution time comparison with and without the synthesized accelerator

Figure 6.6 shows the measured execution time and speedup with different layout densities N. Note that for a very small N, the speedup gets degraded since the communication time dominates the computation time on the FPGA. For a moderate N, we can achieve a speedup around 15X even with the communication overhead between the CPU and the hardware accelerator.

The acceleration on FPGA also provides significant power and energy savings. According to Altera Quartus II PowerPlay analysis tool, the synthesized hardware block consumes 6,954 mW, which is 10X smaller than the power consumption of the AMD Opteron processor (about 70 W). Considering the 15X performance speedup, we can achieve a 150X energy saving over the CPU.

Acknowledgments The authors would like to thank Xilinx for providing the MPEG-4 decoder example, XtremeData for lending the XD1000 development platform, and Yi Zou at UCLA for sharing the lithographic simulation result.

References

1. *SystemC Synthesizable Subset (Draft 1.1.18)*, 2004. Open SystemC Initiative. http://www.systemc.org
2. *IEEE 1666^{TM}–2005 Standard for SystemC*, 2005. IEEE and OCSI. http://www. systemc.org
3. *XD1000^{TM} FPGA Coprocessor Module for Socket 940*, 2006. XtremeData Inc. http://www.xtremedatainc.com
4. *H100 Series FPGA Application Accelerators*, 2007. Nallatech. http://www. nallatech.com
5. Cong, J., Fan, Y., Han, G., Jiang, W., and Zhang, Z. (2006). Platform-Based Behavior-Level and System-Level Synthesis. In *Proc. IEEE International SOC Conference*, pages 199–202
6. Cong, J., Fan, Y., and Jiang, W. (2006). Platform-Based Resource Binding Using a Distributed Register-File Microarchitecture. In *Proc. International Conference on Computer-Aided Design*, pages 709–715
7. Cong, J. and Zhang, Z. (2006). An Efficient and Versatile Scheduling Algorithm Based on SDC Formulation. In *Proc. Design Automation Conference*, pages 433–438
8. Ghenassia, F. (2005). *Transaction-Level Modeling with SystemC: TLM Concepts and Applications for Embedded Systems*. Springer, Berlin Heidelberg New York
9. Schumacher, P., Denolf, K., Chilira-RUs, A., Turney, R., Fedele, N., Vissers, K., and Bormans, J. (2005). A Scalable, Multi-Stream MPEG-4 Video Decoder for Conferencing and Surveillance Applications. In *Proc. IEEE International Conference on Image Processing*, pages II: 886–889
10. Wakabayashi, K. (2004). C-Based Behavioral Synthesis and Verification Analysis on Industrial Design Examples. In *Proc. ASPDAC*, pages 344–348

Chapter 7
"All-in-C" Behavioral Synthesis and Verification with CyberWorkBench
From C to Tape-Out with No Pain and A Lot of Gain

Kazutoshi Wakabayashi and Benjamin Carrion Schafer

Abstract This chapter introduces the benefits of C language-based behavioral synthesis design methodology over traditional RTL-based methods for System LSI, or SoC designs. A comprehensive C-based tool flow, based on CyberWorkBench^{TM} (CWB), developed during the last 20 years at NEC's R&D laboratories is introduced. This includes behavioral synthesis and formal verification and hardware–software co-simulation of entire complex SoC. First we introduce the "all-in-C" concept based on CWB.

Then we discuss the behavioral synthesis for various types of circuits and examine the advantages of behavioral synthesis on the hand of commercial ICs. We show that currently entire SoCs are created using this flow in a fraction of the time taken by traditional approaches.

Behavioral IP and C-based configurable processor synthesis and automatic architecture exploration is explained next. At the end we demonstrate a real world example of a mobile phone SoC where most of the modules are synthesized from C descriptions using CWB.

Keywords: Behavioral synthesis, Control and data intensive flows, All-in-C, Behavioral C level formal verification, Hardware-software co-simulation, Automatic system exploration, Behavioral IP, Configurable processor

7.1 Introduction

The design productivity gap problem is becoming more and more serious as VLSI systems become larger. In the mid-1980s, gate-level design shifted to register transfer level (RTL) design for designs that typically exceeded 100K gates (we assume a hundred thousand gates is the upper limit for hand coded modules to be designed in several months).

Currently, several million gates circuits are commonly used just for random logic parts of a design, which equate to more than several hundreds thousand lines of RTL

P. Coussy and A. Morawiec (eds.) *High-Level Synthesis.*

code. It is therefore needed to move the design abstraction one more level in order to cope with this increasing complexity. Behavioral synthesis is a logic way to go as it allows "less detailed design description" and "higher reusability".

A higher level of abstraction description requires smaller code and provides faster simulation times. For example a one million gates circuit requires about 300K lines of RTL (Verilog or VHDL) code, but only around 40K lines of C code. The RTL simulation of 300K lines, we observed in [1], is on average 10–100 times slower than the 40K lines of equivalent behavioral code (it is important to note that in order to benefit from higher level of abstraction the entire design needs to be modeled at the behavioral level).

It is sometimes claimed that behavioral synthesis is only useful for dataflow intensive circuits, but not for control dominated circuits. We believe that behavioral synthesis *can* and *should* be used for all hardware modules in order to truly benefit from it. We will demonstrate this by an example of a real complex SoC design where all custom design modules, except the analog ones, have been designed using behavioral synthesis. NEC Electronics adopted behavioral synthesis as standard design methodology since 2003 and taped out since then several hundreds million Dollars worth of "C-based" chips every year.

Since the benefits of behavioral synthesis are palpable through multiple commercial chip successes, Behavior Synthesis, or High Level Synthesis, is gaining acceptance within the design community, especially in Japanese industries. Various commercial chips for printers, mobile phones, set-top-boxes and digital cameras are designed using behavioral synthesis these days. ANSI-C is the preferred programming language for behavioral synthesis because embedded software is often described in C and design tools like compilers, debuggers, libraries and editors are easily available and there is a big amount of legacy code.

In this paper, we first provide an overview of our C-based design flow where we compare the efficiency and simulation performance against pure RTL as well as co-simulating it with embedded software. We show the advantages of C-based behavioral IPs over RTL IPs and how application specific processors can benefit from it. We present a hardware architecture explorer at the behavioral level allowing a fast and easy way to study the area, performance and power trade-offs of different designs automatically. Finally we demonstrate on a real complex design, how behavioral synthesis can be used for any hardware module (data and control intensive).

7.2 C-Based Design Flow

We have been developing C-based behavioral synthesis called "Cyber" since the late 1980s [2] and developing C-based verification tools such as formal verification and simulation around Cyber during the last 10 years [3]. All these tools are integrated into an IDE, where designers execute these tools upon the C-source code. We named this IDE tool suite "CyberWorkBench™".

7.2.1 Basic Concept of CyberWorkBench

The main idea behind CyberWorkBench is an "all-in-C" approach. This is built around two principal ideas (1) "all-modules-in-C" and (2) "all-processes-on-C".

(1) *All-modules-in-C*: means that all modules in a VLSI design, including control intensive circuits and data dominant circuits, should be described in behavioral C language. Our system supports legacy RTL or gatenetlist blocks as black boxes, which are called as C functions. At the same time it allows designers to create all new parts in C, although this is not recommended as the designer will need to use two different programming languages and RTL parts will slow down the simulation.

(2) *All-processes-on-C*: means that synthesis and verification (including debugging) tasks should be done at the C source code. As an example we can compare this with a software compiler. In a software compiler, a designer does not have to debug the generated machine language (or, assembler language) directly. Similarly, in behavioral synthesis, a designer should not have to debug the generated RTL code. Our CWB environment allows a designer to debug the original C source code and the CWB model checker allows designer to write properties or assertions directly on the C source code.

7.2.2 Design Flow Overview

CWB targets general LSI systems which normally contain several CPUs or DSPs, dedicated hardware modules and some pre-designed or fixed RTL- or gate level IP modules, which are directly connected or through buses.

Initially, each dedicated hardware module such as an ECC encryption module is described in behavioral C. Once its functionality is verified using the C simulator and debugger, the hardware module is synthesized with our behavioral synthesizer. Configurable processors are also synthesized from their C description in our environment. Legend RTL modules are described as function, and handled as a black box. The CPU bus and bus interface circuits are automatically generated using a CPU bus library. After synthesizing and verifying each hardware module, our design environment allows designers to create a cycles-accurate simulation model for the entire system including CPUs, DSPs and custom hardware modules. With this simulation model, designers can verify both functionality and performance of their hardware design as well as the embedded software run on the CPU, DSP and/or generated configurable processors. Behavioral synthesis is quick enough to allow designers to repeatedly modify and synthesis the hardware modules and embedded software. The behavioral C source code can also be debugged with our formal verification, property/assertion model checker tool. Global properties and in-context (immediate) assertions are described for/in the C source code. The equivalence between behavioral C and generated RTL can be verified both in dynamic and static

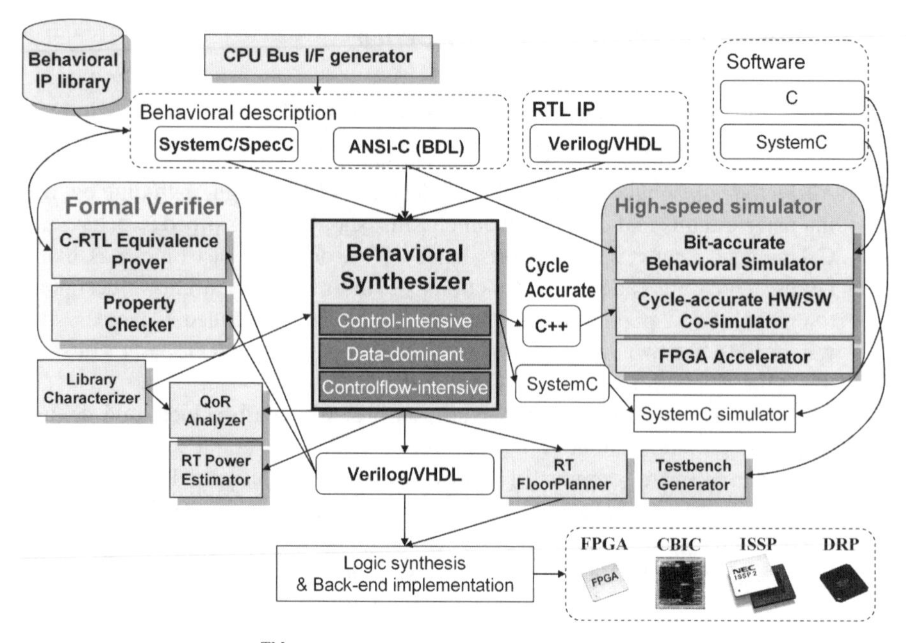

Fig. 7.1 CyberWorkBench^{TM} design flow

way, as described later. Currently, the architectural level parallelization is left to the designer. The designer partitions the C source code into individual hardware modules and embedded software based on the performance result of the cycle simulation or FPGA emulation.

7.2.2.1 Synthesis Flow

Our design flow is shown in Fig. 7.1. A hardware design in extended ANSI-C (called "BDL", or "Cyber-C") [4], or SystemC is synthesized into synthesizable RTL with our "Cyber" behavioral synthesizer [1] with a set of design constraints such as clock frequencies, number and kind of functional units and memories. Usually RTL is handled as a black box, but if necessary, the RTL can also be fed to the behavioral synthesizer. The behavioral synthesizer can insert extra registers to speed up the original RTL and generate new RTL of smaller delay. It also generates a cycle accurate simulation models in C++ or SystemC. The behavioral synthesis can therefore be considered as a Verilog, VHDL, C, C++, and SystemC unification step.

The "Library Characterizer" generates delay and area information of the functional units and memories on a particular technology or FPGA.

A Behavioral IP library, called "Cyberware", is also included in the synthesis environment. Any part of the behavioral IP can be encrypted for security purposes.

Wire delays of global wires between modules need to be analyzed carefully since those delays can be significant when the connected modules are placed far away. Our "RTL FloorPlanner [3]" takes the RTL modules generated by the behavioral synthesizer. Accurate timing information is extracted from the floorplanner and fed back to the behavioral synthesizer. The behavioral synthesizer reads the timing information and re-schedules the C code considering the timing information.

7.2.2.2 Verification Flow

The functionality of the hardware described in C can be verified at the behavioral level, while performance and timing are verified at the cycle-accurate level (or RTL) through simulation. Debugging the generated RTL is however not an easy task since C variables are shared in a register, and various optimizations are applied. We therefore provide a behavioral C source code debugger linked to our cycle-accurate simulation and FPGA emulation tool. After verifying each hardware module, the entire SoC is simulated in order to analyze the performance and/or to find inter-modules problems such as low performance through bus collision, or inconsistent bit orders between modules. Since such entire chip performance simulation is extremely slow in RTL-based HW-SW co-simulation, CWB generates cycle accurate C++ simulation models which can run up to hundred times faster than RTL models. Our HW-SW co-simulator [3] uses the generated cycle-accurate model for this purpose. The simulator allows designers to simulate and debug both hardware and software at the C source code level at the same time. If any performance problems are found, designers can change the hardware-software partitioning or algorithm directly at the C level, and can then repeat the entire chip simulation. This flow implies a much smaller and therefore faster re-design cycle than in a conventional RTL methodology. The C description is the only initial and final SoC description language of the entire design. This entire chip simulation can be further accelerated using an FPGA emulation board [5]. A "Testbench Generator" helps designers to run an RTL simulation with test patterns for behavioral C simulation faster and easier. Its inputs are test patterns for the C simulation and output a Verilog and/or VHDL testbench, which generates stimulus for the RTL simulation. It also creates a script to run commercial simulators to feed the behavioral test patterns and check the equivalence of outputs patterns between the behavioral and RTL simulation.

Another important feature of CWB is the formal verification tool, which is tightly linked to the behavioral synthesizer. With the behavioral synthesis information the formal verification tools can handle larger circuits than usual RTL tools and have C-source level debugging capability even though the model checker works on the generated RTL model. "C-RTL equivalence prover" checks the functional equivalence between a behavioral (un-timed or timed) C description and the generated RTL, using information of the optimizations performed such as loop unrolling, loop merge and array expansion performed by the behavioral synthesis. Without such information, the equivalence check is almost impossible for large circuits.

Designers can specify assertions or properties at the behavioral C level, similar to our cycle accurate simulator. Such behavioral level properties/assertions are converted into RTL ones automatically, and are passed to our RTL model checker.

CWB generates a power enhanced RTL model which estimates the power consumed by the design. A set of power libraries for different technology are provided and used with the generated RTL estimates that power for the selected technology.

A "QoR" synthesis report of the generated circuit shows a quick overview of the design quality. The report file includes area, number of states, critical path delay, number of wires and routability. This information is used for quick micro-architectural exploration as well as system architectural exploration. The system architecture explorer automatically generates different hardware architectures based on the preferences and constraints entered by the user (area, latency, power) at the C level. The designer can analyze the different generated architectures and finally choose the one that meets the design constraints at the smallest cost.

7.3 Behavioral Synthesis

To support the "all-modules-in-C" paradigm presented before, our behavioral synthesizer must cope with three types of circuits: (i) data-dominated, (ii) control-dominated, and (iii) control-flow intensive (CFI) ones. Data-dominated descriptions have many arithmetic operations and less control structures (e.g. only one loop), while control-dominated descriptions have many control-flow operations such as I/O activity in every cycle. A CFI description has a mix of arithmetic operations and control-flow constructs such as loops, conditional operations, jumps ('goto' statements) and functions. Our synthesizer has three types of synthesis engines in order to support these varieties of circuit types: (i) automatic scheduling for CFI and data-flow circuits, (ii) fixed scheduling for control-dominated circuits, and (iii) pipeline scheduling for automatic pipelining or loop folding. Figure 7.2 shows a block diagram of CWB's behavioral synthesizer. CWB supports various C-based language (e.g. BDL, SystemC, SpecC), and RTL as an input description. BDL is directly translated into our tree-structured Control Flow Graph (tCFG) [4], which is a kind of abstract structured expressing control structure of the behavior. Since SystemC and SpecC have different synthesis semantics than BDL, our "Parser/Translator" translates them into BDL semantics and generates the tCFG. In the same way, Verilog-HDL or VHDL is translated into the tCFG. A unique Control Data Flow Graph [2] is then created from the tCFG. All synthesis tasks are performed on those two data structures.

Control dominated circuits such as PCI I/F, DMA controller, DRAM controller, bus bridge, etc, require cycle-by-cycle behavioral description. For this type of circuits, specifying timing constraints for all inputs and outputs is a tough and complex job. Our extended C language called BDL can describe clock boundaries in a behavioral description, and is able to express very complex timing behaviors concisely. Such descriptions are synthesized with a "fixed scheduling" engine, which is fit for

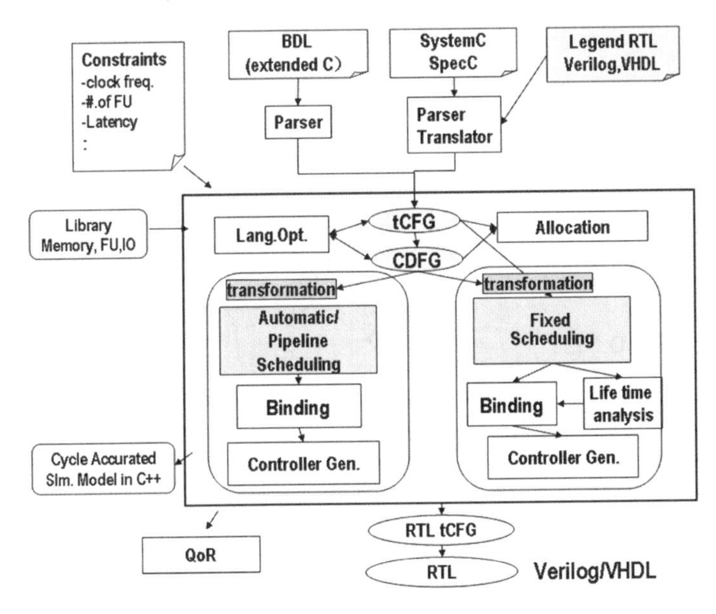

Fig. 7.2 Configuration of Cyber Behavior Synthesis

complex control sequence with exceptional tasks with strict timing constraints. For the circuits, which require fixed sequential communication protocols but all other computations can be freely scheduled, "automatic scheduling" engine is used for synthesis.

For CFI circuit synthesis, the "automatic scheduling" engine is used. The quality of the synthesis is affected by the control flow structure, not just by the data flow. A smart scheduling algorithm is designed to overcome the effects of the programming style. For instance, Fig. 7.3 shows an example of global parallelization among multiple data-dependent conditional branches. These two branches cannot be parallelized in the form given in Fig. 7.3a, because of the control dependency between them. However, if the conditional operations "if (F1)" and "if (F2)" are transformed while scheduling, then they can be parallelized as shown in Fig. 7.3b. This implies that the scheduler will have to modify the control logic in order to obtain circuits with less latency while maintaining the data-flow intact.

Merging two branches into a single one using CDFG transformations is not as effective because the procedure is complex and the merging does not always lead to better results. In contrast, our approach uses a systematic scheduling algorithm without CDFG transformations. In other words, our scheduler schedules all operations in several basic blocks and several branches at the same time in a unique way, as if they were all operations in a single basic block. Our approach handles many other types of speculations, global parallelization with a method called "Generalized Condition Vector [6]", which is extended version of "Condition Vector [2]".

The "Pipeline scheduling" engine generates pipelined circuits from the initial C code with stall signals, which have various "Data Initial Intervals (DII. It also

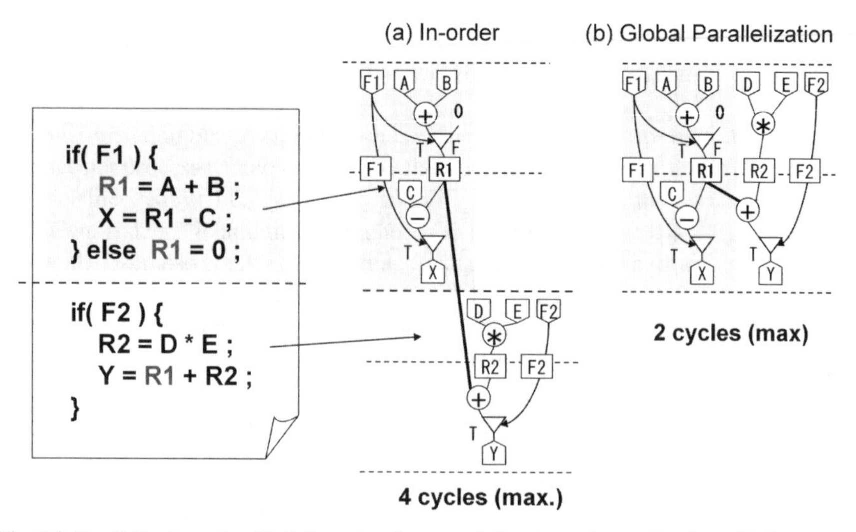

Fig. 7.3 Parallelization of multiple branches for control-flow intensive applications (CFI)

speeds up loop execution by folding loop bodies like software loop pipelining. Global parallelization capabilities are very important even for loop pipelining. Loop carry variables that will be read in the next loop iteration should be scheduled into the states within the given DII cycles sequence. Parallelization beyond control dependencies is one key technique to make loop pipelining possible with a small DII.

7.4 Behavioral Synthesis Advantages Over Conventional Flows

The next sections describes in detail some of the advantages of behavioral synthesis over conventional RTL methodologies like hardware-software co-design, source code re-usability, application specific processor optimizations and automatic architecture exploration.

7.4.1 Shorter Design Period and Less Design Cost

Since C-based behavioral synthesis automates the functional design of hardware, it shortens the design cycle and at the same time shortens the design time of embedded software. Figure 7.4 shows the design cycle of two designs. The first uses the traditional RTL-based design flow and the second the proposed C-based design flow. The total design period and design men-month for the RTL-based design is larger than the C-based one, even though the gate size for RTL design (200K) is one third of that

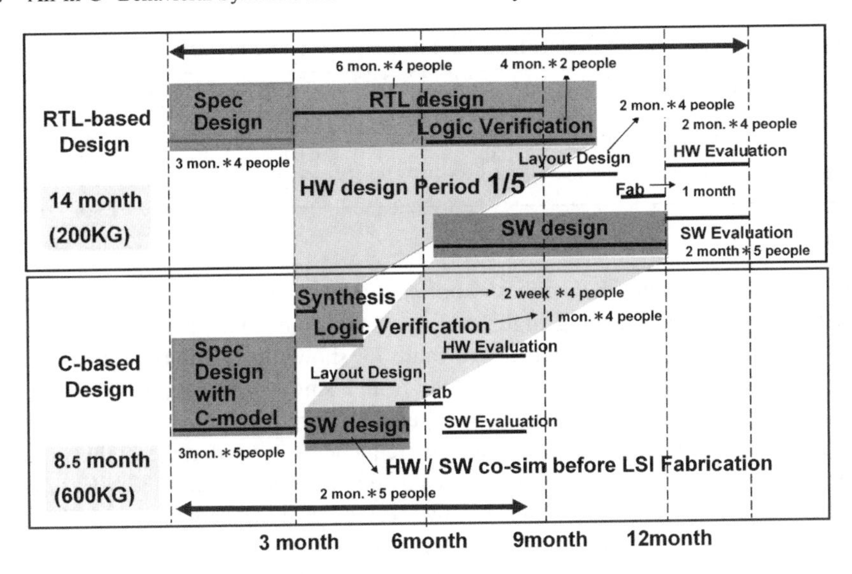

Fig. 7.4 Comparison of design periods with C-based and RTL-based design

for the C-based (600K) one. The hardware design period of the C-based design is 1.5 months, much shorter than the RTL-based design which takes 7 months. It needs to be stressed that the software design in the C-based design takes only 2 months while it takes 6 months for the RTL-based. This is due to the fact that the embedded software can be debugged before the IC fabrication using the hardware-software co-simulator. In RTL design, the software is usually verified on the evaluation board since RTL co-simulation is too slow even for this size of circuits. Lastly, C-based design allows very quick generation of simulation models for embedded software at a very early stage, allowing hardware and software to be concurrently designed both in C.

7.4.2 Source Code Reusability and Behavioral IPs

Another important aspect of C-based behavioral design is the high-reusability of behavioral models; we call this "behavioral IPs" or "Cyberware". An RT level reusable module, called "RTL-IP", can be successfully used for circuits of fixed performance such as bus interface circuits. However, RTL-IPs for general functional circuits such as encryption can only be used for a specific technology, since the RTL-IP's "performance" is hard to adapt for newer technologies. For instance, an encryption RTL-IP at 200 Mbps is difficult to be "upgraded" to perform encryptions at 800 Mbps, because the RTL-IP structure is fixed and the logic synthesis tool is not able to reduce its delay by a forth. On the contrary, a behavioral IP is more flexible and more reusable than RTL-IPs, since it can change its structure

Table 7.1 BS broadcast descrambler behavioral IP comparison

Clock frequency (MHz)	Generated gate size	Generated RTL size	Performance (Mbps)
33	57 KG	7.0 KL	80
54	42 KG	5.9 KL	80
108	26 KG	2.5 KL	80

and behavior allowing the synthesis tool can generate circuits of different performances by simply changing high level synthesis constraints such as number of functional units and clock frequencies. Table 7.1 shows how various circuits of different "clock-frequency" can be generated from a single behavioral IP. This IP is a BS broadcast descramblers (Multi2). All generated circuits satisfy the required performance (more than 80 Mbps) at various frequencies. Note that the highest clock circuit (108 MHz) uses less number of gates than the slow circuit (33 MHz). This never happens in RTL-IPs, which follow the area-delay tradeoff relation of logic synthesis. However, it is natural that a behavioral synthesizer generates a smaller circuit of higher clock frequency for the same performance, since less parallel operations are necessary to achieve the same performance at higher clock frequency.

Another important aspect is that for behavioral IPs it is much easier to modify their "functionality" and "interface" than for RTL-IPs. We designed two types of "Viterbi" decoders for mobile phone and satellite communications. The two required different Bit Error Rate, which is defined by several parameters such as encode rate and constraint bit length. Changing these parameters requires significant modification of the RTL-IP; however, only slight modification is necessary for the behavior IP.

Lastly it has to be noted that behavioral IPs sometimes generates smaller circuits than RTL IPs as behavioral synthesis shares registers and functional units for sequential algorithms such as the Viterbi decoder, but recent RTL designers do not share registers since such time multiplexed sharing makes RTL simulation and debug very difficult.

7.4.3 Configurable Processor Synthesis

Since chip fabrication cost has risen considerably, SoC are becoming as flexible as possible. For this purpose, recent SoC usually have several configurable processors besides a main CPU. These configurable processors should be small, have a high performance and low power consumption for a specific application. Such a configurable processor is also called Application Specific Instruction set Processor (ASIP). ASIPs employ custom instruction-sets to accelerate some applications. There are several commercial ASIPs, such as Xtensa [7] from Tensilica and Mep [8] from Toshiba. Their base-processor and co-processors for adding instructions are described in RTL and they are logic synthesized. In CWB we provide ASIP's base

Table 7.2 Behavioral base-band DSP synthesis results

MIPS(clock)	STB stream 72(108 MHz)	Base-band DSP 15(15 MHz)	Application DSP 60(60 MHz)
#.of Inst.	Base: 81 +Adding: 24	Base: 17 +Adding: 17	Base: 65 +Adding: 21
Gate size	43K	20K	120K
Behavior	2.1KL	1.3KL	2.5KL
Generated RTL	13.0KL	11.4KL	26.0KL
Man-power	1.5 m-m	0.5 m-m	0.8 m-m

Table 7.3 Behavioral configurable processor synthesis

	Behavioral C-based	Manual RTL
Code size	1.3 KL (1/7.6)	9.2 KL
Simulation	61.0 Kc/s(203×) Pentium3@1 GHz	0.3 Kc/s UltraSparc-II@450 MHz
Gate size	19 KG	18 KG

processor and supplementary instructions that are described fully in behavioral C, which are behavioral synthesized. This allows the base-processors and the addition of instructions to share functional units. This sharing leads to much smaller circuits than the conventional RTL-based ASIPs. For an ASIP base-processor, we added 24 instructions suitable for stream processing, such as CRC calculation, with only 25% area increase (34KG to 42KG) due to the of FU sharing.

C-based ASIPs are more flexible than RTL-based ones in terms of public register number, pipeline stages or interrupt policy. In Table 7.2, the synthesis results of three ASIPs are presented. All ASIPs were relatively small, but had enough performance to run the specific application due to the addition of custom instructions. All C-based ASIP designs required only as one tenth man-power of the RTL-based designs.

Table 7.3 shows comparison of C-based and manual RTL design for a configurable DSP design. RTL design flow. The two designs had comparable gate size and delay (RTL design is slightly better). The code efficiency of C-based design flow is shown to be 7.6 compared to the RTL design flow and a simulation speed-up of approximate 200, which leads to high reliability. We believe such advantages are much more important than slight area loss.

7.4.4 Automatic Architecture Exploration

Behavioral synthesis allows the creation of multitude hardware architecture for a unique C design. The user can specify a set of constraints which all architectures have to meet (e.g. area, latency, power) and a set of different architectures that meets those constraints will automatically be generated. The area-performance-power trade-offs can be easily analyzed and the architecture that meets the constraints with the lowest cost can be chosen by the designer. This task is extremely time consuming if it is done at the RTL level as every single architecture requires a major re-work in the RTL code including component types and number of component instantiations. At the behavioral level this can be done by exploring the C code "attributes" of the most significant C code operations (those that will have the highest impact on the final architecture) like functions (e.g. inline expansion, sub-routine), loops (loop merge, unroll, unroll x-times, unroll completely) and mapping arrays as wired logic, registers or memories. Another aspect that is explored is the "global" synthesis options. What kind of scheduling policy is performed such as speculative scheduling, ASAP, ALAP scheduling of inputs and outputs, and which optimization algorithms (e.g. area-, latency-, delay-oriented) should be performed during behavioral synthesis. The third exploration step involves the maximum number of functional units available. This has a significant effect on the scheduler and therefore on the final design. To facilitate the trade-off analyzes the different architectures are displayed as a graph in the IDE's GUI as shown on Fig. 7.5.

The exploration engine is based on a weighted probabilistic search algorithm, where the target options (area and performance) entered by the user are the probabilities that a specific synthesis option or attribute is selected. Each possible synthesis option and attribute has therefore been previously characterized in a library depending on its "usual" contribution to increase performance or area. A unique list of new attributes and synthesis options is generated for each new architecture, avoiding repetition of two equal designs.

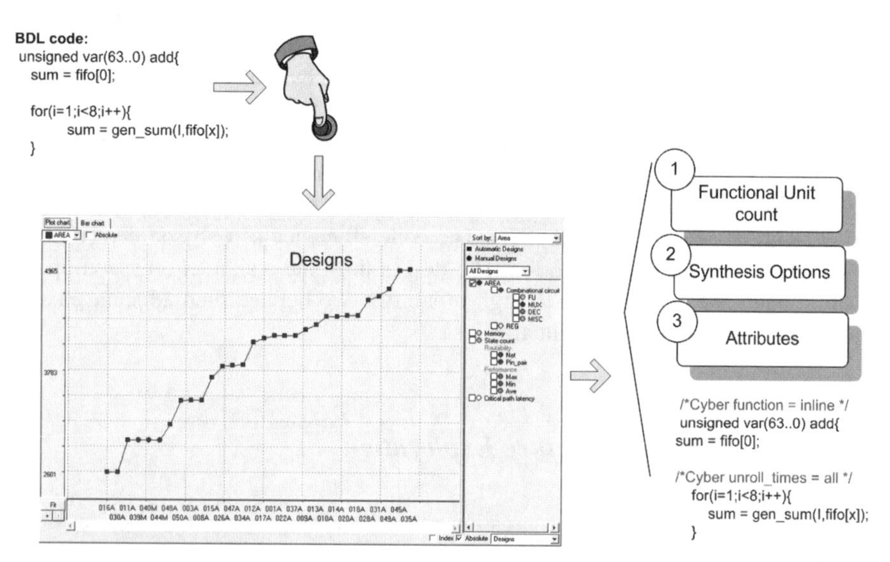

Fig. 7.5 Automatic architectures exploration

Table 7.4 AES core system exploration example

Design	Gates	Registers	Muxes	States	Delay (ns)
1	223,973	59,336	135,891	37	2.06
2	304,203	68,774	186,964	62	1.78
3	80,892	29,940	36,265	61	2.74
4	283,687	8,774	184,015	64	1.78
5	244,997	53,150	173,175	67	2.30

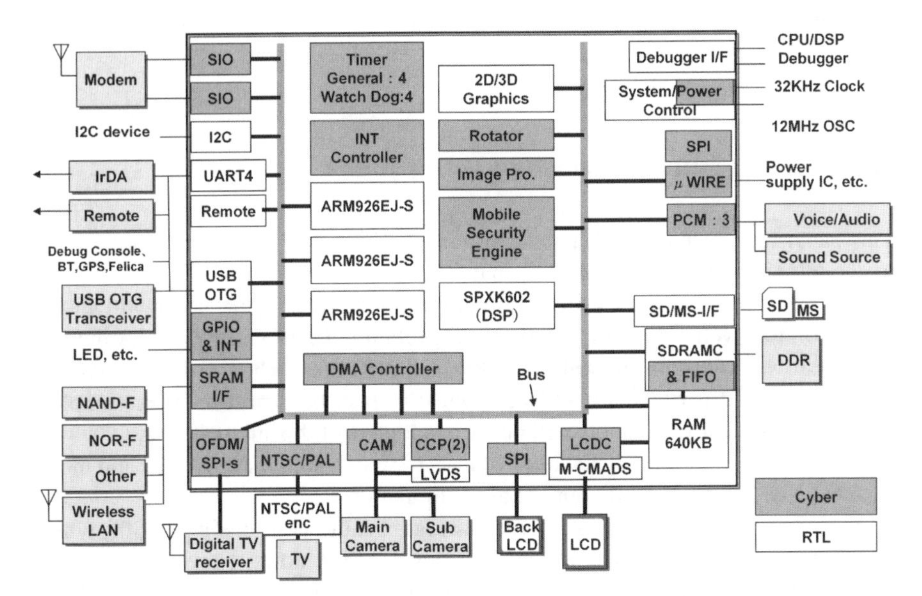

Fig. 7.6 Behavioral design flow design example used in a cell phone SoC (gray boxes design using Cyber)

Table 7.4 shows an example of the architecture exploration of an AES core function which has about 800 lines of C code. The system explorer generates a user defined number of unique architectures (five in this case) based on the target selected by the user (e.g. minimize area, maximize performance).

7.5 System VLSI Design Example Using C-Based Behavioral Synthesis

Figure 7.6 shows a design example of a real complex SoC used at NECs cell phones generated with our behavioral synthesizer. This SoC is called MP211, or Medity [9], which has three ARM cores, one DSP, several dedicated hardware engines and various applications of mobile phone such as audio and video processing, voice recognition, encryption, Java and so on.

Wide ranges of circuits including control dominated circuits and data-intensive circuits were successfully implemented. The grey boxes (including bus) indicate modules that have been synthesized from C descriptions with the proposed behavioral synthesizer, while the white boxes are IP cores given in RTL format (some are legacy RTL components and some are commercial ones). All newly developed modules are designed with our C-based design flow. This example clearly illustrates that our C-based environment is able to design entire SoC designs, and not only algorithmic modules. C-based design flow became a standard ASSP development flow since 2003 at NEC, and several billon dollars worth of ICs have been taped out since.

7.6 Summary and Conclusions

This paper introduced the advantages of behavioral synthesis over traditional RTL methodologies in system LSI design on the hand CyberWorkBench. Faster development time, hardware-software co-simulation and development, easier and faster verification as well as automatic system exploration are some of these. Although many hardware designs are still very skeptical regarding behavioral synthesis the facts show that it is necessary and will sooner or later be a must in every complex hardware design flow. Winners will be early adopters of this methodology.

Currently, we are using behavior synthesis for most of our new designs and more system LSIs are verified with our C-based simulation.

Behavior synthesis tool is as mature as logic synthesis in the late 1980s, when designers started to use them widely RTL level design flows. However, it is taking time to make designers adopt this new design paradigm shifting from RTL "structural" domain thinking to "behavioral" domain thinking. Education and training on behavioral thinking for RTL designers is a crucial and difficult task.

Acknowledgments The authors would like to acknowledge the work of everyone at EDA R&D center, Central Research Laboratories at NEC Corporation, and NEC Information Systems Ltd., NEC Electronics Corp. NEC-HCL-ST for all their work developing CyberWorkBench and designing various chips with it.

References

1. H. Kurokawa, Y. Ikegami, H. Otsubo, K. Asao, K. Kirigaya, K. Misumi, S. Takahashi, T. Kawatsu, K. Nitta, K. Ryu, K. Wakabayashi, M. Tomobe, W. Takahashi, A. Mukaiyama, T. Takenaka, "Study and Analysis of System LSI Design Methodologies Using C-Based Behavioral Synthesis," IEICE Trans. Fundamentals, Vol. E85-A, 2002
2. K. Wakabayashi, "Cyber: High Level Synthesis System from Software into ASIC," Kluwer, Dordecht, pp. 127–151, 1991

3. K. Wakabayashi and T. Okamoto, "C-Based SoC Design Flow and EDA Tools: An ASIC and System Vendor Perspective," IEEE Trans. Comput. Aided Design Integr. Syst., Vol. 19, No. 12, pp. 1507–1522, 2000
4. N. Kobayashi, K. Wakabayashi, H. Tanaka, N. Shinohara, T. Kanoh, "Design Experiences with High-Level Synthesis System Cyber I and Behavioral Description Language BDL," Proceedings of Asia-Pacific Conference on Hardware Description Languages, Oct. 1994
5. Y. Nakamura, K. Hosokawa, I. Kuroda, K. Yoshikawa, T. Yoshimura, "A Fast Hardware/Software Co-Verification Method for System-On-a-Chip by Using a C/C++ Simulator and FPGA Emulator with Shared Register Communication", pp. 299–304, DAC, 2004
6. K. Wakabayashi, "Unified Representation for Speculative Scheduling: Generalized Condition Vector", IEICE Trans. Fundamentals, Vol. E89-A, VLSI Design and CAD Algorithm, pp. 3408–3415, 2006
7. Xtensa, http://www.tensilica.com
8. Mep, http://www.mepcore.com/english/
9. S. Torii, S. Suzuki, H. Tomonaga, T. Tokue, J. Sakai, N. Suzuki, K. Murakami, T. Hiraga, K. Shigemoto, Y. Tatebe, E. Ohbuchi, N. Kayama, M. Edahiro, T. Kusano, N. Nishi, "A 600 MIPS 120 mW 70 μA Leakage Triple-CPU Mobile Application Processor Chip", pp. 136–137, ISSCC, 2005

Chapter 8
Bluespec: A General-Purpose Approach to High-Level Synthesis Based on Parallel Atomic Transactions

Rishiyur S. Nikhil

Abstract Bluespec SystemVerilog (BSV) provides an approach to high-level synthesis that is general-purpose. That is, it is widely applicable across the spectrum of data- and control-oriented blocks found in modern SoCs. BSV is explicitly parallel and based on atomic transactions, the best-known tool for specifying complex concurrent behavior, which is so prevalent in SoCs. BSV's atomic transactions encompass communication protocols across module boundaries, enabling robust scaling to large systems and robust IP reuse. The timing model is smoothly refinable from initial coarse functional models to final production designs. A powerful type system, extreme parameterization, and higher-order descriptions permit a single parameterized source to generate any member of a family of microarchitectures with different performance targets (area, clock speed, power); here, too, the key enabler is the control-adaptivity arising out of atomic transactions. BSV's features enable design by refinement from executable specification to final implementation; architectural exploration with early architectural feedback; early fast executable models for software development; and a path to formal verification.

Keywords: High level synthesis, Atomic transactions, Control adaptivity, Transaction-level modeling, Design by refinement, SoC, Executable specifications, Parameterization, Reuse, Virtual platforms

8.1 Introduction

SoCs have large amounts of concurrency, at every level of abstraction – at the system level, in the interconnect, and in every block or subsystem. The complexity of SoC design is a direct reflection of this heterogeneous concurrency. Tools for high-level synthesis (HLS) attempt to address this complexity by automating the creation of concurrent hardware from high-level design descriptions.

P. Coussy and A. Morawiec (eds.) *High-Level Synthesis.*

At first glance, it may seem surprising that C, a sequential language, is being used successfully in some tools for such a highly concurrent target. However, a deeper understanding of the technology resolves the apparent contradiction. It turns out that certain loop-and-array computations for signal-processing algorithms such as audio/video codecs, radios, filters, and so on, can be viewed as equivalent parallel computations. Their mostly homogeneous and well-structured concurrency can be automatically parallelized and hence converted into parallel hardware.

Unfortunately, traditional (C-based) HLS technology does not address the many parts of an SoC that do not fall into the loop-and-array paradigm – processors, caches, interconnects, bridges, DMAs, I/O peripherals, and so on. One of Bluespec's customers estimated that 90% of their IP portfolio will not be served by C-based synthesis. These components are characterized by heterogeneous, irregular and complex parallelism for which the sequential computational model of C is in fact a liability. High-level synthesis for these components requires a fundamentally different approach.

In contrast, Bluespec's approach is fundamentally parallel, and is based first on *atomic transactions*, the most powerful tool available for specifying complex concurrent behaviors. Second, Bluespec has mechanisms to compose atomic transactions across module boundaries, addressing the crucial but often underestimated complexity that many control circuits fundamentally must straddle module boundaries. Handling this fundamental non-modularity smoothly and automatically is key to system integration and IP reuse. Third, it has a precise notion of mapping atomic transactions to synchronous logic, and can do so in a "refinable" way; that is, it can be refined from an initial coarse timing to the final desired circuit timing. Fourth, it is based on high-level types and higher-order programming facilities more often found in advanced programming languages, delivering succinctness, parameterization, reuse and control adaptivity. Finally, all this is synthesizable, enabling design by refinement, early estimates of architectural quality, early and fast emulation on FPGA platforms for embedded software development, and early and high-quality hardware for final implementations. In this chapter, we provide an overview of this "whole-SoC" design solution, and describe its growing validation in the field.

8.2 Atomic Transactions for Hardware

In many high-level specification languages for complex concurrent systems, such as Guarded Commands [6], Term Rewriting Systems [2, 10, 23], TLA+ [11], UNITY [4], Event-B [17] and others, the concurrent behavior of a system is expressed as a collection of rewrite rules. Each rule has a *guard* (a boolean predicate on the current state), and an *action* that transforms the state of the system. These rules can be applied in parallel, that is, any rule whose guard is true can be applied at any time. The only assumption is that each rule is an *atomic transaction* [12, 16], that is, each rule observes and delivers a consistent state, relative to all the other rules. This formalism is popular in high-level specification systems because it permits concurrent behavioral descriptions of the highest abstraction, and it simplifies establishment of correctness with both informal and formal reasoning, because atomicity directly supports the concept of reasoning with *invariants*. It is also universally applicable to all kinds of concurrent computational processes, not just "data parallel" applications. Atomic transactions have been in widespread use for decades in database systems and distributed systems, and recently there has been a renewed spurt of interest even for traditional software because of the advent of multithreaded and multicore processors [8, 22].

When viewed through the lens of atomicity, it suddenly becomes startlingly clear why RTL is so low-level, fragile, and difficult to reuse. The complexity of RTL is fundamentally in the *control logic* that is used to orchestrate movement of data and, in particular, for access to shared resources – arbitration and flow control. In RTL, this logic must be designed explicitly by the designer from scratch in every instance. This is tedious by itself and, because it is ad hoc and without any systematic discipline, it is also highly error-prone, leading to race conditions, interface protocol errors, mistimed data sampling, and so on – all the typical difficult-to-find bugs in RTL designs. Further, this control logic needs to be redesigned each time there is a small change in the specification or implementation of a module.

Another major problem affecting RTL design arises because atomicity – consistent manipulation of shared state – is fundamentally *non-modular*, that is, you cannot take two modules independently verified for atomicity and use them as black boxes in constructing a larger atomic system. Textbooks on concurrency usually illustrate this with the following simple example: imagine you have created a "bank account" module with transactions *withdraw*() and *deposit*(), and you have verified their correctness, that is, that each transaction performs its read-modify-write atomically. Now imagine a larger system in which there are concurrent activities that are attempting to perform *transfer*() operations between two such bank account modules by withdrawing from one and depositing to the other. Unfortunately there is no guarantee that the *transfer*() operation is atomic, even though the *withdraw*() and *deposit*() transactions, which it uses, are atomic. Additional control structure is needed to ensure that *transfer*() itself is atomic. The problem gets even more complicated if the set of shared resources is dynamically determined; if concurrent activities have to *block* (wait) for certain conditions before they can proceed; and if concurrent activities have to make *choices* reactively based on current availability of shared resources. This issue of non-compositionality is explored in more detail in [8] and although explained there in a software context, it is equally applicable to hardware modules and systems. Atomicity requires control logic, and that control logic is non-modular.

This leads precisely to the core reason why Bluespec SystemVerilog [3] dramatically raises the level of abstraction – automatic synthesis of all the complex control logic that is needed for atomicity.

In addition, Bluespec contributes the following:

- Provision of compositional atomic transactions within the context of a familiar hardware design language (SystemVerilog [9])
- Definition of precise mappings of atomic transactions into clocked synchronous hardware

- An industrial-strength synthesis tool that implements this mapping, that is, automatically transforms atomic transaction-based source code into RTL
- Simulation tools based on atomic transactions

The synthesis tool produces RTL that is competitive with hand-coded RTL, and the simulator executes an order of magnitude faster than the best RTL simulators (see Sect. 8.9).

We first illustrate the impact of supporting atomicity with a small example, and then with a larger one. We realize that the small example may seem too low level and narrow for a discussion on High Level Synthesis, but it is eye-opening to realize how much complexity in RTL can be attributed to atomicity concerns, even with such a small example. Ultimately, atomic transactions prove their value when you scale to larger systems (because atomicity is not too difficult to implement manually in the small).

Consider the situation in the figure below. Three concurrent activities A, B and C periodically update the registers x and y. Activity A increments x when condA is true, B decrements x and increments y when condB is true, and C decrements y when condC is true. Let us also specify that if both condB and condC are true, then C gets priority over B, and similarly that B gets priority over A (Fig. 8.1).

The following Verilog RTL is one way to express this behavior. (There are several alternate styles in which to write the RTL, but every variation is susceptible to the same analysis below).

```
always @(posedge CLK) begin
  if (condC)
    y <= y - 1;
  else if (condB) begin
    y <= y + 1; x <= x - 1;
  end;
  if (condA && (!condB || condC))     // SchedA
    x <= x + 1;
end
```

The conditional statements and their boolean expressions represent *control logic* that governs what each register is updated with, and when. Note in particular the last conditional expression, which is flagged with the comment SchedA. A naïve coder might have just written (condA && !condB), reflecting the priority of B over

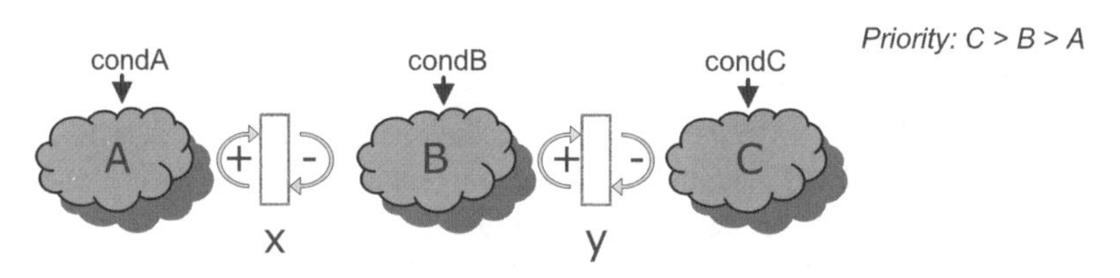

Fig. 8.1 Small atomicity example – consistent access to multiple shared resources

A for updating x. But here the designer has exploited the following transitive chain of reasoning: if condC is true, then B cannot update x even if condB is true because B must update x and y together and C has priority over B for updating y. Therefore, it is now ok for A to update x.

Said another way, the competition for resource y shared between atomic transactions B and C can affect the scheduling of the atomic transaction A because of the competition between A and B for another shared resource, x. In microcosm, this transitive effect also illustrates why atomicity is fundamentally non-modular; that is, the control structures for managing consistent access to shared resources require a non-local view.

Next, we show how the same problem is solved using Bluespec SystemVerilog (BSV).

```
rule rA (condA);
    x <= x + 1;
endrule

rule rB (condB);
    y <= y + 1; x <= x - 1;
endrule

rule rC (condC);
    y <= y - 1;
endrule

(* descending_urgency = "rC, rB, rA")
```

Each rule represents an atomic transaction. It has a *guard*, which is a boolean condition indicating a necessary (but not sufficient) condition for the rule to fire. It has a *body*, or *action*, which is a logically instantaneous state transition (this can be composed of more than one sub-action, all of which happen in parallel, as in rule rB,). The final line expresses, declaratively, the desired priority of the rules. The textual ordering of the rules and the final phrase is irrelevant, and the textual ordering of the two actions in the body of rule rB is also irrelevant; in this sense, it is a highly *declarative* specification of the solution. From this specification, the Bluespec compiler (synthesis tool) produces RTL equivalent to that shown earlier; that is, it produces all the control logic that had to be designed and written explicitly in RTL, taking into account all the scheduling nuances discussed earlier, including transitive effects.

The reason a rule's guard is necessary but not sufficient for its firing is precisely because of contention for shared resources. For example, condB is necessary for rB, but not sufficient – the rule should not fire if condC is true.

To drive home the importance of this automation, imagine what modifications would be needed in the code under the following changes in the specification:

- The priority is changed to A > B > C, or B > A > C. In each case the RTL design needs an almost complete rethink and rewrite, because the control logic changes drastically and this must be expressed in the RTL. In the BSV code, however, the only change is to the priority specification, and the control logic is regenerated automatically.
- Activity B only decrements x if y is even. In the RTL code, the decrement of x can easily be wrapped with an "if (even(y)..." condition. But now consider the condition SchedA for the x increment. It changes to the following:

```
if (condA && (!(condB && even(y)) || condC))
  x <= x + 1;
```

  In other words, A has access to x if condC is true (as before, because then C has priority for y and so B cannot run anyway), or else if B is not competing for x; that is, it is not the case that condB is true and y is even.

We can see that the control logic for managing competing accesses to shared resources gets more and more messy and complex, even in such a small example. There is even some repetition in the control expressions, such as the tests for condB and even(y), leading to the possibility of cut-and-paste errors. The complexity increases when the set of shared resources demanded by an atomic transaction is dynamic or data dependent, as in the last bullet, where B competed for x with A only if y was even. A small slip-up in writing one of those complex access conditions results in a race condition, or a protocol error, or dropping a value, or writing a wrong value into a register – all the common bugs that plague RTL design.

For a larger example, consider a packet switch (perhaps in an SoC interconnect) that has N input ports and N output ports. Consider that not all inputs may need to be connected to all outputs, and vice versa. Consider that at the different points in the switch where packets merge to a common destination, different arbitration policies may be specified. Consider that for each incoming packet, the set of resources needed is dependent on the contents of the packet header (destination buffers, unicast vs. multicast, certain statistics to be counted, and so on). When coding in RTL, the control logic for such a switch is a nightmare. With BSV rules, on the other hand, the behavior can be elegantly and correctly captured by a collection of atomic transactions, where each transaction encapsulates all the actions needed for processing packets from a particular input – all the control logic to manage all the shared resources in the switch is automatically synthesized based on atomicity semantics.

In summary, much of the complexity of coding in RTL, much of the complexity in debugging RTL, and much of its fragility against change or reuse arises from the ad hoc treatment of concurrent access to shared resources, that is, the lack of a discipline of atomicity. Further, decades of experience with multithreaded software shows clearly that a discipline of atomicity cannot be imposed merely by programming conventions or style – it needs to be built into the semantics of the language, and it needs to be built into implementations – simulation and synthesis tools (see also [13] and [22]). For this reason, much of this critique also applies to SystemC, which has atomic primitives but not atomic transactions. By making atomic transactions part of the semantics and automating the generation of control logic thereby implied, BSV dramatically simplifies the description and implementation of complex hardware systems.

8.3 Atomic Transactions with Timing, and Temporal Refinement

Atomic transactions are of course an old idea in computer science [12]. In BSV, uniquely, they are additionally mapped into synchronous time and this, in turn, provides the basis for automatic synthesis into synchronous digital hardware. In pure rule semantics [2, 4, 10, 23], one simply executes one enabled rule at a time, and hence rules are trivially atomic. In BSV, we have a notion of a global clock (BSV actually has powerful facilities for multiple clock domains, but this is not necessary for the current discussion). In each "clock cycle", BSV executes a subset of the enabled rules – the subset is chosen based on certain practical hardware constraints. The BSV synthesis tool compiles parallel hardware for these rules, but it is always *logically equivalent* to a serialized execution of the subset. Thus, the parallel hardware is true to pure rule semantics, and hence preserves atomicity and correctness.

Every BSV program has this model of computation, whether it represents an early, coarse, functional model or a final, silicon-ready, production implementation. An early functional model may lump all of the computation into a single rule or just a few rules. Its execution can be imagined to be governed by a clock with a long time period (in general we may not care much about this "clock" at the stage). The designer splits rules into finer, smaller rules according to architectural considerations such as pipelining, or concurrency, or iteration, and so on. These later refinements may be imagined to execute with a faster, finer clock, and permit more concurrency because of the finer grain. Thus, the process of design involves not only a refinement of functionality, but also a refinement of time, from the early, coarse, possibly highly uneven clock (untimed) of an early model to the final, full speed, evenly-spaced synchronous clock of the delivered digital hardware. At every step of refinement, the designer can measure latencies and bandwidths, and identify bottlenecks with respect to the current granularity of rule contention. This is a much more disciplined, realistic and accurate modeling of time compared to the typically ad hoc mechanisms often used in so-called PVT models (Programmer's View plus Timing).

The mapping of a logical ordering of rules into clock cycles can be viewed as a kind of *scheduling*. BSV does this scheduling automatically, with occasional high-level guidance from the designer in the form of assertions about the desired schedule. There is a full theory of how such schedules can be specified formally to control precisely how rules are mapped into clocks [19]. Because these scheduling specifications are about timing, they are also known as "performance specifications".

8.4 Atomic Transactional Module Interfaces

It is widely accepted that RTL's signal-level interfaces or SystemC's sc_signal level interfaces are very low-level. In SystemC modeling, and in SystemVerilog test-benches, there is a trend towards so-called "transactional" interfaces, which use an object-oriented "method calling" style for inter-module communication. This is certainly an improvement, but without atomicity, they are severely limited. Many interface protocol issues can be traced once again to the lack of a discipline for atomicity.

Consider a simple FIFO, with the usual *enqueue*() and *dequeue*() methods. In general, we cannot enqueue when a FIFO is full, nor dequeue when it is empty. In a hardware FIFO, there is also a concept of simultaneity, namely "in the same clock" (we ignore for now the situation of multiple clock domains), and in this context we can ask the question: "Can one enqueue and dequeue simultaneously, under what conditions, and with what meaning?"

One can imagine three different kinds of FIFOs, all of which have exactly the same set of hardware signals at their interface. Assume all the FIFOs allow simultaneous enqueues and dequeues in the non-boundary conditions, that is, when it is neither full nor empty. The interesting differences are in the boundary conditions:

- The *naïve* FIFO allows only dequeue if full, and only enqueue if empty. The reason for the FIFO name is that this is typically the first FIFO designed by an inexperienced designer!
- The *pipeline* FIFO, the most common kind, allows only enqueue if empty, but allows a simultaneous enqueue and dequeue if full. The reason for the name is that when full, it behaves like a pipeline buffer, that is, a new element can simultaneously arrive while the oldest value departs.
- The *bypass* FIFO allows only dequeue if full, but allows a simultaneous enqueue and dequeue if empty. The reason for the name is that when empty, a new value can arrive via the enqueue operation and "bypass" through the FIFO to depart immediately via the dequeue operation.

(Of course, one can imagine a fourth FIFO that has both pipeline and bypass behavior, but it is not necessary for this discussion.) To illustrate the ad hoc nature of how this is typically specified, a certain commercial IP vendor's data sheet for a pipeline FIFO covers several pages. On one page it states, "An error occurs if a push [enqueue] is attempted while the FIFO is full". On another page it states, "Thus, there is no conflict in a simultaneous push and pop when the FIFO is full". These partially contradictory specifications are only given informally in English.

These nuances are not academic. Although these three FIFOs have *exactly* the same RTL signals at its module interface, the *control logic* in a client module governing access to such a FIFO is different for each of the different types of FIFO. *Every* instance of this FIFO imposes a verification obligation on the designer of the client module to ensure that the operations are invoked correctly, particularly at the boundary conditions.

What has all this got to do with atomic transactions? In BSV, interface methods like *enqueue* and *dequeue* are parameterized, invocable, shareable components of atomic transactions. In other words, an atomic transaction in a client module may invoke the *enqueue* or *dequeue* operation (using standard object-oriented syntax), and those operations become part of the atomic transaction. If in the current clock the *enqueue* operation is not ready (perhaps because the FIFO is full), the atomic transaction containing the *enqueue* operation cannot execute. Thus, one can think of every method as having a condition and an action (just like a *rule*), and its condition and action become part of the overall condition and action of the invoking rule. Methods are also shareable. For example, many rules may invoke the *enqueue* method of a single FIFO. This, too, plays a role in atomic semantics because in any given clock cycle, only one of the rules can be invoke the shared method, so if a particular rule is inhibited for this reason, its other actions should also be inhibited on that clock (because its actions must be atomic).

Because of atomicity (and its related concept of *serializability*), there is a precise and well-defined concept of "logically before" and "logically after", when rules and methods are scheduled simultaneously, that is, within the same clock. Given any two rule executions R1 and R2, either R1 happens before R2 (logically), or it happens after. This concept directly gives us a formal way to express the differences between the three kinds of FIFOs. The following table summarizes the terminology, focusing only on the boundary conditions:

	When empty	When full
Naïve FIFO	*enqueue*	*dequeue*
Pipeline FIFO	*enqueue*	*dequeue* < *enqueue*
Bypass FIFO	*enqueue* < *dequeue*	*dequeue*

In the left-hand column (when empty) the Bypass FIFO allows both operations "simultaneously", but it is logically as if the *enqueue* occurred before the *dequeue*. In the logical ordering, the *enqueue* is ok when the FIFO is empty, and then the dequeue is ok because logically the FIFO is no longer empty, and, further, it receives the freshly enqueued value. Similarly, in the right-hand column (when full) the Pipeline FIFO allows both operations "simultaneously", but it is logically as if the *dequeue* occurred before the *enqueue*. In the logical ordering, the *dequeue* is ok when the FIFO is full, and then the *enqueue* is ok because logically the FIFO is no longer full. The oldest value departs and a new value enters.

This discussion gives a flavor of how Bluespec extends atomicity semantics into inter-module communication, and uses these semantics to capture formally the "scheduling" properties of the interface methods; in short, the protocol of the interface methods. Given a BSV module, the tool automatically infers properties like those shown in the table. Then, for every instance of these FIFOs, the tool produces the correct external control logic, by construction. The verification obligation on the RTL designer's shoulders, mentioned earlier, is eliminated completely.

Although transactional interfaces exist in SystemC and in SystemVerilog (and may not always be synthesizable), it is their atomicity semantics in Bluespec that gives them tremendous *compositional* power (scalability of systems) and full synthesizability.

8.5 A Strong Datatype System and Atomic Transactional Interfaces

It is well acknowledged that C has a weak type system. C++ has a much stronger type system, but it is not clear how much of it can be used in the synthesizable subsets of existing tools. Advanced programming languages like Haskell and ML have even stronger type systems. The type systems themselves provide abstraction (abstract types), parameterization and reuse (polymorphism and overloading). Type checking in such systems is a form of strong static verification.

Bluespec's type system strengthens the SystemVerilog type system to a level comparable to C++ and beyond (in fact it is strongly inspired by Haskell). As an example of this, we show how it is used to provide very high level interfaces and connections.

We start with an extremely simple interface:

```
interface Put#(t);
    method Action  put  (t x);
endinterface
```

This defines a new *interface type* called Put#(). It is polymorphic; that is, it is parameterized by another type, t. It contains one method, put(), which takes an argument x of type t and is of type Action. Action is the abstract type of things that go into atomic transactions (rules and methods); that is, atomic transactions consist of a collection of Actions. The method expresses the idea of communicating a value (x) into a module and possibly affecting its internal state. In C++ terminology, interfaces are like virtual classes and polymorphism is the analog of template classes. Unlike C++, however, BSV's polymorphic interfaces, modules and functions can be separately type-checked fully, whereas in C++ template classes can be fully type-checked only after the templates have been instantiated.

Similar to Put#(), we can also define Get#():

```
interface Get#(t);
    method ActionValue#(t) get();
endinterface
```

The get() method takes no argument, and has type ActionValue#(t); that is, it returns a value of type t and may also be an Action – it may also change the state of the module. It expresses the idea of retrieving a value from a module.

Interface types can be nested, to produce more complex interfaces. For example:

```
interface Client#(reqT, respT);
    interface Get#(reqT) request;
    interface Put#(respT) response;
endinterface

interface Server#(reqT, respT);
    interface Put#(reqT) request;
    interface Get#(respT) response;
endinterface
```

A Client#() interface is just one where we get requests and put responses, and a Server#() interface is just the inverse. Now consider a cache between a processor and a memory. Its interface might be described as follows:

```
interface Cache#(memReq, memResp);
    interface Server#(memReq, memResp) toCPU;
    interface Client#(memReq, memResp) toMem;
endinterface
```

The cache interface contains a Server#() interface towards the CPU, and a Client#() interface towards the memory. It is parameterized (polymorphic) on the types of memory requests and memory responses.

In this manner, it is possible to build up very complex interfaces systematically, starting with simpler interfaces. Polymorphism allows heavy reuse of common, standard interfaces (and many are provided for the designer in Bluespec's standard libraries).

Next, we consider user-defined *overloading*. Many pairs of interfaces are natural "duals" of each other. For example, a module with a Get#(t) interface would naturally connect to a module with a Put#(t) interface, provided t is the same. Similarly, Client#(t1,t2) and Server#(t1,t2) are natural duals. And this is of course an open-ended collection – AXI masters can connect to AXI slaves (provided they agree on address widths, data widths, and other polymorphic parameters), OCP masters to OCP slaves, my-funny-type-A to my-funny-type-B, and so on.

Of course, a connection is, in general, just another module. It could be as simple as a collection of wires, but connecting some interfaces may need additional state, internal state machines and behaviors, and so on.

BSV has a powerful, user-extensible *overloading* mechanism in its type system, patterned after Haskell's overloading mechanism, which allows us to define a single "design pattern" called *mkConnection*(i1, i2) to connect an interface of type i1 to an interface of type i2, for suitable pairs of types i1 and i2, such as Get#(t) and Put#(t). Note: many languages provide some limited overloading, typically of binary infix operators, but what is being overloaded here is a *module*. In BSV, any kind of elaboration value can be overloaded – operators, functions, modules, rules, and so on.

As a consequence, the complete top-level structure of a CPU-cache-memory system can be expressed succinctly and clearly with no more than a few lines of code:

```
module mkSystem;
    Client#(MReq, MResp)    cpu <- mkCPU;
    CacheIfc#(MReq, MResp) cache <- mkCache;
    Server#(MReq, MResp)    mem <- mkMem;
    mkConnection (cpu, cache.toCPU);
    mkConnection (cache.toMem, mem);
endmodule
```

In the first line mkCPU instantiates a CPU module which yields a Client interface that we call cpu. Similarly the next two lines instantiate the cache and the memory. The fourth line instantiates a module that establishes the cpu-to-cache connection, and the final line instantiates a module that establishes the cache-to-memory connection. Note that the two instances of mkConnection may be used at different types; overloading resolution will automatically pick the required mkConnection module.

The final feature of BSV's type system we wish to mention in this section is one that deals with the *sizes* of entities, and the often complex relationships that exist between sizes. For example, a multiplication operation may take operands of width m and n, and return a result of width $m + n$. These are directly expressible in Bluespec's type system as three types Int#(m), Int#(n) and Int#(mn) along with a *proviso* (a constraint) that $\mathrm{m} + \mathrm{n} = \mathrm{mn}$. Another example is a buffer whose size is K, with the implication that a register that indexes into this buffer must have width log(K). These constraints can be used in many ways. First, they can be used as pure constraints that are checked statically by the compiler. But, in addition, they can be *solved* by the Bluespec compiler to derive some sizes from others. For example, in designing a module containing a buffer of size K, it can derive the size of its index register, log(K), or vice versa. These features are extremely useful in designing hardware, particularly for fixed-point arithmetic algorithms, where each item is precisely sized to the correct width and all constraints between widths are automatically checked and preserved by the compiler.

8.6 Control-Adaptive Architectural Parameterization and Elaboration

In BSV, one can abstract out the concept of a "functional component" as a reusable building block. Then, separately, one can express how to compose these functional components into microarchitectures, such as combinational, pipelined, iterative, or concurrent structures. For example, a function of ActionValue type in BSV expresses a piece of sequential behavior. A function of type Rule expresses a complete piece of reactive behavior, in fact a complete reactive atomic transaction. All these components are "first class" data types, so one can build and manipulate "collections" such as lists and vectors of ActionValues, Rules, Modules, and so on.

Second, BSV has some powerful "generate" mechanisms that allow one to compose microarchitectures flexibly and succinctly. For example, the microarchitectural structure can be expressed using conditionals, loops, and even recursion. These can manipulate lists of rules, interfaces, modules, ActionValues, and so on, in order to *programmatically* construct modules and subsystems.

Third, BSV has very powerful parameterization. One can write a single piece of parameterized code that, based on the choice of parameters, results in different microarchitectures (such as pipelined vs. concurrent vs. iterative, or varying a pipeline pitch, or using alternative modules, and so on.).

Finally, and most important, what makes all this flexibility work is the control-adaptivity that arises out of the core semantics of atomic transactions. Each change in microarchitecture from these capabilities of course needs a corresponding change in the control logic. For example, if two functional components are composed in a pipelined or concurrent fashion, they may conflict on access to some shared resource, whereas when composed iteratively, they may not – these require different control logics. When designing with RTL, it is simply too tedious and error-prone to even contemplate such changes and to redesign all this control logic from scratch. Because BSV's synthesis is based on atomic semantics, this control logic is resynthesized automatically – the designer does not have to think about it.

For example, in a mathematical algorithm, many sections of the code represent N-way 'data parallel' computations, or 'slices'. We first abstract out this slice function, and then we can write a single parameterized piece of code that chooses whether to instantiate N concurrent copies of this slice, or N/2 copies to be used twice, or N/4 copies to be used four times, and so on. Similarly, each of these slices could be pipelined, or not. BSV automatically generates all the intermediate buffering, muxing and control logic needed for this.

So, the designer can rapidly adjust the microarchitecture in response to timing, area and power estimation results from actual RTL-to-netlist synthesis, and converge quickly on an optimized design. The baseline atomicity semantics of BSV is key to preserving correctness and eliminating the effort that would be needed to redesign the control logic. Reference [5] presents a detailed case study of an 802.11a (WiFi) transmitter design in BSV using these techniques, including a somewhat counter-intuitive result about which micro-architecture resulted in the least-power implementation. In other words, without the kind of architectural flexibility described in this section, the designer's intuition may have led to a dramatically sub-optimal implementation.

8.7 Some Comparisons with C-Based HLS

Having described the various features of the BSV approach, we can now make some brief comparisons with classical C-based High Level Synthesis.

In classical C-based HLS, the design-capture language is typically C (or C++). To this are added proprietary "constraints" that specify, or at least guide, the synthesis tool in microarchitecture selection, such as loop unrolling, loop fusion, number of resources available, technology library bindings, and so on. The synthesis tool uses these constraints and knowledge about a particular target technology and technology libraries to produce the synthesized output.

Since the reference semantics for C and C++ are sequential, what C-based HLS tools do is a kind of automatic parallelization; that is, by analyzing and transforming the intermediate form of Control/Data Flow Graphs (CDFGs), they relax the reference sequential semantics into an equivalent parallel representation suitable for hardware implementation. In general, this kind of automatic parallelization is only successful on well-structured loop-and-array computations, and is not applicable to more heterogeneous control-dominated components such as processors, caches, DMAs, interconnect, I/O devices, and so on. Even for loop-and-array computations, it is rare that an off-the-shelf C code results in good synthesis; the designer often must spend significant effort "restructuring" the C code so that it is more amenable to synthesis, often undoing many common C idioms into more analyzable forms, such as converting pointer arithmetic into array indexing, elimination of global variables so that the data flow is more apparent, and so on. Reference [20] describes in detail the kinds of source-level transformations necessary by the designer to achieve good synthesis, and reference [7] describes in more generality the challenge of getting good synthesis out of C sources.

As described in the previous section on "Control-Adaptive Architectural Parameterization and Elaboration", in BSV the microarchitecture is specified precisely in the source, but with such powerful generative and parameterization mechanisms that a single source can flexibly represent a rich family of microarchitectures, within which different choices may be appropriate for different performance targets (area, clock speed, power). Further, the structure can be changed quickly and easily without compromising correctness or hardware quality, in order quickly to converge to a satisfactory implementation. Thus, BSV provides synthesis from very high level descriptions but, paradoxically, the microarchitecture is precisely specified in the parameterized program structure.

Experience has shown that with these capabilities, the BSV approach, although radically different, easily matches the productivity and quality of results of classical C-based HLS for well-structured loop-and-array algorithmic codes. But unlike C-based synthesis, BSV is not limited to such computations – its explicit parallelism and atomic transactions make it broadly suitable to all the different kinds of components found in SoCs, whether data- or control-oriented.

BSV synthesis is currently technology neutral – it does not try to perform technology-specific optimizations or retimings (BSV users rely on downstream tools to perform such technology-specific local retiming optimizations).

These properties of BSV also provide a certain level of transparency, predictability and controllability in synthesis; that is, even though the design is expressed at a very high level, the designer has a good idea about the structure of the generated

RTL (the synthesis tool is also heavily engineered to produce RTL that is not only highly readable, but where the correspondence to the source is evident).

Although, as we have discussed, BSV is universal and can be applied to design all kinds of components in an SoC, there is no reason why BSV cannot be used in conjunction with classical C-based HLS. Indeed, one of Bluespec's customers has implemented a complex "data mover" for multiple video data formats, where some of the sources and destinations of the data are "accelerators" for various video algorithms that are implemented using another C-based synthesis tool.

8.8 Additional Benefits

The features of BSV we have described provide a number of additional benefits that we explore in this section.

Design-by-refinement: Because of the control-adaptiveness of BSV, that is, the automatic reconstruction of control circuits as microarchitecture changes, BSV enables repeated incremental changes to a design without damaging correctness. A common practice is to start by producing a working skeleton of a design, literally within hours or days, by using the powerful parameterized interfaces and connections already defined in Bluespec's standard libraries, such as Client and Server and mkConnection. This initial approximation already defines the broad architecture of the design, and the broad outlines of the testbench. Then, repeatedly, the designer adds or modifies detail, either to increase functionality or to adjust the microarchitecture for the existing functionality. At every step, the design is recompiled, resimulated, and tested – verification is deeply intertwined with design, instead of being a separate activity following the design.

Because the concept of mapping atomic transactions to synchronous execution is present from the beginning, the methodology also involves a refinement of timing. The first, highly approximate and incomplete model itself has a notion of clocks, and hence abstract timing measurements of latency and throughput can begin immediately. Bottlenecks can be identified and resolved through microarchitecture refinement.

As this refinement proceeds, since everything is synthesizable to RTL from the beginning, one may also periodically run RTL-to-netlist synthesis and power estimation tools to get an early indication of whether one is approaching silicon area, clock speed and power targets.

Thus the whole process has a smooth trajectory from high level models to final implementation, without any disruptive transitions in methodology, and with no late surprises about meeting latency, bandwidth or silicon area and clock speed targets. Early BSV models can thus also be viewed as executable specifications.

Early fast simulation on FPGAs: Because synthesis is available from the very earliest approximate models in the above refinement methodology, many BSV users are able quickly to run their models on FPGA platforms and emulators. Note, the microarchitecture may be nowhere near the final version, and its FPGA implementation may run at nowhere near the clock speed of the final version, but it can still provide, effectively, a simulator that is much faster than software simulation.

This capability can more rapidly identify microarchitectural problems, and can provide a fast "virtual platform" early to the software developers.

Formal specification and verification: In the beginning of Sect. 8.2 we mentioned several well-known formal specification languages that share the same basic computational model as BSV – a collection of rewrite rules, each of which is an atomic transaction, that collectively express the concurrent behavior of a system. As such, the vast theory in that field is in principle directly applicable to BSV. In practice, some individual projects have been done in this area with BSV, notably processor microarchitecture verification [1], systematic derivation of processor microarchitectures via transformation [15], and the verification of a distributed, directory-based cache-coherence protocol [21]. We expect that, in the future, BSV tools will incorporate such capabilities, including integration with formal verification engines.

8.9 Experience and Validation, and Conclusion

Bluespec SystemVerilog is an industrial-strength tool, with research roots going back at least 10 years, and production-quality implementations going back at least 7 years. It also continues to serve as a fertile research vehicle for Bluespec and its university partners. Many large designs (from 100 Ks to millions of gates) have been implemented in Bluespec, and some of them are in silicon in delivered products today.

Measured over several dozens of medium to large designs, BSV designs have routinely matched hand-coded RTL designs in silicon area and clock speed. In a few instances, BSV has actually done much better than hand-coded RTL because BSV's higher-level of abstraction permitted the designer clearly to see a better architecture for implementation, and BSV's robustness to change allowed modifications to the design accordingly.

Bluesim, Bluespec's simulator, is capable of executing an order of magnitude faster than the best RTL simulators. This is because the simulator is capable of exploiting the semantic model of BSV, where atomic transactions are mapped into clocks, to produce significant optimizations over RTL's fine-grained event-based simulation model.

Of course BSV has proven excellent for highly control-oriented designs like processors, caches, DMA controllers, I/O peripherals, interconnects, data movers, and so on. But, interestingly, it has also had excellent success on designs that were previously considered solely the domain of classical High Level (C-based) Synthesis. These designs include, as examples:

- OFDM transmitter and receiver, parameterized to cover 802.11a (WiFi), 802.16 (WiMax), and 802.15 (WUSB). Reference [5] describes the 802.11a transmitter part. This BSV code is available in open source, courtesy of MIT and Nokia [18]

- H.264 decoder [14]. This code is capable of decoding 720 p resolution video at 75 fps in .18 um technology (about the same computational effort as 1,080 p at 30 fps). This BSV code is available in open source, courtesy of MIT and Nokia [18]
- Components of an H.264 encoder (customer proprietary)
- Color correction for color images (customer proprietary)
- MIMO decoder in a wireless receiver (customer proprietary)
- AES and DES (security)

Thus, BSV has been demonstrated to be truly general-purpose, applicable to the broad spectrum of components found in SoCs. In this sense it can truly be seen as a high level, next generation tool for whole-SoC design, in the same sense that RTL was used in the past.

To date, the concept of High Level Synthesis has been almost synonymous with classical C-based automatic synthesis. This, in turn, has limited its applicability only to certain components of modern SoCs, those based on structured loop-and-array computations. We hope this chapter will serve to raise awareness of a very unusual alternative approach to high level synthesis that is potentially more promising for the general case and applicable to whole SoCs.

Acknowledgments The original ideas in synthesizing rules (atomic transactions) into RTL were due to James Hoe and Arvind at MIT. Lennart Augustsson augmented this with ideas on composing atomic transactions across module boundaries, strong type checking, and higher-order descriptions. Subsequent development of BSV, since 2003, is due to the team at Bluespec, Inc.

References

1. Arvind and X. Shen, *Using Term Rewriting Systems to Design and Verify Processors*, IEEE Micro 19:3, 1998, pp. 36–46
2. F. Baader and T. Nipkow, *Term Rewriting and All That*, Cambridge University Press, Cambridge, 1998, 300 pp
3. Bluespec, Inc., *Bluespec SystemVerilog Reference Guide*, www.bluespec.com
4. K.M. Chandy and J. Misra, *Parallel Program Design: A Foundation*, Addison-Wesley, Reading, MA, 1988, 516 pp
5. N. Dave, M. Pellauer, S. Gerding and Arvind, *802.11a Transmitter: A Case Study in Microarchitectural Exploration*, in *Proc. Formal Methods and Models for Codesign (MEMOCODE), Napa Valley, CA, USA*, July 2006
6. E.W. Dijkstra, *A Discipline of Programming*, Prentice-Hall, Englewood Cliffs, NJ, 1976
7. S.A. Edwards, *The Challenge of Hardware Synthesis from C-Like Languages*, in *Proc. Design Automation and Test Europe (DATE), Munich, Germany*, March 2005
8. T. Harris, S. Marlow, S. Peyton Jones and M. Herlihy, *Composable Memory Transactions*, in *ACM Conf. on Principles and Practice of Parallel Programming (PPoPP'05)*, 2005
9. *IEEE Standard for SystemVerilog – Unified Hardware Design, Specification, and Verification Language*, IEEE Std 1800-2005, http://standards.ieee.org, November 2005
10. J. Klop, *Term Rewriting Systems*, in *Handbook in Computer Science*, S. Abramsky, D.M. Gabbay and T.S.E. Maibaum, editors, Vol. 2, Oxford University Press, Oxford, 1992, pp. 1–116

11. L. Lamport, *Specifying Systems: The TLA+ Language and Tools for Hardware and Software Engineers*, Addison-Wesley Professional (Pearson Education), Reading, MA, 2002
12. B. Lampson, *Atomic Transactions*, in *Distributed Systems – Architecture and Implementation, An Advanced Course*, Lecture Notes in Computer Science, Vol. 105, Springer, Berlin Heidelberg New York, 1981, pp. 246–265
13. E.A. Lee, *The Problem with Threads*, IEEE Comput 39:5, 2006, pp. 33–42
14. C-C. Lin, *Implementation of H.264 Decoder in Bluespec System Verilog*, Master's Thesis, Department of Electrical Engineering and Computer Science, Massachusetts Institute of Technology, MA, February 2007. Available as CSG Memo-497 at http://csg.csail.mit.edu/pubs/publications.html
15. M. Lis, *Superscalar Processors Via Automatic Microarchitecture Transformation*, Master's Thesis, Department of Electrical Engineering and Computer Science, Massachusetts Institute of Technology, MA, May 2000
16. N. Lynch, M. Merritt, W.E. Weihl and A. Fekete, *Atomic Transactions*, series in *Data Management Systems*, Morgan Kaufman, San Mateo, CA, 1994, 476 pp
17. C. Métayer, J.-R. Abrial and L. Voisin, *Event-B Language*, rodin.cs.ncl.ac.uk/deliverables/D7.pdf, May 31, 2005, 147 pp
18. MIT Open Source Hardware Designs, http://csg.csail.mit.edu/oshd
19. D.L. Rosenband and Arvind, *Hardware Synthesis from Guarded Atomic Actions with Performance Specifications*, in *Proc. ICCAD, San Jose*, November 2005
20. G. Stitt, F. Vahid and W. Najjar, *A Code Refinement Methodology for Performance-Improved Synthesis from C*, in *Proc. Intl. Conference on Computer Aided Design (ICCAD), San Jose*, November 2006
21. J.E. Stoy, X. Shen and Arvind, *Proofs of Correctness of Cache-Coherence Protocols*, in *Formal Methods for Increasing Software Productivity (FME2001)*, Lecture Notes in Computer Science, Vol. 2021, Springer, Berlin Heidelberg New York, 2001, pp. 43–71
22. *Transactional Memory Online*, online bibliography for literature on transactional memory, www.cs.wisc.edu/trans-memory/biblio
23. Terese, *Term Rewriting Systems*, Cambridge University Press, Cambridge, 2003, 884 pp

Chapter 9
GAUT: A High-Level Synthesis Tool for DSP Applications

From C Algorithm to RTL Architecture

Philippe Coussy, Cyrille Chavet, Pierre Bomel, Dominique Heller, Eric Senn, and Eric Martin

Abstract This chapter presents GAUT, an academic and open-source high-level synthesis tool dedicated to digital signal processing applications. Starting from an algorithmic bit-accurate specification written in C/C++, GAUT extracts the potential parallelism before processing the allocation, the scheduling and the binding tasks. Mandatory synthesis constraints are the throughput and the clock period while the memory mapping and the I/O timing diagram are optional. GAUT next generates a potentially pipelined architecture composed of a processing unit, a memory unit and a communication with a GALS/LIS interface.

Keywords: Digital signal processing, Compilation, Allocation, Scheduling, Binding, Hardware architecture, Bit-width, Throughput, Memory mapping, Interface synthesis.

9.1 Introduction

The technological advances have always forced the IC designers to consider new working practices and new architectural solutions. In the SoC context, the traditional design methodology, relying on EDA tools used in a two stages design flow – a VHDL/Verilog RTL specification, followed by logical and physical synthesis – is no more suitable. However, the increasing complexity and the data rates of Digital Signal Processing (DSP) applications still require efficient hardware implementations. Indeed, concerning DSP applications, pure software solutions based on multi-processor architectures are not acceptable, and optimized hardware accelerators or coprocessors – composed of a set of computing blocks communicating through point-to-point links – are still needed in the final architecture. Thus SoC embedded DSP cores will need new ESL design tools in order to raise the specification abstraction level up to the "algorithmic one". Algorithmic descriptions enable an IC designer to focus on functionality and target performances rather than debugging

P. Coussy and A. Morawiec (eds.) *High-Level Synthesis.*

RTL. Designers will spend more time exploring the design space with multiple "what if" scenarios. They will obtain a range of implementation alternatives, from which they will select the architecture providing the best power/speed/gate count trade-off.

This chapter presents GAUT which is an open-source HLS tool dedicated to DSP applications [1]. Starting from an algorithmic bit-accurate specification written in C/C++, a throughput constraint (Initiation Interval) and a clock period, the tool extracts the potential parallelism before processing the selection, the allocation, the scheduling and the binding tasks. GAUT generates a potentially pipelined architecture composed of a processing unit, a memory unit and a communication unit. Several RTL VHDL models for the logic synthesis and SystemC CABA (Cycle Accurate Bit Accurate) and TLM-T (Transaction Level Model with Timing) are automatically generated with their respective test benches.

The chapter is organized as follow: Sect. 9.2 introduces our design flow and presents the targeted architecture. Section 9.3 details each step of our high-level synthesis flow. In Sect. 9.4, experimental results are provided.

9.2 Overview of the Design Environment

High-level synthesis enables the (semi) automatic search for architectural solutions that respect the specified constraints while optimizing the design objectives. To be efficient, the synthesis must rely on a design method which takes into account the specificity of the application fields. We have focused on the domain of real-time digital signal processing and we have formalized a dedicated design approach for this type of application where the regular and periodic data-intensive computations dominate.

GAUT [1] takes as input a C description of the algorithm that has to be synthesized. The mandatory constraints are the throughput (specified through an initiation interval which represents the constant interval between the start of successive iterations) and the clock period. Optional design constraints are the memory mapping and I/O timing diagram. The architecture of the hardware components that GAUT generates is composed of three main functional units: a processing unit PU, a memory unit MEMU and a Communication & Interface Unit COMU (see Fig. 9.1). The PU is a datapath composed of logic and arithmetic operators, storage elements, steering logic and a controller (FSM). Storage elements of the PU can be strong semantic memories (FIFO, LIFO) and/or registers. The MEMU is composed of memory banks and their associated controllers. The COMU includes a synchronization processor and an operation memory which allow to have a GALS/LIS (Globally Asynchronous Locally Synchronous/Latency Insensitive System) communication interface.

As described in Fig. 9.2, GAUT first synthesizes the Processing Unit. Then it generates the Memory Unit and the Communication Unit. During the design of the PU, GAUT initially selects arithmetic operators and after targets their best use according to the design constraints and objectives. Then GAUT processes the registers and

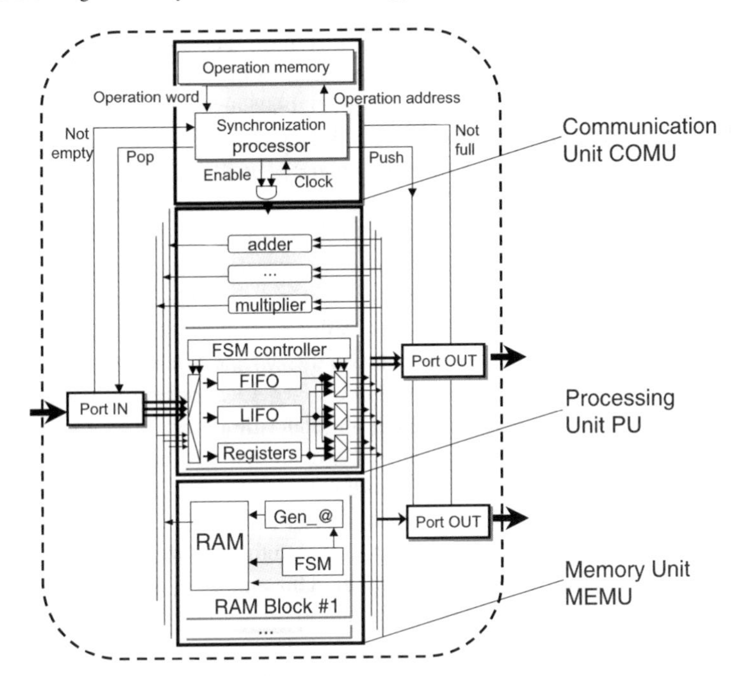

Fig. 9.1 Target architecture

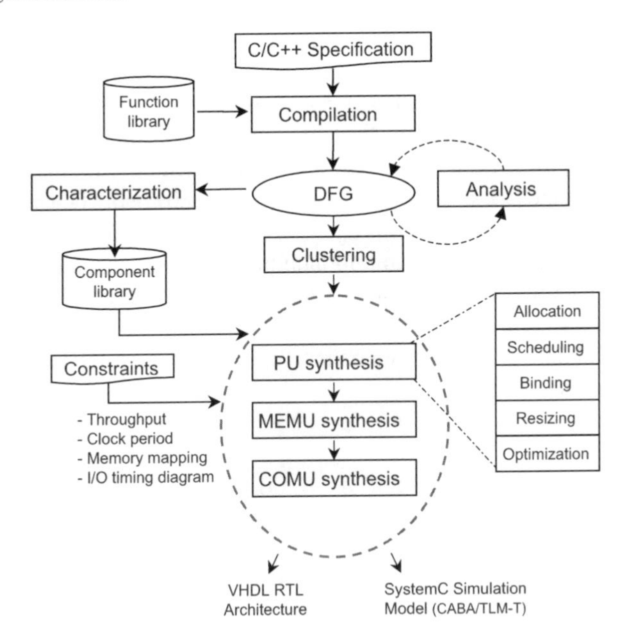

Fig. 9.2 Proposed high-level synthesis flow

memory banks, which are part of the memory unit. The register's optimization, which is done before the memory optimization, is based on prediction techniques. The communication paths will then be optimized, followed by the optimization of the address generators of the memory banks dedicated to the application being considered. The communication interface is generated next by using the I/O timing behavior of the component. To validate the generated architecture, a test bench is automatically generated to apply stimulus to the design and to analyze the results. The stimulus can be incremental, randomized or user defined values allowing automatic comparison with the initial algorithmic specification (i.e. the "golden" model). The processing unit can be verified alone. In this case, the memory and communication units are generated as VHDL components whose behavior is described as a Finite State Machine with Data path. GAUT generates not only VHDL models but also scripts necessary to compile and simulate the design with the Modelsim simulator. It can also compare the results of two simulations (produced by different timing behaviors (I/O, pipeline...)). Both "Cycle Accurate, Bit Accurate" (CABA) and "Transaction-Level Model with Timing" (TLM-T) simulation models are generated which allow to integrate the components into the Soclib platform [1]. GAUT also addresses the design of multi-mode architectures (see [3] for details).

9.3 The Synthesis Flow

9.3.1 The Front End

The input description is a C/C++ function where Algorithmic C™ class library from Mentor Graphics [5] is used. This allows the designer to specify signed and unsigned bit-accurate integer and fixed-point variables by using *ac_int* and *ac_fixed* data types. This library, like SystemC [6], hence SystemC [6], hence provides fixed-point data-types that supply all the arithmetic operations and built-in quantization (rounding, truncation...) and overflow (saturation, wrap-around...) functionalities. For example, an *ac_fixed <5,2,true,AC_RND,AC_SAT>* is a signed fixed-point number of the form bb.bbb (five bits of width, two bits integer) for which the quantization and overflow modes are respectively set to 'rounding' and 'saturation'.

9.3.1.1 Compilation

The role of the compiler is to transform the initial C/C++ specification into a formal representation which exhibits the data dependencies between operations. The compiler of GAUT derives gcc/g++ 4.2 [7] to extract a data flow graph (DFG) representation of the application annotated with the bit-width information (the code optimizations performed by the compiler will not be presented in this paper). For the quantization/overflow functionality of a fixed-point variable, the compiler generates dedicated operation nodes in the DFG. As described later, this allows to share (i.e. reuse) (1) arithmetic operators between bit-accurate integer operations and fixed-point operations and (2) quantization/overflow operators between fixed-point operations. Timing performance optimization is addressed through the operator chaining.

As detailed in [7], the gcc/g++ compiler includes three main components: a front end, a middle end and a back end. The front end performs lexical, syntactical and semantic analysis on the code. The middle end operates code optimizations on the internal representation named "GIMPLE". The back end performs hardware dependent optimizations and finally generates assembly language. The source file is processed in four main steps: (1) the C preprocessor (cpp) expands the preprocessor directives; (2) the front end constructs the Abstract Syntax Tree (AST) for each function of the source file. The AST tree is next converted into a CDFG-like unified form called GENERIC which is not suitable for optimization. The GENERIC representation is lowered into a subset called GIMPLE form; (3) false data dependencies are eliminated with Static Signal Assignment and various scalar optimizations (dead code elimination, value range propagation, redundancy elimination). Loop optimizations (loop invariant, loop peeling, loop fusion, partial loop unrolling) are applied; (4) finally the GIMPLE form is translated into the GAUT internal representation.

9.3.1.2 Bit-Width Analysis

The bit-width analysis which next operates on the DFG is based on the two following steps:

- *Constant bit-width definition*: the compiler carries out a DFG representation where the constants are represented by nodes with a 16, 32 or 64 bit size. This first analysis step defines for each constant the exact number of bits needed to represent its value. We use the simple following formula for unsigned and signed values:
$$Number\ of\ bits = \lfloor \log_2 |Value| \rfloor + 1 + S_{igned}.$$
- *Bit-width and value range propagation*: infers the bit-width of each variables of the specification by coupling work from [9] and [10]. A bit-width analysis is hence performed to optimize the word-length of both the operations and the variables. This step performs a forward and a backward propagation of both the value ranges and the bit-width information to figure out the minimum number of bits required.

9.3.1.3 Library Characterization

Library characterization uses a DFG, a technological library and a target technology (typically the FPGA model). This fully automated step, based on commercial logic synthesis tools like ISE from Xilinx and Quartus from Altera, produces a library of time characterized operators to be used during the following HLS steps. The technological library provides the VHDL behavioral description of operators and the DFG

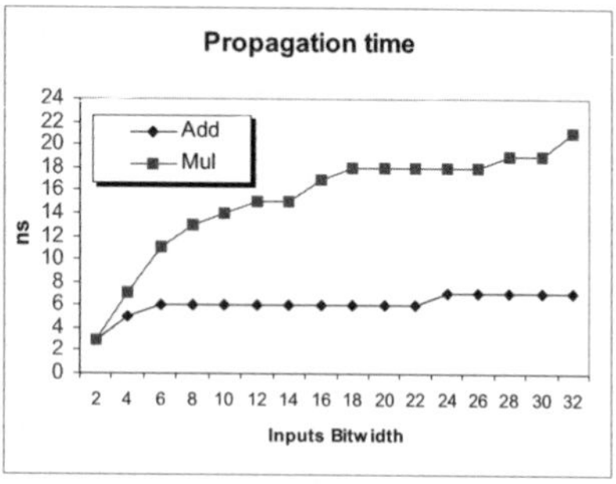

Fig. 9.3 Propagation time vs. bit-width for addition-subtraction and multiplication operations

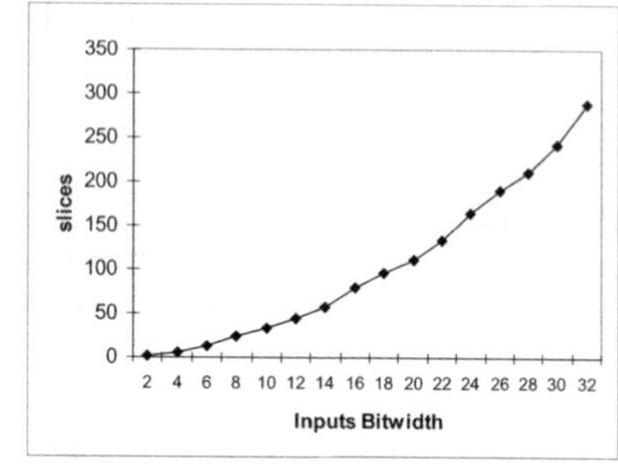

Fig. 9.4 Multiplier area vs. bit-width

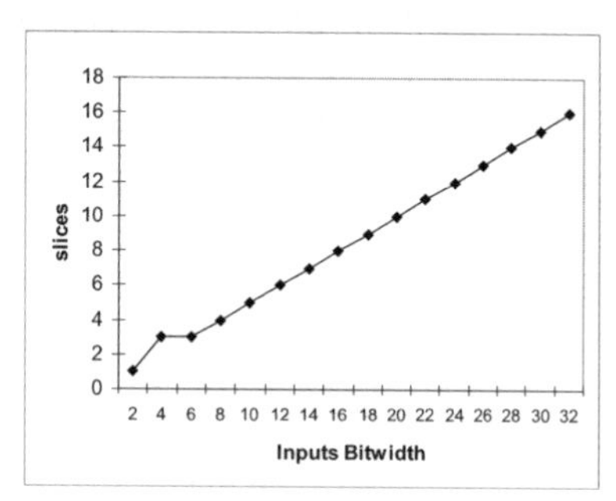

Fig. 9.5 Adder area vs. bit-width

provides the set of operations to be characterized with their bit-width information. The characterization step synthesizes each operator from the technological library which is able to realize one operation of the DFG. It next retrieves synthesis results in terms of logical cell number and propagation time to generate a characterized operator library. Figures 9.3–9.5 present results provided by the characterization step.

9.3.1.4 Operation Clustering

For clustering operations we propose to combine the computational function and the operation delay. This allows to indirectly consider operation's bit-width since the propagation time of an operator depends on its operand's size. In order to maximize the use of operators, one operation that belongs to a cluster $C1$ with a propagation time $t1$ can be assigned to operators allocated for a cluster $C2$ if the propagation time $t2$ is greater than $t1$.

9.3.2 Processing Unit Synthesis

The design of the Processing Unit (PU) integrates the following tasks: resource selection and allocation, operation scheduling, and binding of operations onto operators. First, GAUT executes the allocation task, and then executes the scheduling and the assignment tasks (see Figs. 9.2 and 9.6).

```
Inputs:
   DFG, timing constraint and resource allocation
Output:
   A scheduled DFG
Begin
   cstep=0;
   Repeat until the last node is scheduled
       Determine the ready operations RO;
       Compute the operations mobility;
       While there are RO
        If there are available resources
          Schedule the operation with the highest priority;
          Remove resource from available resource set;
           If the current operation belongs to a chaining pattern
             Update the ready operations RO;
             If there are available resources
                Schedule the operations corresponding to the pattern;
                 Remove resources from available resource set;
             End if
           End if
        Else
          If the operations can be delayed
             Delay the operations;
          Else
             Allocate resources (FUs);
             Schedule the operations;
          End if
        End if
       End while
       Bind all the scheduled operations;
       cstep++;
End
```

Fig. 9.6 Pseudo code of the scheduling algorithm

9.3.2.1 Resource Allocation

Allocation defines the type and the numbers of operators needed to satisfy the design constraints. In our approach, in order to respect the throughput requirement specified by the designer, allocation is done for each a priori pipeline stage. The number of a priori pipeline stage is computed as the ratio between the minimum latency, *Latency*, of the DGF (i.e. the longest data dependency path in the graph) and the Initiation Interval *II* (i.e. the period at which the application has to (re)iterate): $\lceil Latency/II \rceil$. Thus we compute the average parallelism of the application extracted from the DFG dated by an As Soon As Possible (ASAP) unconstrained scheduling. The average parallelism is calculated separately for each *type* of operation and for each pipeline stage *s* of the DGF, comprising the set of the date operations belonging to $[s.II, (s+1).II]$. The average number of operators, for a given operation type *type*, that is allocated to an a priori pipeline stage is defined as follow:

$$avr\_opr(type) = \left\lceil \frac{nb\_ops(type)}{\left\lfloor \frac{II}{T(opr)} \right\rfloor * \left\lceil \frac{Tclk}{II(opr)} \right\rceil} \right\rceil$$

with *Tclk* the clock period, *nb_ops*(*type*) the number of operators of type *type* that belong to the current pipeline stage, *T*(*opr*) the propagation time of the operator and *II*(*opr*) the iteration period of pipelined operators.

This first allocation is considered as a lower bound. Thus, during the scheduling phase, supplementary resources can be allocated and pipeline stages may be created if necessary. This is done subsequently to operation scheduling on the previously allocated operators.

9.3.2.2 Operation Scheduling

The classical "list scheduling" algorithm relies on heuristics in which the ready operations (operations to be scheduled) are listed by priority order. An operation can be scheduled if the current cycle is greater than or equal to its earliest time. Whenever two ready operations need to access the same resource (this is a so-called resource conflict), the operation with the highest priority is scheduled. The other is postponed.

Traditionally, bit-width information is not considered and the priority function depends on the mobility only. The operation mobility is thus defined as the difference between the As Late As Possible (ALAP) time and the current c-step (see Fig. 9.6). In order to optimize the final architecture area, we modified the classical priority function to take into account the bit-with of the operations in addition to the mobility. Hence, the priority of an operation is a weighted sum of (1) its timing priority (i.e. the inverse of its mobility) and (2) the inverse of the over-cost inferred by the pseudo assignment of the largest operator (returned by the *maxsize* function) with the operation.

$$\text{Priority} = \frac{\alpha}{mobility} + \frac{1-\alpha}{over\,\text{cost}(operation, \max size(operator))},$$

$$over\,\text{cost}\,(ops, opr) = Min \left\{ \begin{pmatrix} \frac{opr_{in1} - ops_{in1}}{opr_{in1}} + \frac{opr_{in2} - ops_{in2}}{opr_{in2}} \end{pmatrix}, \begin{pmatrix} \frac{opr_{in2} - ops_{in1}}{opr_{in2}} + \frac{opr_{in1} - ops_{in2}}{opr_{in1}} \end{pmatrix} \right\}.$$

The *overcost* function return the lowest sum of gradients of operation input's bit-width and of operator input's bit-width. This means that for a same mobility, the priority will be given to the operation that best minimizes the over-cost. For different mobility, the user defined factor α allows to increase the priority of an operation O_1 having more mobility than an operation O_2 if *overcost*(O_1) is less than *overcost*(O_2). In the over-cost computation, the reuse of an operator (already used) is avoided through a pseudo-assignment made during the scheduling. A pseudo-assignment is a preliminary binding which allows to remove the largest operator from the available resource set.

Once the operations can be no more scheduled in the current cycle, the resource binding is performed.

Operation Chaining

To respect the specified timing constraints (latency or throughput) while optimizing the final area, operator chaining can be used. In our approach, the candidate for chaining are identified by using templates in a library. Through a dedicated specification language, the user defines chaining patterns with their respective maximum delays. These latency constraints are expressed in number of clock cycles which allows to be bit-width independent in the pattern specification.

In order to allow the sharing of arithmetic operators between bit-accurate and/or fixed-point operations, the compiler generates for fixed-point operations two nodes in the DFG: one node for the arithmetic operation and one other for the quantization/overflow functionality.

Figure 9.7a depicts a fixed-point dedicated operator where the computational part is merged with the quantization/overflow functionality. This kind of operator architecture neither allows to share the arithmetic logic nor the quantization/overflow

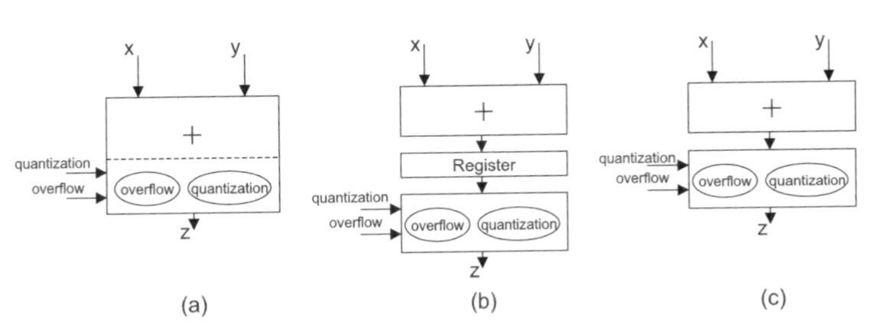

Fig. 9.7 (**a**) Monolithic fixed-point operator, (**b**) "Unchained" fixed-point operator and (**c**) Chained fixed-point operator

part between bit-accurate and/or fixed-point operations Fig. 9.7b shows the resulting architecture when the compiler generates dedicated nodes for a fixed-point operation and when chaining is not used. Figure 9.7c presents an architecture where the arithmetic part and the quantization/overflow functionality have been chained by coupling both the compiler results and a fixed-point templates.

9.3.2.3 Resource Binding

The assignment of an available operator with a candidate operation has to respond to the minimization of interconnections (steering logic) between operators and to the minimization of the operator's size. Given the set of allocated Functional Units FUs, our binding algorithm assigns all the scheduled operations of the current step (see Fig. 9.6). The pipeline control of each operator is managed by a complementary priority on assignment. When an operator is allocated, but not yet used, its priority for assignment is primarily inferior to that of an already bound operator.

The first step consists in constructing a bipartite weighted graph $G = (U, FU(V), E)$ with:

- U, the set of operations in *c-step* S_k of the DFG
- $FU(V)$, the set of available FUs in *c-step* S_k that can implement at least one operation from V
- E, the set of weighted edges $(U, FU(V))$ between a pair of operations $u \in U$ and a functional unit $fu(v)$ where $v \in V$

The edge weight w_{uv} is given by the following equation:

$$w_{u,v} = \beta * con(u,v) + (1-\beta) * dist(u,v),$$

where:

- $con(u,v)$ is the maximum number of existing connections between $fu(v)$ and each FUs assigned to the set of predecessors of u
- *dis(u,v)* is the reciprocal of the positive difference between bit-widths of u and v operands
- β is user defined factor which allow minimizing either steering logic area or computational area

The second step consists in finding the maximal weighted edge subset by using the maximum weighted bipartite matching (MWBM) algorithm described in [8].

Assuming:

- The scheduling and binding of the operations of the DFG in Fig. 9.8a on *c-step1* and *c-step2*, has been already done
- The operations O_1 and O_4 have been scheduled in *c-step3*
- Allocated operators are SUB_1, SUB_2 and ADD_1
- O_9, O_1 have been bound to SUB_1
- O_3, O_0 have been bound to ADD_1

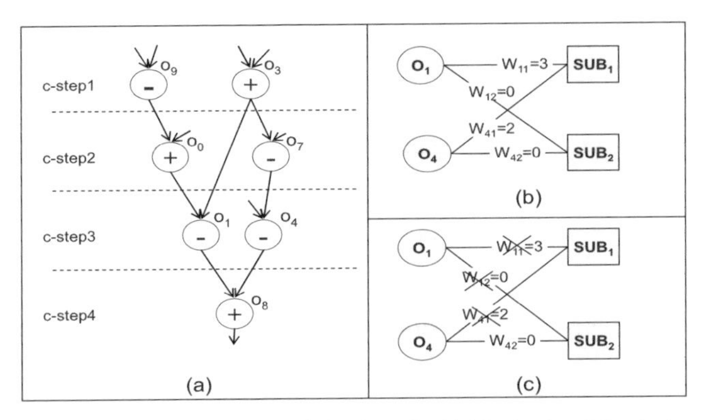

Fig. 9.8 (**a**) DFG example, (**b**) Bipartite weighted graph, (**c**) Maximal weighted edge matching

We will focus on O_1 and O_4 binding. Our algorithm first constructs the bipartite weighted graph (Fig. 9.8b) taking β equal to 1 for the sake of simplicity (i.e. only steering logic is considered). Afterwards, the MBWM algorithm is applied to identify the best edges.

Thus, operation O_1 is assigned to SUB_1 thanks to the edge weight $w_{11} = 3$. Nodes connected to w_{11} are then removed from the bipartite graph and so forward (Fig. 9.8c). In other word, connection between ADD_1 (FU bound to O_1 predecessor) and SUB_1 is maximized thereby the creation of multiplexers is avoided. Thus the final architecture has been optimized.

9.3.2.4 Operator Sizing

In this design step the operators have to be sized according to the operations which have been assigned on. In order to get correct computing results, the width of the operator inputs/outputs have to be greater or equal to the width of the operation variables. Operation variables can have different sizes which can greatly impact the propagation time and the area of the operator.

The input's width of an operator is used to be the maximum of all its inputs as described in the available literature (see [9] and and [11] for example). This computing method increases considerably the final area (see Figs. 9.4 and 9.9 and [12]). However, an operator can have different input width. Thus, the operator sizing task can optimize the final operator area by (1) computing the maximum width for each input respectively (Fig. 9.9b) or (2) computing the optimal size for each input by considering commutativity (Fig. 9.9c). However swapping inputs can infer steering logic.

Let's consider a multiplier that executes two operations O_1 and O_2. Their respective input widths are ($in_1 = 8$, $in_2 = 4$) and ($in_1 = 3$, $in_2 = 9$) and output width is 12. Figure 9.9 shows respectively for each approach the synthesis results we obtained by using a Xilinx Virtex2 xc2v8000 -4 FPGA device and the ISE 8.2 logic synthesis tool. Considering different widths for each input can thus reduce the operator area.

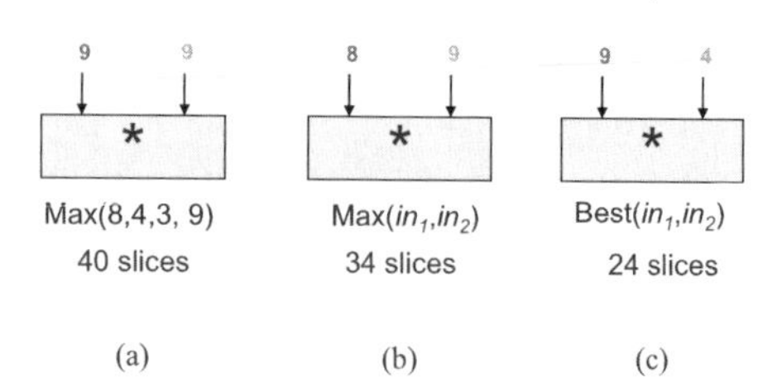

Fig. 9.9 Operator area vs. sizing approaches

9.3.2.5 Storage Element Optimization

Because currently there is no feed-back loop in the design flow, the registers optimization has to be done during the conception of the processing unit. The choice of the location of an unconstrained variable (user can define the location of variables) in a register or in a memory, has to be done according to the minimization of two contradictory cost criteria:

- The cost of a register is higher than the cost of a memory point.
- The cost to access data in a register is lower than the cost to access data in memory (because of the necessity to compute the address).

Two criteria are used to choose the memorization location of the data:

- A variable whose life time is inferior to a locality threshold is stored in a register.
- The location of memorization depends on the class of the variable.

Data are classified into three categories:

- Temporary processing data (declared or undeclared).
- Constant data (read-only).
- Ageing data (which serves to express the recursivity of the algorithm to be synthesized, via their assignment after having been utilized).

The optimal storage of a given data element depends upon its declaration and its life time. It can be either stored in a memory bank of the MEMU or in a storage element of the processing unit PU. The remaining difficulty lies in selecting an optimal locality threshold which results in minimizing the cost of the storage unit. The synthesis tool leaves the choice of the value of the locality threshold up to the user. In order to help the designer, GAUT proposes a histogram of the life time of the variables, normalized by the utilization frequency, which is calculated from the scheduled DFG.

The architecture of the processing unit is composed of a processing part and a memory part (i.e. memory plan) and the associated control state machine FSM (Fig. 9.1). The memory part of the datapath is based on a set of strong semantic memories (FIFO, LIFO) and/or registers. Spatial adaptation is performed by an interconnection logic dealing with data dispatching from operators to storage elements, and from storage elements to operators. Timing adaptation (data-rates, different input/output data scheduling) is realized by the storage elements. Once the location of data has been decided, the synthesis of the storage elements located in

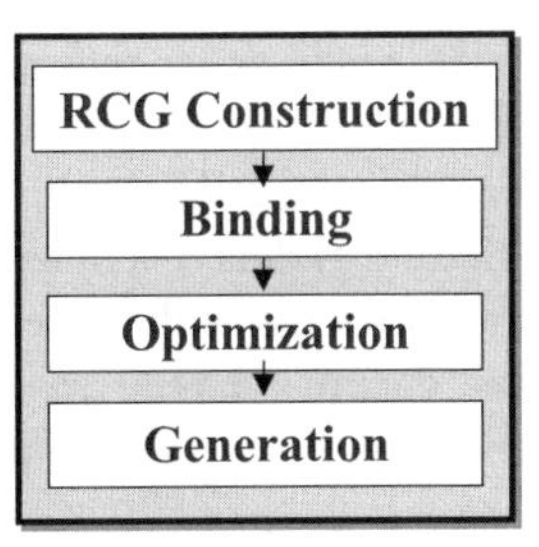

Fig. 9.10 Four-step flow

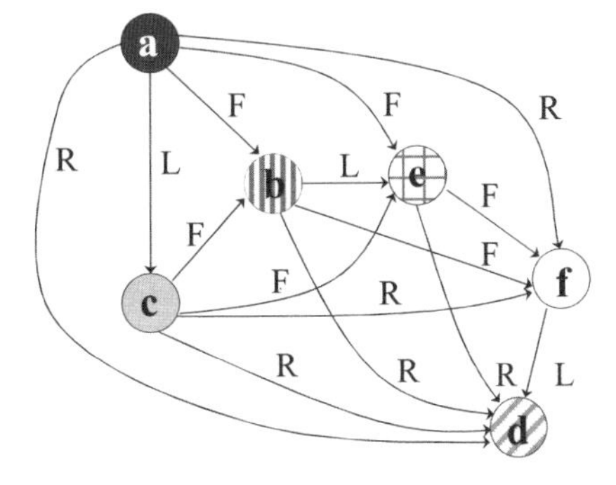

Fig. 9.11 Resource compatibility graph

the PU is done. This design step inputs data lifetimes resulting from the scheduling step and spatial information resulting from the binding step of the DFG. The spatial information is the source and destination for each data. First, we formalize both timing relationships between data (thanks to data lifetimes) and spatial information through a Resource Compatibility Graph RCG. This formal model is then used to explore the design space. We named timing relationships and spatial information as Communication Constraints.

This synthesis task is based on a four-step flow: (1) Resource Compatibility Graph (RCG) construction, (2) Storage resource binding, (3) Architecture optimization and (4) VHDL RTL generation (see Fig. 9.10). During the first step of the component generation, a Resource Constraints Graph is generated from the communication constraints. The analysis of this formal model allows both the binding of data to storage elements (queue, stack or register), and the sizing of each storage element. This first architecture is then optimized by merging storage elements that have non-overlapping usage time frames.

Formal model: In order to explore the design space of such a component, the first step consists in generating a Resource Compatibility Graph, from the communication constraints. This RCG specifies through formal modeling the timing relationship between data that have to be handled by the datapath architecture.

The vertex set $V = \{v_0, \ldots, v_n\}$ represents data, the edge set $E = \{(v_i, v_j)\}$ represents the compatibility between the data. A tag $t_{ij} \in T$ is associated with each edge (v_i, v_j). This tag represents the compatibility type between the two data (i and j), T = {Register R, FIFO F, LIFO L}, e.g. Fig. 9.11.

In order to assign compatibility tags to edges, we need to identify the timing relationship that exists between two data. For this purpose we defined a set of rules based on functional properties of each storage element (FIFO, LIFO, Register). The lifetime of data a is defined by $\Gamma(a) = [\tau_{min}(a), \tau_{max}(a)]$ where $\tau_{min}(a)$ and $\tau_{max}(a)$ are respectively the date of the write access of a into the storage element, and the last date of the read access to a. $\tau_{first}(a)$ is the first read access to a, τ_{Ria} is the ith read access to a, with $first \leq i \leq max$.

***Rule 1**: Register compatibility*
$If\ (\tau_{min_b} \geq \tau_{max_a})$ then we create a "Register" tagged edge.
***Rule 2**: FIFO compatibility*
$If\ [(\tau_{min_b} > \tau_{min_a})$ and $(\tau_{fisrt_b} > \tau_{max_a})$ and $(\tau_{min_b} < \tau_{max_a})]$ then we create a "FIFO" tagged edge.
***Rule 3**: LIFO compatibility*
$If\ [[(\tau_{min_b} > \tau_{min_a})$ and $(\tau_{first_a} > \tau_{max_b})]$ or $[(\tau_{Ri_a} < \tau_{min_b} < \tau_{max_b} < \tau_{Ri+1_a})]]$ then we create a "LIFO" tagged edge.
***Rule 4**: Otherwise, No edge – No compatibility.*

An analysis of the communication constraints enables the RCG generation. The graph construction supposes edge creation between data, respecting a chronological order (τ_{min}). If n is the number of data to be handled, the graph may contain: $n(n-1)/2$ edges, $O(n^2)$.

Storage element binding: The second step consists in binding storage elements to data thanks to the timing relations modeled by the RCG.

Resource identification: The second step consists in binding storage elements to data by using the timing relations modeled by the RCG. The aim is to identify and to bind as many FIFO or LIFO structures as possible on the RCG.

Theorem 1. *If a is FIFO compatible with b and b is FIFO compatible with c, then a is transitively FIFO (or Register) compatible with c.*

As a consequence of Theorem 1, a FIFO compatible datapath, PF, is by construction equivalent to a FIFO compatibility clique (i.e. the data of the PF path can be stored in the same FIFO).

Theorem 2. *If a is LIFO compatible with b and b is LIFO compatible with c, then a is transitively LIFO compatible with c.*

As a consequence of Theorem 2, a LIFO compatible datapath, P_L, is by construction equivalent to a LIFO compatibility clique (i.e. the data of the P_L path can be stored in the same LIFO).

Resource sizing: The size of a LIFO structure equals the maximum number of data stored by a LIFO compatible data path. So, we have to identify the longest LIFO compatibility path P_L in a LIFO compatibility tree, and then the number of vertices in P_L from the longest LIFO path in the tree equals the maximum number of data that can be stored in it.

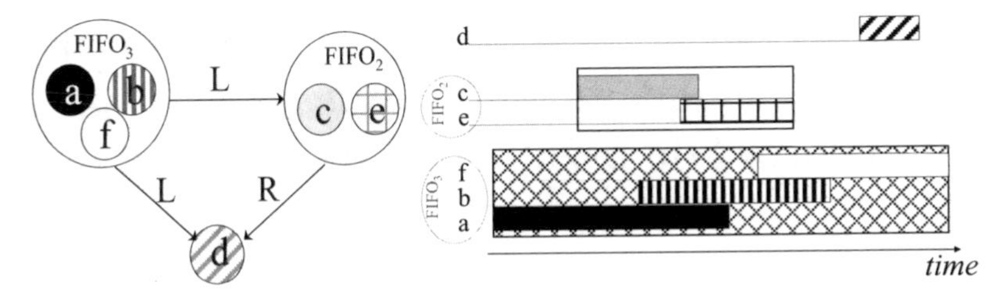

(a) Resulting hierarchical graph (b) Resulting constraints

Fig. 9.12 A possible binding for graph

The size of a FIFO is the maximum number of data (of the considered path) stored at the same time in the structure. In fact, the aim is to count the maximum number of overlapped data (respecting I/O constraints) in the selected path P. These sizes can be easily extracted from our formal model.

Resource binding: Our greedy algorithm is based on user plotted metrics (minimal amount of data to use a FIFO or a LIFO, average use factor, FIFO/LIFO usage priority factor...) to bind as many FIFO or LIFO structures as possible on the RCG. A two-steps flow is used: (1) identification of the best structure, (2) merging all the concerned data in a hierarchical node.

Each node represents a storage element, as shown on Fig. 9.12a (e.g. data a, b and f are merged in a three-stages FIFO). We say *hierarchical node* because merging a set of data in a given node, supposes adding information that will be useful during the optimization step: the lifetime of this structure (i.e. the time interval during which this structure will be used. e.g. Fig. 9.12b).

Let $\mathrm{P} = \{v_0, \ldots, v_n\}$ be a compatible data path,

- If P is a FIFO compatible path, the structure lifetime will be $[\tau_{min_{v_0}}, \tau_{max_{v_n}}]$.
- If P is a LIFO compatible path, the structure lifetime will be $[\tau_{min_{v_0}}, \tau_{max_{v_0}}]$.

Storage element optimization: The goal of this final task is to maximize storage resource usage, in order to optimize the resulting architecture by minimizing the number of storage elements and the number of structures to be controlled. To tackle this problem, we built a new hierarchical RCG by using the merged nodes, and their lifetimes. In order to avoid any conflict, the exploration algorithm of the optimization step will only search for Register compatibility path, between same type vertices. When two structures of the same type are Register compatible, they can be merged.

Let $\mathrm{P} = \{v_0 \ldots v_n\}$ be a Register compatible data path,

- The lifetime of the resulting hierarchical merged structure will be $[\tau_{min_{v_0}}, \tau_{max_{v_n}}]$ $\mathrm{U} \ldots \mathrm{U}$ $[\tau_{min_{v_n}}, \tau_{max_{v_n}}]$.

The algorithm is very similar to the one used during binding step. When there is no more merging solution, the resulting graph is used to generate the RTL VHDL

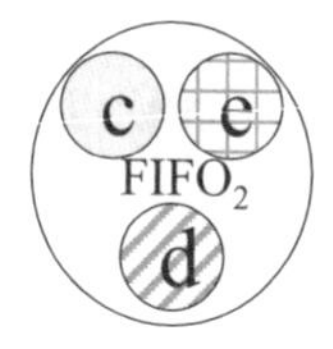

Fig. 9.13 Optimization of Fig. 9.11 graph

architecture. Figure 9.13 is a possible architectural solution for the Resource Compatibility Graph presented in Fig. 9.11. Here, the resulting architecture consist in a three-stages FIFO that handles three data, and a two-stages FIFO that handles three data: one memory place has been saved.

9.3.3 Memory Unit Synthesis

In this section, we present two major features of GAUT, regarding the memory system. First the data distribution and placement are formalized as a set of constraint for the synthesis. We introduce a formal model for the memory accesses, and an accessibility criterion to enhance the scheduling step. Next, we propose a new strategy to implement signals described as ageing vectors in the algorithm. We formalize the maturing process and explain how it may generate memory conflicts over several iterations of the algorithm. The final Compatibility Graph indicates the set of valid mappings for every signal. Our scheduling algorithm exhibits a relatively low complexity that allows to tackle complex problems in a reasonable time.

9.3.3.1 Memory Constrained Scheduling

In our approach the data flow graph DFG first generated from the algorithmic specification is parsed and a memory table is created. This memory table is completed by the designer who can select the variable implementation (memory or register) and place the variable in the memory hierarchy (which bank). The resulting table is the memory mapping that will be used in the synthesis. It presents all the data vertices of the DFG. The data distribution can be static or dynamic.

In the case of a static placement, the data remains at the same place during the whole execution. If the placement is dynamic, data can be transferred between different levels in the memory hierarchy. Thus, several data can share the same location in the circuit memory. The memory mapping file explicitly describes the data transfers to occur during the algorithm execution.

Direct Memory Address (DMA) directives will be added to the code to achieve these transfers. The definition of the memory architecture will be performed in the first step of the overall design flow. To achieve this task, advanced compilers such as Rice HPF compiler, Illinois Polaris or Stanford SUIF could be used [14]. Indeed, these compilers automatically perform data distribution across banks, determine

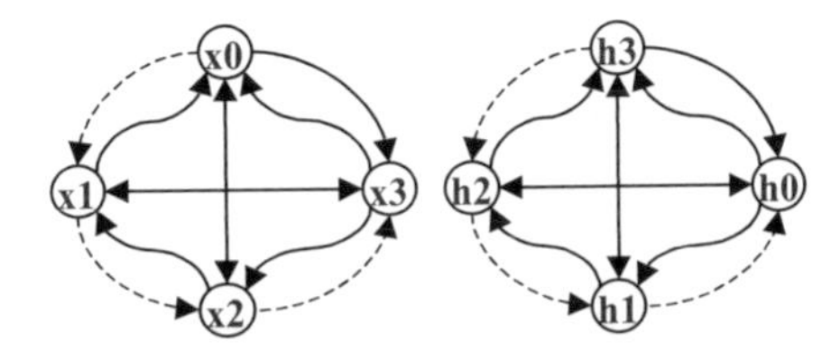

Fig. 9.14 Memory constraint graph

which access goes to which bank, and then schedule to avoid bank conflicts. The Data Transfer and Storage Exploration (DTSE) method from IMEC and the associated tools (ATOMIUM, ADOPT) are also a good mean to determine a convenient data mapping [15].

We modified the original priority list (see Sect. 9.3.2.2) to take into account the memory constraint: an accessibility criterion is used to determine if the data involved by an operation is available, that is to say, if the memory where it is stored is free. Operations are still listed according to the mobility and bit-width criterion, but all operations that do not match the accessibility criterion are removed. Every operation that needs to access a busy memory will not be scheduled, no matter its priority level. Fictive memory access operators are added (one access operator per access port to a memory). The memory is accessible only if one of its access operators is idle. Memory access operators are represented by tokens on the Memory Constraint Graph (MCG): there are as many tokens as access ports to the memory or bank. Figure 9.14 shows two MCG, for signal samples x[0] to x[3] stored in bank 1, and coefficients h[0] to h[3] stored in bank 2 (in the case of a four points convolution filter for instance).

If one bank is being accessed, one token is placed on the corresponding data. Only one token is allowed for a one port bank. Dotted edges indicate which following access will be the faster. In the case of a DRAM indeed, slower random accesses are indicated with plain edges and faster sequential accesses with dotted edges. Our scheduling algorithm will always favor fastest sequences of accesses whenever it has the choice.

9.3.3.2 Implementing Ageing Vector

Signals are the input and output flows of the applications. A mono-dimensional signal x is a vector of size n, if n values of x are needed to compute the result. Every cycle, a new value for x ($x[n+1]$) is sampled on the input, and the oldest value of x ($x[0]$) is discarded. We call x an ageing, or maturing, vector or data. Ageing vectors are stored in RAM. A straightforward way to implement, in hardware, the maturing of a vector, is to write its new value always at the same address in memory, at the end of the vector in the case of a 1D signal for instance. Obviously, that involves to shift every other values of the signal in the memory to free the place for the new value. This shifting necessitates n reads and n writes, which is very time and power consuming. In GAUT, the new value is stored at the address of the oldest one in the

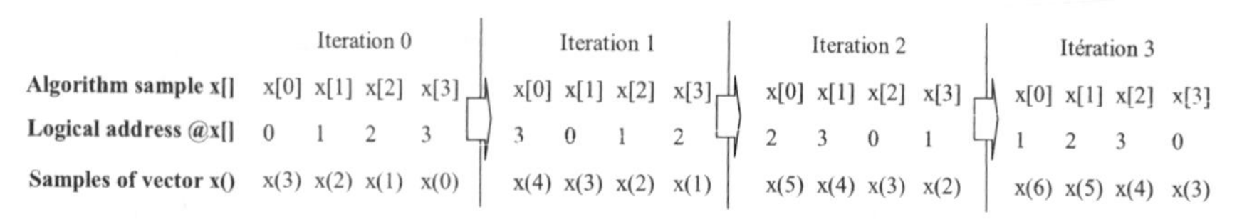

Fig. 9.15 Logical addresses evolution for signal x

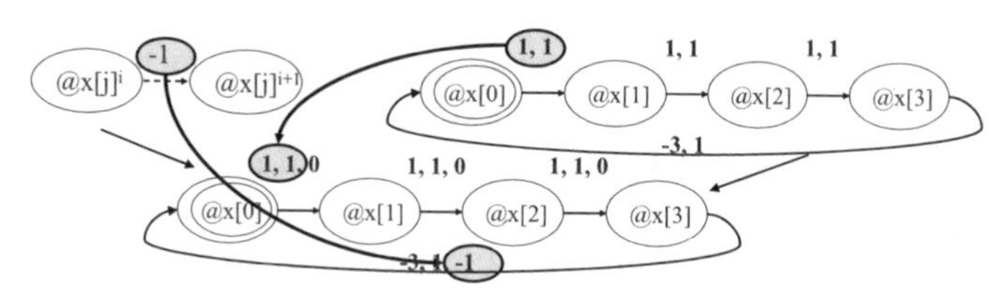

Fig. 9.16 LAG, AG and USG

vector. Only one write is needed. Obviously, the address generation is more difficult in this case, because the addresses of the samples called in the algorithm change from one cycle to the other. Figure 9.15 represents the evolution of the addresses for a $L = 4$ points signal x from one iteration to the other.

The methodology that we propose to support the synthesis of these complex logical address generators is based on three graphs (see Fig. 9.16). The *logical address graph* (LAG) traces the evolution of the logical addresses for a vector *during the execution of one iteration* of the algorithm. Each vertices correspond to the logical address where samples of signal x are to be accessed. Edges are weighted with two numbers. The first number, f_{ij}, indicates how the logical address evolves between two successive accesses to vector x. $f_{ij} = (j-i)\%L$ (% indicates the modulo). The second number $g_{i,j}$ indicates the number of iteration between those two successive accesses.

To actually calculate the evolution of logical addresses of x from one iteration to the other, we must take into account the ageing of vector x. We introduce the *ageing factor* k as the difference between the logical address of element $x[i]$ at the iteration o and the logical address of element $x[i]$ at the iteration $o+1$, so that:

$$@\mathrm{x[j]}^{\mathrm{i}+1} = (@\mathrm{x[j]}^{\mathrm{i}} - k)\%L.$$

In our example, $k = 1$. The *Ageing Graph* (Fig. 9.16) is another representation of this equation. We finally combine the LAG and the ageing factor to get the *Unified Sequences Graph* (USG) (Fig. 9.16). A detailed definition of those three graphs may be find in [16].

By moving a token in the USG, and by adding to the first logical address for x the value of weight $f_{i,j}$ minus the ageing factor k, we get the address sequence for x during the complete execution of the algorithm. Then, the corresponding address generator is generated.

If a pipelined architecture is synthesized, the ageing factor k is multiplied by the number of pipeline slices, and as many tokens as pipeline slices are placed and moved in the USG. Of course, as much memory locations as supplemental tokens in the USG must be added to guarantee data consistency. Concurrent accesses to elements of vector x may appear in a pipelined architecture. While moving tokens in the USG, a *Concurrent Accesses Graph* is constructed. This graph is finally colored to obtain the number of memory banks needed to support access concurrency.

9.3.4 Communication and Interface Unit Synthesis

9.3.4.1 Latency Insensitive Systems

Systems on a chip (SoCs) are the composition of several sub-systems exchanging data. SoC size increase is such that an efficient and reliable interconnection strategy is now necessary to combine sub-systems and preserve, at an acceptable design cost, the speed performances that the current very deep sub-micron technologies allow [20]. This communication requirement can be satisfied by a LIS communication network between hardware components. The LIS methodology enables to build functionally correct SoCs by (1) promoting pre-developed components intensive reuse (IPs), (2) segmenting inter-components interconnects with relay stations to break critical paths and (3) bringing robustness to data stream latencies to components by encapsulating them into synchronization wrappers. These encapsulated blocks are called "patient processes". Patient processes [21] are a key element in the LIS theory. They are suspendable synchronous components (named *pearls*) encapsulated into a wrapper (named shell) which function is to make them insensible to the I/O latency and to drive the clock. The decision to drive or not the component's clock is implemented with combinatorial logic. The LIS approach relies on a simplifying, but restricting, assumption: a component is activated only if all its inputs are valid and all its outputs are able to store a result produced at the next clock cycle. Now, it is frequent that only a subset of the inputs and outputs are necessary to execute one step of computation in a synchronous block.

To limit the patient process sensitivity to a subset of the inputs and outputs, in [22] authors suggest to replace the combinatorial logic that drives the clock by a Mealy type FSM. This FSM tests the state of only the relevant inputs and outputs at each cycle and drives the component clock only when they are all ready. The major drawbacks of FSMs are their difficult synthesis and large silicon size when communication scenarios are long and complex like for computing intensive digital signal processing applications. To reduce the hardware cost, in [23] the component activation static schedule is implemented with shift registers which contents drive the component's clock. This approach relies on the hypothesis that there are no irregularities in the data streams: it is never necessary to randomly freeze the components.

9.3.4.2 Proposed Approach

As (1) LIS methodology lacks the ability to dynamically sense I/O subsets, (2) FSMs can become too large as communication bandwidth does, and (3) shift registers based synchronization targets only extremely rapid environments, we propose to encapsulate hardware components into a new synchronization wrapper model which area is much less than the FSM-based wrappers area, which speed is enhanced (mostly thanks to area reduction) and synthesizability is guaranteed whatever the communication schedule is.

The solution we propose is functionally equivalent to the FSMs. This is a specific processor that reads and executes cyclically operations stored in a memory. We name it a "synchronization processor" (SP). Figure 9.1 specifies the new synchronization wrapper structure with our SP.

The SP communicates with the LIS ports with FIFO-like signals. These signals are formally equivalent to the *voidin/out* and *stopin/out* of [19] and *valid*, *ready* and *stall* of [22]. Number of input and output ports can be any. It drives the component's clock with the *enable* signal. The SP model is specified by a three states FSM: a reset state at power up, an operation-read state, and a free-run state. This FSM is concurrent with the component and contains a data path: this a "concurrent FSM with data path" (CFSMD). Operation's format is the concatenation of an input-mask, an output-mask and a free-run cycles number. The masks specify respectively the input and output ports the FSM is sensible to. The run cycles number represents the number of clock cycles the component can execute until the next synchronization point. To avoid unnecessary signals and save area, the memory is an asynchronous ROM (or SRAM with FPGAs) and its interface with the SP is reduced to two buses: the *operation address* and *operation word*. The execution of the program is driven by an operation "read-counter" incremented modulo the memory size.

9.4 Experiments

Design synthesis results for Viterbi decoders are presented in this section. Results are based on a Virtex-E FPGA technology from the hardware prototyping platform that we used and that we present first.

9.4.1 The Hardware Platform

The Sundance platform [24] we used as an experimental support is composed of the last generation of C6x DSPs and Virtex FPGAs. Communications between different functional blocs are implemented with high throughput SDB links [24]. We have automated the generation of communication interface for software and hardware components which frees the user from designing the communication interfaces. At the hardware level the communication between computing nodes is handled by four-phases handshaking protocols and decoupling FIFOs. The handshaking protocols synchronize computing with communication and the FIFOs enable to store data in order to overcome potential data flow irregularities. Handshaking protocols are used either to communicate seamlessly between hardware nodes or between hardware and software nodes. Handshaking protocols are automatically refined by the GAUT tool to fit with the selected (SDB) inter-node platform communication interfaces (bus width, signal names, etc). To end the software code generation, platform specific code has to be written to ensure the inter processing elements communication. The communication drivers of the targeted platform are called inside the interface functions introduced in the macro-architecture model through an API mechanism. We provide a specific class for each type of link available on the platform.

9.4.2 Synthesis Results

The Viterbi algorithm is applicable to a variety of decoding and detection problems which can be modeled by a finite-state discrete-time Markov process, such as convolutional and trellis decoding in digital communications [25]. Based on the received symbols, the Viterbi algorithm estimates the most likely state sequence according to an optimization criterion, such as the a posteriori maximum likelihood criterion, through a trellis which generally represents the behavior of the encoder. The generic C description of the Viterbi algorithm allowed us to synthesize architectures using different values for the following functional parameters: state number and throughput. A part of synthesis results that have been obtained is given in Fig. 9.17. For each generated architecture, the table presents the throughput constraint and the complexity of both the algorithm (number of operations) and the generated architecture (amount of logic elements).

In the particular case of the DVB-DSNG Viterbi decoder (64 states) different throughput constraints (from 1 to 50 Mbps) have been tested. Figure 9.18 present the synthesis results.

State number	8	16	32	64	128
Throughput (Mbps)	44	39	35	26	22
Number of operations	50	94	182	358	582
Number of logic elements	223	434	1130	2712	7051

Fig. 9.17 Synthesis results for different Viterbi decoders

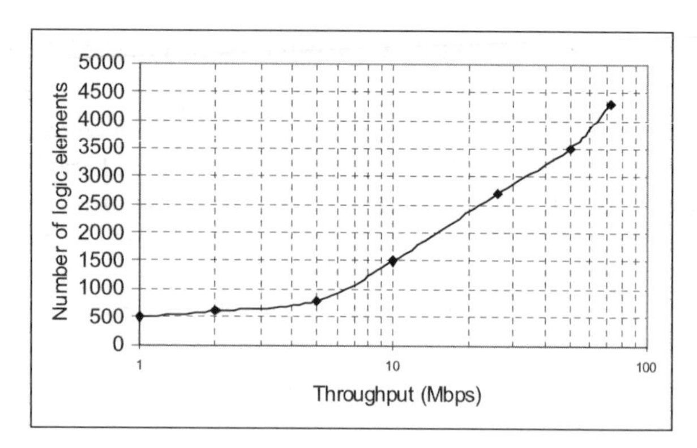

Fig. 9.18 Logic size for different throughputs

9.5 Conclusion and Perspectives

In this chapter, we presented GAUT [1], which is an academic and open source high-level synthesis tool dedicated to digital signal processing applications. We described the different tasks that compose the datapath synthesis flow: compilation, operator characterization, operation clustering, resource allocation, operation scheduling and binding. Memory and communication interface synthesis has also been described in this chapter.

Current work targets the area optimization of the architecture generated by GAUT through an approach based on iterative refinement. The integration of Control and Data Flow Graph CDFG model to be used as internal representation is also in progress. The loop transformations will be addressed during the compilation step thanks to the features provided by the last versions of the gcc/g++ compiler. An approach to map data in memories will be proposed to limit the access conflicts. Automatic algorithm transformation will be addressed through Taylor Expansion Diagram. Multi-clock domain synthesis is also currently considered. Starting from an algorithmic specification and design constraints a Globally Asynchronous Locally Synchronous GALS architecture will be automatically synthesized. This will allow to design high-performance and low-power architectures.

Acknowledgments Authors would like to acknowledge all the GAUT contributors and more specifically Caaliph Andriamisaina, Emmanuel Casseau, Gwenolé Corre, Christophe Jego, Emmanuel Juin, Bertrand Legal, Ghizlane Lhairech and Kods Trabelsi.

References

1. http://web.univ-ubs.fr/gaut
2. B. Ramakrishna Rau, "Iterative modulo scheduling: an algorithm for software pipelining loops", In *Proceedings of the 27th annual international symposium on Microarchitecture*, pp. 63–74, November 30–December 02, 1994, San Jose, CA, United States

3. C. Chavet, C. Andriamisaina, P. Coussy, E. Casseau, E. Juin, P. Urard and E. Martin, "A design flow dedicated to multi-mode architectures for DSP applications", In *Proceedings of the IEEE International Conference on Computer Aided Design ICCAD*, 2007
4. http://soclib.lip6.fr
5. www.mentor.com
6. www.systemc.org
7. http://gcc.gnu.org
8. Z. Galil, "Efficient algorithms for finding maximum matching in graphs", *ACM Computing Survey*, Vol. 18, No. 1, pp. 23–38, 1986
9. S. Mahlke, R. Ravindran, M. Schlansker, R. Schreiber and T. Sherwood, "Bitwidth cognizant architecture synthesis of custom hardware accelerators", *IEEE Transactions on Computer-Aided Design of Circuits and Systems*, Vol. 20, No. 11, pp. 1355–1371, 2001
10. C. Andriamisaina, B. Le Gal and E. Casseau, "Bit-width optimizations for high-level synthesis of digital signal processing systems", In *SiPS'06, IEEE 2006 Workshop on Signal Processing Systems*, Banff, Canada, October 2006
11. J. Cong, Y. Fan, G. Han, Y. Lin, J. Xu, Z. Zhang and X. Cheng, "Bitwidth-aware scheduling and binding in high-level synthesis", In *Proceedings of ASPDAC*, Computer Science Department, UCLA and Computer Science and Technology Department, Peking University, 2005
12. N. Herve et al., "Data wordlength optimization for FPGA synthesis", In *IEEE Workshop on Signal Processing Systems Design and Implementation*, pp. 623–628, 2005
13. A. Baganne, J.-L. Philippe and E. Martin, "A formal technique for hardware interface design", *IEEE Transactions on Circuits and Systems*, Vol. 45, No. 5, 1998
14. P. Panda et al., "Data and memory optimization techniques far embedded systems", *Transactions on Design Automation of Electronic Systems*, Vol. 6, No. 2, pp. 149–206, 2001
15. F. Catthoor, K. Danckaert, C. Kulkami and T. Omns, "*Data Transfer and Storage (DTS) architecture issues and exploration in multimedia processors*", Marcel Dekker, New York, 2000
16. G. Corre, E. Senn, P. Bornel, N. Julien and E. Martin, "Memory accesses management during high level synthesis", In *Proceedings of International Conference on Hardware/Software Codesign and System Synthesis, CODES+ISSS, 2004*, pp. 42–47, 2004
17. P. Bomel, E. Martin and E. Boutillon, "Synchronization processor synthesis for latency insensitive systems", In *Proceedings of the Conference on Design, Automation and Test in Europe*, Vol. 2, pp. 896–897, 2005
18. P. Coussy, E. Casseau, P. Bomel, A. Baganne and E. Martin, "A formal method for hardware IP design and integration under I/O and timing constraints", *ACM Transactions on Embedded Computing Systems*, Vol. 5, No. 1, pp. 29–53, 2005
19. L. P. Carloni, K. L. McMillan and A. L. Sangiovanni-Vincentelli, "Theory of latency-insensitive design," *IEEE Transactions on Computer Aided Design of Integrated Circuits and Systems*, Vol. 20, No. 9, p. 18, 2001
20. International Technology Roadmap for Semiconductors ITRS, 2005 editions
21. L. P. Carloni and A. L. Sangiovanni-Vincentelli, "Coping with latency in SoC design," *IEEE Micro, Special Issue on Systems on Chip*, Vol. 22, No. 5, p. 12, 2002
22. M. Singh and M. Theobald, "Generalized latency-insensitive systems for single-clock and multi-clock architectures," In *Proceedings of the Design Automation and Test in Europe Conference (DATE'04)*, Paris, February 2004
23. M. R. Casu and L. Macchiarulo, "A New Approach to Latency Insensitive Design," In *Proceedings of the Design and Automation Conference (DAC'04)*, San Diego, June 2004
24. Sundance Mu1ti-Processor Technology, http://www.sundance.com
25. A. J. Viterbi, "Error bounds for convolutional codes and an asymptotically optimum decoding algorithm", *IEEE Transactions on Information Theory*, Vol. IT-13, pp. 260–269, 1967

Chapter 10
User Guided High Level Synthesis

Ivan Augé and Frédéric Pétrot

Abstract The User Guided Synthesis approach targets the generation of coprocessor under timing and resource constraints. Unlike other approaches that discover the architecture through a specific interpretation of the source code, this approach requires that the user guides the synthesis by specifying a draft of its data-path architecture. By providing this information, the user can get nearly the expected design in one shot instead of obtaining an acceptable design after an iterative process. Of course, providing a data-path draft limits its use to circuit designers.

The approach requires three inputs: The first input is the description of the algorithm to be hardwired. It is purely functional, and does not contain any statement or pragma indicating cycle boundaries. The second input is a draft of the data-path on which the algorithm is to be executed. The third one is the target frequency of the generated hardware.

The synthesis method is organized in two main tasks. The first task, called *Coarse Grain Scheduling*, targets the generation of a fully functional data-path. Using the functional input and the draft of the data-path (*DDP*), that basically is a directed graph whose nodes are functional or memorization operators and whose arcs indicate the authorized data-flow among the nodes, this task generates two outputs:

- The first one is a RT level synthesizable description of the final coprocessor data-path, by mapping the instructions of the functional description on the *DDP*.
- The second one is a coarse grain finite state machine in which each operator takes a constant amount of time to execute. It describes the flow of control without knowledge of the exact timing of the operators, but exhibits the parallelism among the instruction flow.

The data-path is synthesized, placed and routed with back-end tools. After that, the timings such as propagation, set-up and hold-times, are extracted and the second task, called *Fine Grain Scheduling*, takes place. It basically performs the retiming of the Coarse Grain finite state machine taking into account the target frequency and the fine timings of the data-path implementation.

P. Coussy and A. Morawiec (eds.) *High-Level Synthesis.*

Compared to the classical High Level Synthesis approaches, the User Guided Synthesis induces new algorithmic problems. For *Coarse Grain Scheduling*, it consists of finding whether an arithmetic and logic expression of the algorithmic input can be mapped on the given draft data-path or not, and when several mappings are found, to choose the one that maximizes the parallelism and minimizes the added resources. So the *Coarse Grain Scheduling* can be seen as a classical compiler, the differences being firstly that the target instruction set is not hardwired in the compiler but described fully or partially in the draft data-path, and secondly that a small amount of hardware can be added by the tools to optimize speed.

For *Fine Grain Scheduling*, it consists of reorganizing the finite state machine to ensure that the data-path commands are synchronized with the execution delays of the operators they control. The fine grain scheduling also poses interesting algorithmic problems, both in optimization and in scheduling.

Keywords: Behavioral synthesis, FSM retiming, Design space exploration, Scheduling, Resource binding, Compilation

10.1 Introduction

10.1.1 Enhanced Y Chart

The Y chart representation proposed by Gajski [10] can not accurately represent the recent synthesis and high level synthesis tools. So we have enhanced it as shown Fig. 10.1. In the enhanced Y chart, a control flow level is inserted between the system and data flow levels. It corresponds to the coprocessor synthesis and allows to distinguish co-design from high level synthesis. In the structural view, a coprocessor is usually a data-path controlled by a FSM. There are two possible types of description in the behavioral view. The synchronized description is more or less a Register Transfer Language where cycle boundaries are explicitly set by the language. The non-synchronized description is based on imperative languages such as C, PASCAL, and in this type of description, the cycle boundaries are not given. As shown in the Fig. 10.1, High Level Synthesis consists of making a structural description of a circuit from a non-synchronized description of a coprocessor. A usual approach [6, 14, 24] is to generate the synchronized description of the coprocessor (plain arrow in Fig. 10.1) and to submit it to a CAD frameworks having an efficient RTL synthesis tool.

10.1.2 UGH Overview

The multiple arrows noted 1.*a*, 1.*b* and 1.*c* on the Fig. 10.2a describe the User Guided High level synthesis tool (UGH). It starts from a non-synchronized description of an algorithm and generates a structural description of the coprocessor (arrow

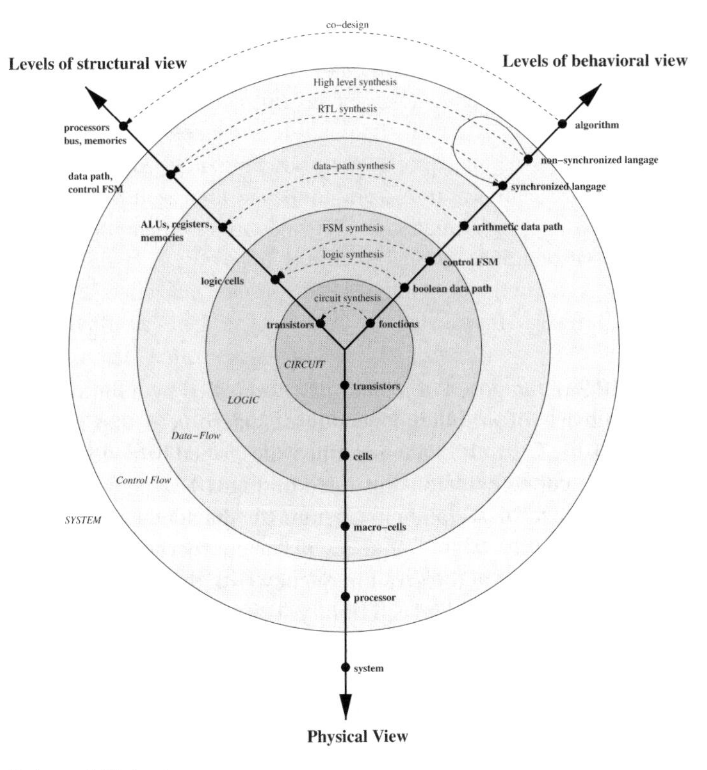

Fig. 10.1 Enhanced Y chart

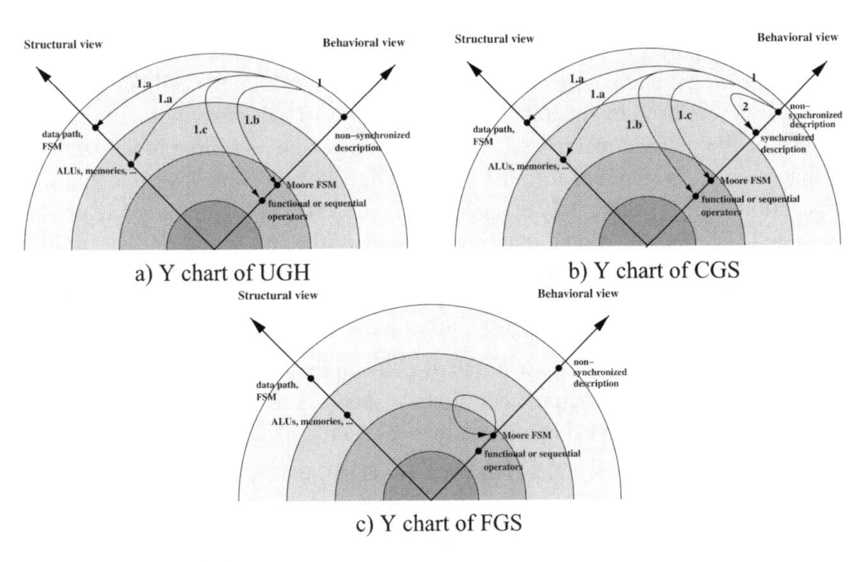

Fig. 10.2 Y charts of UGH tools

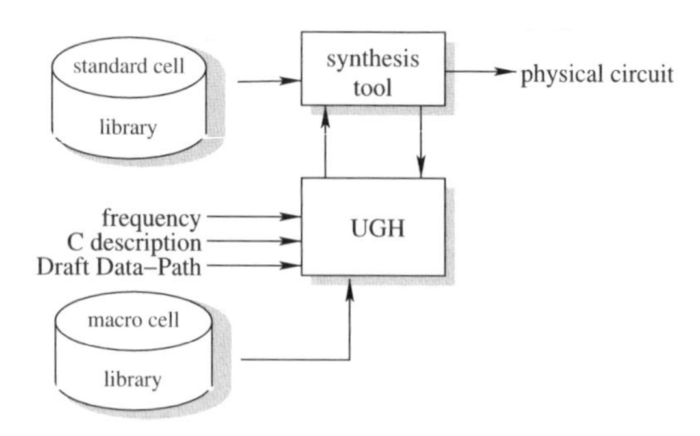

Fig. 10.3 User view

1.a on the Fig. 10.2a) composed of a data-path controlled by a finite state machine. The data-path consists of an interconnection of macro-cells that are described in data flow (arrow 1.b). The finite state machine is described behaviorally (arrow 1.c).

For describing a coprocessor, as illustrated on Fig. 10.3, the user inputs of UGH are the non-synchronized description in *C* language, the synthesis constraints (Draft Data Path or DDP) and the target frequency of the coprocessor. UGH produces the coprocessor circuit running at the given frequency with the help of a logic and FSM synthesis framework and a standard cell library. The main features of UGH reside in:

Macro-cell library. UGH synthesis process is based on a macro-cell library containing both functional cells such as *and*, *adder*, *ALU*, *multiplier*, ... and sequential ones such as *DFF*, *input/output ports*, *RAMs*, *register files*, ... A macro-cell is generic, the parameters being the bit size and various others ones depending on its type. For instance, for a *DFF* a parameter indicates if it has a synchronous reset, for a *RAM* parameters indicate whether there is a single address port for read and write or not.

A macro-cell is a complex object with different views. The functional view describes the operation, or the operations for multi-operation cells such as ALU (plus, minus, logic and, ...) or register (write, read, set, ...), the cell performs. The synthesis view is used to generate the synthesizable VHDL description of a specific instance of the generic macro-cell. The scheduling view is a time modelization of the macro cell. As opposed to most current High Level Synthesis tools that use a propagation delay, it accurately and generically describes the timing behavior of the macro-cell. For the functional macro-cells, these rules are based on the minimum and maximum propagation times between every output ports and the input ports. For the sequential macro-cells, the rules are more complex and take into account propagation, setup and hold times.

Every *C* operator has at least a macro-cell implementing it, but some specific macro-cells such as input/output operations or special shift operations (see 10.3.3.1) can be explicitly invoked using a procedure call in the *C* description.

Design space exploration. Design space exploration is a crucial problem of high level synthesis. The HLS tools usually propose an iterative approach to explore the design space. The user runs the synthesis, the result being the FSM graph and various cross-reference tables (between states and source statements, between cells and source statements, . . .). Then, using pragma in the source file, the user can force "specific" allocations. He runs again the HLS synthesis to get the new results and so on until he obtains the expected design. This iterative approach is difficult to use primarily because: (1) For large designs the time between iterations is too long. (2) The tables are difficult to interpret. The analysis of the results to set judicious pragmas requires to rebuild the data-path from the cross-reference tables, and this is a very long and tedious work. (3) This latter work must be done again at each iteration, because it is not obvious to predict the effect of a change in a pragma. So the iterative approach is not suited to large designs.

The UGH approach, on the contrary, allows to guide the tool towards the solution in a single step. It is however only aimed at VLSI designers. The designer does not have to change his working habits. He provides a data-path and a FSM, the only difference is that for UGH only a draft of the data-path is needed (see Fig. 10.7 and Sect. 10.3.1) and that the FSM (see Fig. 10.6) is a *C* program. So designers can obtain designs very close to the one they would have by RTL flows, but can easily explore many solutions.

Input frequency. A circuit is most often a piece of a larger system with specifications that determine its running frequency. Most of the HLS tools let the logic synthesis adapt their RTL outputs to the frequency. This approach neither ensures that the circuit can be generated (logic synthesis tools may not respect the clock frequency) nor ensures that the generated circuit is functional at the given clock frequency and even at any frequency if the circuit mixes short and long combinational paths. Furthermore, this approach generates very large circuits when the logic synthesis tools enter into speculative computation techniques. Taking an opposite view, UGH adapts the synthesizable description to the given frequency to guarantee that logic synthesis will be able to produce the circuit and that the circuit will run at the required frequency.

Input/output. Our point of view is that the synthesis of the communications of the coprocessor with the external world is not the purpose of the high level synthesis process. As the imperative languages do, UGH defines input and output primitives mapped to the macro-cells presented in Fig. 10.4a. These macro-cells implement

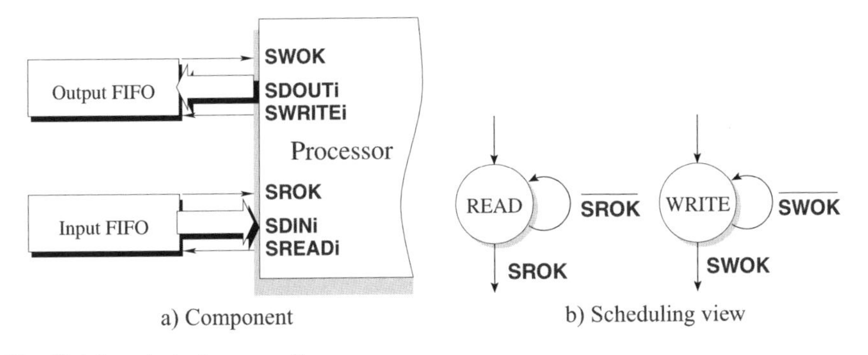

Fig. 10.4 Input/output macro-cells

basic asynchronous communication. From the coprocessor side, the data read action from a FIFO is shown Fig. 10.4b. In the READ state, the coprocessor asserts the SREAD signal and loads the SDIN signals' data into an internal register. If SROK is not asserted, it means the SDIN signals' data are not significant and the state must be run again. Otherwise the value loaded from SDIN is significant, and the producer pops it. The writing action is similar to the reading action. The read and write primitives are blocking. As shown Fig. 10.4a, if the flow of data is bursty, the designer can use hardware FIFO to smooth the transfers.

10.2 User Guided HLS Flow

The synthesis process, presented in the Fig. 10.5, is split into three main steps: The *Coarse Grain Scheduling* (CGS) generates a data-path and a finite state machine from the *C* program and the *DDP*. This finite state machine does not take the propagation delays into account. It is more a finite control step machine which maximizes the parallelism that is possible on the data-path, and we call it CG-FSM.

Then the mapping is performed. Firstly, the generation of the physical data-path is delegated to classical back-end tools (logic synthesis, place and route) using a target cell library. Secondly, the temporal characteristics of the physical data-path are extracted. At this point, the data-path of the circuit is available.

Finally, the *Fine Grain Scheduling* (FGS) retimes for the given frequency the finite control step machine, taking accurately into account the annotated timing delays of the data-path, and produces the finite state machine of the circuit.

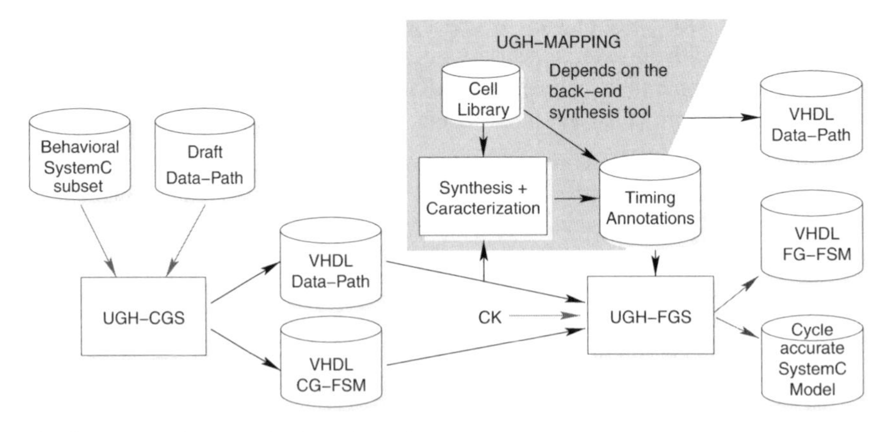

Fig. 10.5 User guided high level synthesis flow

10.3 Coarse Grain Scheduling

The arrow 1 in the Y chart of the Fig. 10.2b represents CGS. It is similar to the UGH arrow but the generated circuit is probably not functional.

CGS can also produce a synchronized description functionally and temporally equivalent to the former (arrow 2 on the Fig. 10.2b). This output is similar to those generated by usual high level synthesis tools and delegates the main work to a RTL synthesis tool.

10.3.1 Inputs

The first input (see Fig. 10.6), is the behavior of the coprocessor given as a *C* program. Most *C* constructs are allowed, but pointers, recursive functions and the use of standard functions (e.g., printf, strcat, ...) is forbidden. Furthermore, all variables must be either global or static unless their assignments can be inlined in the statements that read them. The basic *C* types, `char`, `short`, `int`, `long` and their unsigned version are extended with `int`*N* and `uint`*N*, where *N* is an integer in range 1–128 which defines the bit-size of the type.

The entry point is the `ugh_main` function. The `ugh_inChannel`*N* and `ugh_outChannel`*N* types define communication ports compatible with the hardware FIFO components. The `ugh_read` and `ugh_write` functions generate the state and arcs shown in Fig. 10.4b.

The second input (see Fig. 10.7a) is a simplified structural description of the target data-path called Draft Data-Path (*DDP*). The *DDP* is a directed graph (Fig. 10.7b) whose nodes are functional or memorization operators and whose arcs indicate the authorized data-flow between the nodes. For instance, the 2 arcs that point to the `a` input of the `Subst` node indicate that in the final data-path the bits of this input can be driven by: (a) constants, (b) bits of the `q` port of the `x` register, (c) bits of the `q` port of the `y` register, (d) any bit-wise combination of the former cases. Furthermore, the *DDP* does not express the bit size of the operators associated to the nodes, nor the bit size of the arcs. Notice that specifying the arcs is optional as explained in Sect. 10.3.3.2.

```
#include <ugh.h>
/* communication channels */
ugh_inChannel32  instream;
ugh_outChannel32 outstream;
/* registers */
uint32 x,y;
/* behavior */
void ugh_main()
{
  while (1) {
    ugh_read(instream,&x);
    ugh_read(instream,&y);
    while (x!=y) {
      if (x<y) y = y - x ;
      else     x = x - y ;
    }
    ugh_read(outstream,&x);
  }
}
```

Fig. 10.6 UGH-C for Euclid's GCD algorithm

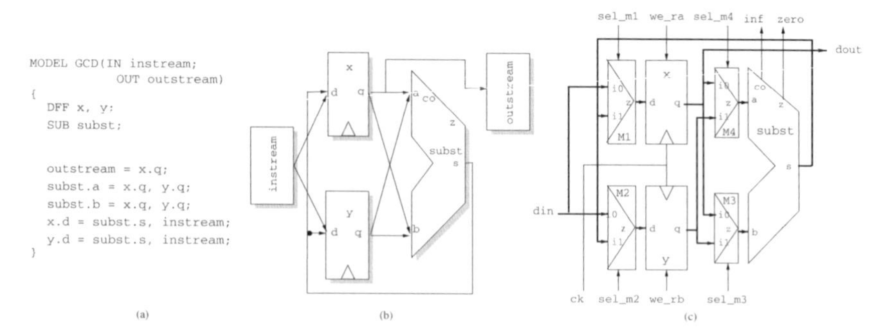

Fig. 10.7 Draft Data-Path of the GCD example

The *C* input and the *DDP* are interdependent. A global or static variable (respectively: array) of the *C* input must correspond to a register (respectively: register file or static RAM) of the *DDP* having the same name. For each statement of the *C* input there must be at least a sub-graph of the directed graph that can execute the statement.

10.3.2 CGS Overview

The *Coarse Grain Scheduling* uses the *C* input and the draft data-path to produce firstly the circuit data-path and secondly a *coarse grain* finite state machine (CG-FSM).

CGS starts with a consistency check. Enough registers must have been instantiated to store all the non-trivial variables. Each statement of the *C* description must correspond to at least one sub-graph of the *DDP*.

Then the binding takes place: Each node of the *DDP* corresponds to a macro-cell of the data-path. Its bit size is deduced from the bit size of the *C* variables, the input connectors of the cells are connected to output connectors either directly or using a multiplexer when inputs are driven by different sources. The resulting data-path of the GCD example is shown in Fig. 10.7c.

Finally the CG-FSM is elaborated, where *coarse grain* means that the operations are only partially ordered like in soft scheduling [26]. This FSM is built using the following timing constraints: multipliers need 2 cycles, adders and subtracters need 1 cycle, and all other functional cells have negligible propagation times.

10.3.3 Features and Algorithms

10.3.3.1 *C* Synthesis Rules

The Table 10.1 summarizes the computation of the size of the physical operators bound to a *C* operator. A *C* operator is used either in an assignment, such as

Table 10.1 Bit-size of hardware operators

C expression	Assignment	Other
$a*b$	s_r	$\min(32, s_a+s_b)$
$a+b$, $a-b$	s_r	$\min(32, \max(s_a, s_b)+1)$
a, $-a$	$\min(s_r, s_a)$	s_a
$a<b$, $a<=b$, ...	$\max(s_a, s_b)+1$	$\max(s_a, s_b)+1$
$a\&b$	$\min(s_r, s_a, s_b)$	$\min(s_a, s_b)$
$a\|b$, a^b, $a==b$, $a!=b$	$\min(s_r, \max(s_a, s_b))$	$\max(s_a, s_b)$
$a!=0\|\|b!=0$, $a!=0\&\&b!=0$	1	1
$a<<b$, $a>>b$	$\max(32, s_a)$	$\max(32, s_a)$

"`r=a*b+c;`" or not, such as "`t[a*b+c];`" or "`if (a<b)`". These cases are different, because the *C* language enforces integer promotion in expression, but in case of assignment the size of the result is known and can be used to minimize the operator size. In the formula, s_a, s_b, s_r correspond respectively to the size of a, b and the expected result.

Following the *C* standard is a bit costly, because only the assignments can be optimized. For example, in "`short t[10],a, b; t[a+b] = 0;`", the + operator must be 17 bit wide, while four bit would be enough. Using 17 bit is compulsory because otherwise the hardware and the *C* will not be equivalent. The shift operator case is quite expensive because in the statement "`char x=1, y=12; x = x << y;`", the *C* standard indicates to promote `x` on 32 bit and set the shift value to `y%32` prior to shift, so `x` is set to 0, while using a 8 bit shifter would lead to set `x` to 16. One can work around this problem by asking explicitly for a 8 bit shifter with the statement "`char x=1, y=12; x = ugh_shift(x,y);`".

10.3.3.2 Path Discovery

The *DDP* can define more or less accurately the data-path. The nodes (functional and sequential macro-cells except of the multiplexers and logic gates) are mandatory but the arcs are optional so the minimal description of the *DDP* presented Fig. 10.7a is:

```
MODEL GCD(IN instream;OUT outstream) {
  DFF x,y;
  SUB subst;
}
```

CGS starts by adding all the missing arcs such as for instance "`instream` → `subst.a`". In the following we call the arcs created by CGS *added* arcs and *user* arcs those defined in the *DDP* description. Note that CGS has an option which disables the addition of arcs to let one defines the data-path very accurately.

As opposed to the other high level synthesis compilers that build a path for each expression, CGS must find for every *C* expression the set of paths in the *DDP* that allow to realize it. It firstly searches paths with only *user* arcs and if such paths are not found then it searches paths mixing *user* and *added* arcs.

Searching paths is quite difficult due to the commutativity, distributivity, associativity, and various equivalences among arithmetic and logic operations. For instance, the expression `a+~b+1` can be mapped on a subtractor and the expression `(a&0xFF)+(b&0xFF)` does not require any & operator. Furthermore, in this last example, an adder with 8 bit inputs is enough, independently of s_a and s_b.

To the best of our knowledge, there is no canonical representation for general arithmetic and logic expressions, identical to what the BDD [2] are for Boolean expressions. In our implementation, the logical masks with constant values are replaced by wiring. The path discovery is done by a brute force algorithm which knows the operators properties and some equivalence rules. After a predefined number of trials, the algorithm indicates that it cannot find a path. If a path really exists, the user must help the tool by indicating it explicitly in the *C* expression. This is done by reordering the operations and adding parenthesis.

10.3.3.3 Scheduling Algorithm with Register Bindings

Firstly as shown Fig. 10.8, one builds a register transfer flow graph (*RFG*) [4] from the *C* statements which represents a Data Flow Graph in which the binding of the variables on the registers has been performed, thus mixing data dependencies and hardware constraints. In such a graph there are three types of relations [20]: the *RaW* (read after write) relation is set when the destination node reads a register written by the source node, the *WaR* (write after read) relation is set when the destination node writes a register read by the source node, the *WaW* (write after write) relation is set when the destination and source nodes write the same register. All these relations indicate that the destination node must be scheduled at least one cycle later than the source node. Secondly, all the execution paths of each register transfer instruction are computed as explained in Sect. 10.3.3.2.

Then the algorithm schedules the register transfer instructions of the RFG using a kind of *list* scheduling [18]. At a given cycle, a node of the RFG can be scheduled if all its predecessors have been scheduled in the previous cycles and if there is

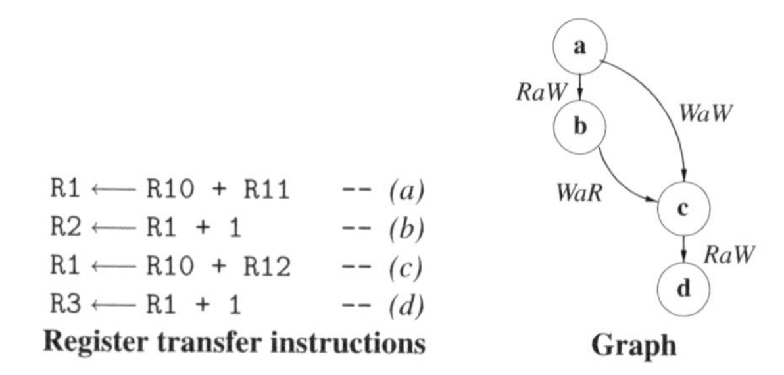

Fig. 10.8 Register transfer flow graph

one free execution path with all its nodes being free. The main objective of the algorithm is to obtain the minimal number of cycles and then to minimize the data-path area. This last objective occurs when there are several free execution paths for an instruction. In this case, the algorithm chooses the path which minimizes the cell bit sizes and/or the multiplexer sizes.

It has been shown in [8] that the *WaW* relation is superfluous, and that the *WaR* relations tend to over-constrain the scheduling. So the idea is to start the scheduling using only the true data dependencies (*RaW*) and to add the *WaR* constraints during the execution of the algorithm to ensure the correctness of the computations. This allows for more scheduling choices and potentially better solutions.

The algorithm is presented in Algorithm 1. This algorithm may deadlock because adding the *WaR* arcs during the scheduling may create cycles in the graph, thus leading to a scheduling that is not compatible with the register bindings. These cycles are due to implicit register dependencies. An algorithm that minimizes these dependencies has been devised, but at worst backtracking must be applied, leading to an exponential computation time. A formal complexity analysis of the scheduling problem with register bindings as we have defined it has been done in [7]. This work proves that it is NP-complete to decide if scheduling a given node first will lead to a deadlock or not.

Nevertheless, the algorithm is usable and fast in practice, even on complex inputs, as it can be seen in Sect. 10.6.

Algorithm 1 RFG scheduling algorithm

Require: $\mathcal{N}$ the set of *RFG* nodes and $\mathcal{R}$ the set of *RaW* arcs
Ensure: $\mathcal{S}$ the set of scheduled nodes

  Let $\mathcal{S}_c$ the set of nodes scheduled at cycle c, $\mathcal{W}$ the set of arcs of type *WaR*, c the current scheduling cycle, υ the current node, u a node, s a successor node of υ, w a node that writes into a register and $a_{(u_1,u_2)}$ an arc from u_1 to u_2.
  $c \leftarrow 0$
  while $\mathcal{N} \neq \emptyset$ **do**
    Choose the best node that does not create a conflict using the select *function that selects the node with the lowest mobility*
    $\upsilon \leftarrow \text{select}(\{u \in \mathcal{N} \text{ such that } \forall w \in (\mathcal{S}_c \cup \mathcal{N}), a_{(w,u)} \notin (\mathcal{R} \cup \mathcal{W})\}$
    if $\upsilon \neq \varnothing$ **then**
      Has a node being chosen for the current cycle ?
      $\mathcal{S}_c \leftarrow \mathcal{S}_c \cup \{\upsilon\}$
      $\mathcal{N} \leftarrow \mathcal{N} \setminus \{\upsilon\}$
      for all $w \in \mathcal{N}$ such that w has the same destination register than u **do**
        for all $s \in \mathcal{N}, s \neq w$ and $a_{(\upsilon,s)} \in \mathcal{R}$ **do**
          $\mathcal{W} \leftarrow \mathcal{W} \cup \{a_{(s,w)}\}$
        end for
      end for
    else
      $c \leftarrow c+1$
    end if
  end while

10.3.3.4 Binding of the Combinational Operators

If the scheduling aims at minimizing the global execution time of the circuit, the combinational operator binding phase aims at minimizing its area once the scheduling known.

In the user guided synthesis, the operator number and kind are known, so the only degrees of freedom concern the number of inputs of the added multiplexers and the bit sizes of the arithmetic and logic operators. Minimizing both has the nice property that it also lowers the operators propagation time.

Under the assumption that there is a multiplexer in front of each combinational operator input, the optimization of the binding phase corresponds to minimize the size of the multiplexers. Each input is connected to a virtual multiplexer with at least one input, and the binding phase chooses at every cycle the binding that minimizes the multiplexer cost, computed as the number of inputs times the number of bits. This simple function allows to rank the solutions correctly, using as cost for the entire data-path the sum of the mux costs.

It has been shown in [4] that a simple exchange of commutative operators operands allows to decrease the number of inputs of the multiplexers by 30%. (More elaborate solutions can reach 40% [5] at a higher computational cost.) For each control step, the set of possible bindings of the operations and their operands on the physical operators and their inputs is built. Starting from an initial binding, we search in the sets a binding that minimizes the cost function, and we apply this binding. This is repeated until there is no binding that minimizes the cost.

10.4 Data-Path Implementation and Analysis

The link of high level synthesis with low level synthesis tools is seldom described in the literature. The synthesis tools most often generate a VHDL standard cell netlist. The circuit is obtained by placing and routing the VHDL netlist. The generated circuit will probably not run at the expected frequency. The main reasons are that the FSM has been constructed with estimated operator and connection delays, and that often the FSM is a Mealy one and its commands may have long delays. Furthermore it is also possible that the circuit does not run at any frequency if it mixes short and long paths. This happens frequently in circuits having both registers and register files.

Of course, these problems also occur with designs done by hand: in that case the designer solves them by adding states to the FSM, adding buffers to speed up or down some paths. This is not easy, and it takes time, but it is possible because he has an intimate knowledge of the design. After high level synthesis, these problems can not be corrected because the designer has lost the knowledge of the design.

From our point of view this mapping phase is an issue that must be dealt with, and not a minor one, because the generated circuit must run as it comes out of the tool. If it is not the case the synthesis tool is simply unusable.

In UGH, the mapping is done in three steps:

1. *Logic synthesis preparation*: The data-path produced by CGS is translated to a synthesizable VHDL description. The data-path is described structurally as an interconnection of UGH macro-cells. Every macro-cell is described as a behavior. Furthermore, a shell script is generated to automatically run the synthesis of each VHDL description using a standard cell library and giving constraints such as maximum fan out for the connectors.
2. *Logic synthesis*: The execution of this script invokes a logic synthesis tool to generate structural VHDL files respecting the given constraints.
3. *Delay extraction*: For each macro-cell instantiated in the CGS data-path, we extract the delays from the corresponding VHDL file produced by the logic synthesis. For that, we have the characteristics of the standard cells and we apply the following rules for computing, in this order, the minimum and maximum propagation times, the setup and hold times:

$$
\begin{aligned}
t_{\min_{I \rightsquigarrow O}} &= \min_{p \in \mathcal{P}_{\mathcal{I} \rightsquigarrow \mathcal{O}}} \sum_{(c_i,c_o) \in p} prop_{\min}(c_i, c_o, l_{c_i}, l_{c_o}) \\
t_{\max_{I \rightsquigarrow O}} &= \max_{p \in \mathcal{P}_{\mathcal{I} \rightsquigarrow \mathcal{O}}} \sum_{(c_i,c_o) \in p} prop_{\max}(c_i, c_o, l_{c_i}, l_{c_o}) \\
t_{\mathrm{setup}_{I \rightsquigarrow CK}} &= \max_{(c_i,c_{ck}) \in \mathcal{C}} \left(t_{\max_{I \rightsquigarrow c_i}} + t_{\mathrm{setup}_{c_i \rightsquigarrow c_{ck}}} - t_{\min c_{ck} \rightsquigarrow CK}\right) \\
t_{\mathrm{hold}_{I \rightsquigarrow CK}} &= \max_{(c_i,c_{ck}) \in \mathcal{C}} \left(t_{\min_{I \rightsquigarrow c_i}} + t_{\mathrm{hold}_{c_i \rightsquigarrow c_{ck}}} - t_{\max c_{ck} \rightsquigarrow CK}\right)
\end{aligned}
$$

In these formulae, I, CK and O represent the macro-cell inputs and outputs, $\mathcal{P}_{x \rightsquigarrow y}$ is the set of paths from the port x to the port y, a path p being a sequence of couples of ports of the same cell. $\mathcal{C}$ is a set of couples of input ports of the same cell having setup and hold times. $prop_{\min}$ and $prop_{\max}$ are the functions characterizing the standard cells taking into account the input and output loads (l_{c_i}, l_{c_o}).

Of course this step may be quite long for large data-paths. For this reason, UGH gives the possibility to bypass the mapping during design tuning and instead uses pessimistic estimated delays.

Currently, this delay extraction is implemented for the Synopsys tools. Furthermore, even though the backend tools use VHDL, they use different VHDL dialects. This requires to adapt the mapping tool to the backend.

10.5 Fine Grain Scheduling

The arrow in the Y chart of the Fig. 10.2c represents FGS. It shows that its job is to retime [21] a FSM.

We illustrate the algorithm on a small example. The Fig. 10.9 presents the inputs of Fine Grain Scheduler: (1) a data-path with known electrical (Fig. 10.9a); (2) the RTL instructions directly extracted from the CG-FSM control-steps

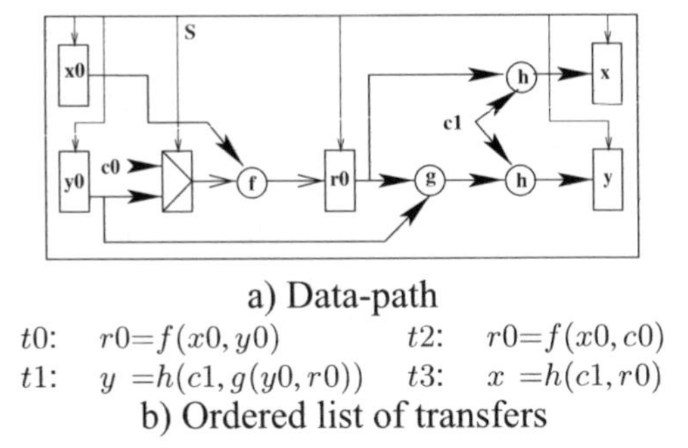

Fig. 10.9 Inputs of the FGS algorithm

(Fig. 10.9b), those are called transfers, and their order matters; (3) a running frequency.

FGS deals with the scheduling of basic-blocks. As a reminder, a basic-block is a sequence of RTL instructions without any control statements, except optionally the last one. Furthermore, in the global program, there is no branch instruction that jumps in the basic block, except at its beginning.

The idea behind FGS is to reorganize the basic-blocks of the CG-FSM, moving instructions from one control-step to either a close control-step or to an added control-step, and then suppressing the useless control-steps.

10.5.1 Definitions

Transfer A transfer is the motion of data from the outputs of a set of registers to the input of a target register.

A transfer t is represented as a DAG, $D^t(V^t, A^t)$, whose vertices are operations and arcs are data dependencies as realized on the data-path. The Fig. 10.10a shows the DAG of the $t0$ transfer of Fig. 10.9. In this DAG, the rectangles represent the output of the control unit (memorized in the micro-instruction register MIR), and the circles represent functional operations. There are three kind of vertices:

COP Concurrent OPerations do not modify the state of the data-path. For instance, changing the selection command of a multiplexer in a control-step only assigns MIR. The next control-step may restore the previous value and so restore the circuit in the previous state. They correspond to a value on bit fields of MIR. Two COPs are equivalent if they match the same bit field

POP Permanent OPerations always perform the same task and are associated to a single functional resource

SOP Sequential OPerations modify the state of the data-path. They perform memorization operation: Once done, the overwritten value is lost. They usually correspond to a data-path register, and a bit field of MIR. Two SOPs are equivalent if they match the same bit field

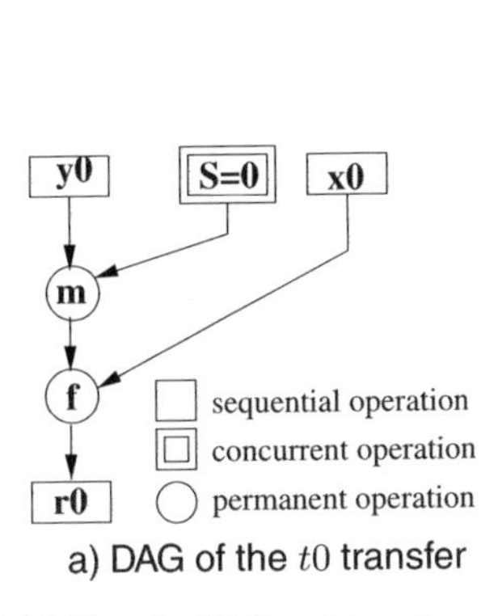

a) DAG of the $t0$ transfer

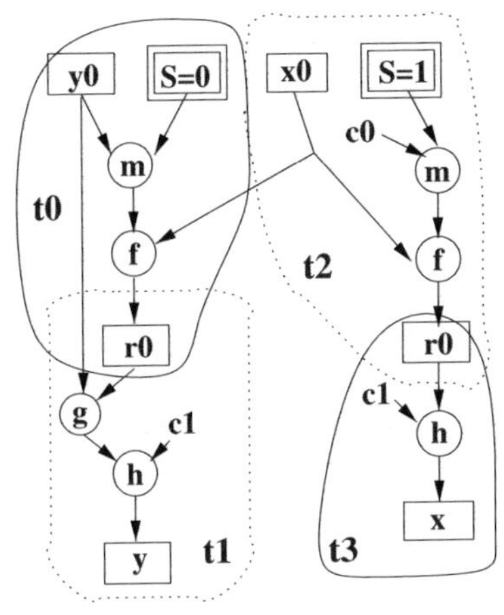

b) transfer graph of Figure 1.9

Fig. 10.10 Transfer DAG and transfer graph

A transfer $D^t(V^t, A^t)$ has the following structural properties:

- V^t_{source} the set of vertices that have no predecessors. V^t_{source} only contains COP and SOP.
- V^t_{sink} the set of vertices that have no successors. $|V^t_{sink}| = 1$ and its element is a SOP.
- $V^t_{operator} = V^t - (V^t_{source} \cup V^t_{sink})$. All elements of $V^t_{operator}$ are POPs.

Transfer Graph A transfer graph is a directed acyclic graph, $D(V,A)$, that represents the set of transfers that occur in the data-path for a given top level FSM transition. The transfer graph is the concatenation of all transfers of the input list in the list order (Fig. 10.9b). The transfer D^t is added to the graph, and the vertices $v \in V^j_{source}$ are merged to the most recently added equivalent vertices. Fig. 10.10b shows the transfer graph resulting of the example of Fig. 10.9.

Characterized Transfer A characterized vertex is a vertex annotated with delays (see Fig. 10.11a).

A POP vertex has a value associated to each couple of incoming and outgoing arcs of the vertex. These values represent the set of propagation times of the corresponding physical cell.

A COP vertex has only one value associated to the outgoing arc, it corresponds to the propagation time from the clock to the MIR output bits associated to the COP.

A SOP vertex has two values associated to each incoming arc and one for each outgoing arc. They represent the set-up and hold times from the input relative to the clock and the propagation time from the clock to the output from the corresponding physical cell.

These values are delays extracted from the physical placed and routed data-path, so wire delays are taken into account.

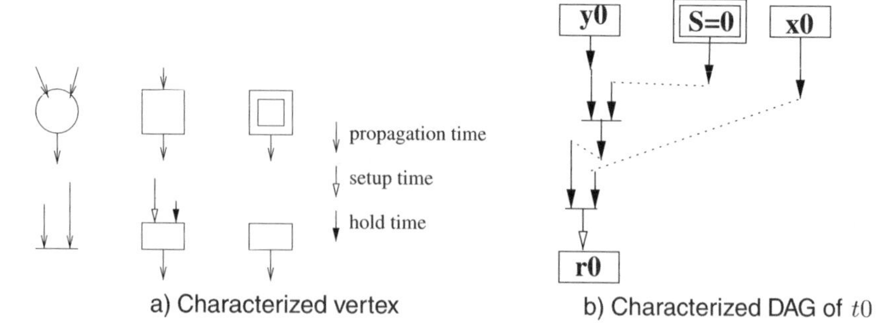

Fig. 10.11 Characterized vertex

The characterized transfer is obtained by replacing the original transfer vertices by characterized vertices. Figure 10.11b shows the characterized transfer of the transfer presented Fig. 10.10a. The values of the characterized vertices are graphically represented by the length of the plain arrows.

Characterized Transfer Graph It is obtained from the transfer graph by replacing transfers with characterized transfers. Nevertheless other arcs must be added to correctly model the behavior of the initial transfer sequence. These arcs implement the *WaR* and *WaW* precedence relations.

- The *RaW* relation denotes the usual data dependencies.
- The *WaR* relation expresses the fact that two equivalent COPs are used with different values. In our example this occurs for $S = 0$ in the $t0$ transfer and $S = 1$ in the $t2$ transfer. S can be set to 1 only when this will not disturb the $t0$ transfer. This gives the arc from the $r0$ hold time to $S = 1$ propagation time.
- The *WaW* relation indicates that two equivalent SOPs are used within two different transfers. In the example, $r0$ is used simultaneously in $t0$ and $t2$. The SOP of $t2$ must be performed after the SOP of $t0$, because the same register cannot be loaded twice in the same cycle.

The resulting graph is plotted in Fig. 10.12a, with the previous relations outlined.

10.5.2 Scheduling of a Basic Block

The scheduling rules are:

R1 Load a given register only once in a given cycle
R2 Loading a register must respect its set-up time
R3 Loading a register must not violate the hold time

The clock period defines a grid on the which the SOPs and the COPs must be *snapped*. A simple ASAP [19] algorithm with the constraint that all arcs point downwards (Fig. 10.12b) produces a scheduling that verifies the scheduling rules. This

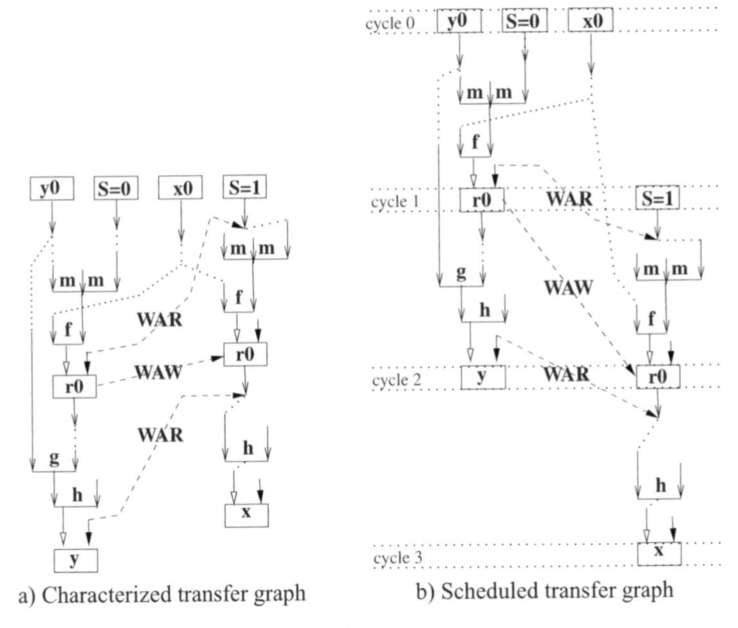

Fig. 10.12 Characterized and scheduled transfer graph

pointing downwards relation is either combinational when concurrent operations are involved or sequential when a permanent operation is involved.

Rule *R1* is enforced by the arcs implementing the *WaW* relations. Rule *R2* is enforced by *RaW* relations (data dependencies: the plain arrows). Rule *R3* is enforced by the the arcs implementing the *WaR* relations.

This scheduling allows all kinds of chaining and especially multi-cycles chaining without intermediate memorizations.

The only delays not taken into account are the propagation times from the FSM state register to the data-path. This is solved because the control unit is a Moore FSM with a MIR that synchronizes the control signals and that we assume that the delays due to routing capacitances between the MIR and the operator command inputs are similar. This can be ensured by increasing the fan-out of the MIR buffers.

10.5.3 Optimization of the WaR *Relations*

The Fig. 10.13 represents the scheduling of our example using a double frequency comparatively to the scheduling given Fig. 10.12b, only the $t0$ transfer and the beginning of $t2$ transfer are represented.

In the FGS implementation, the arcs expressing the *WaR* relation do not start from the hold time of the SOP but from the hold time minus *mt* (see Fig. 10.13a) where *mt* is the minimum propagation time from the COP ($S = 0$) to the SOP ($r0$).

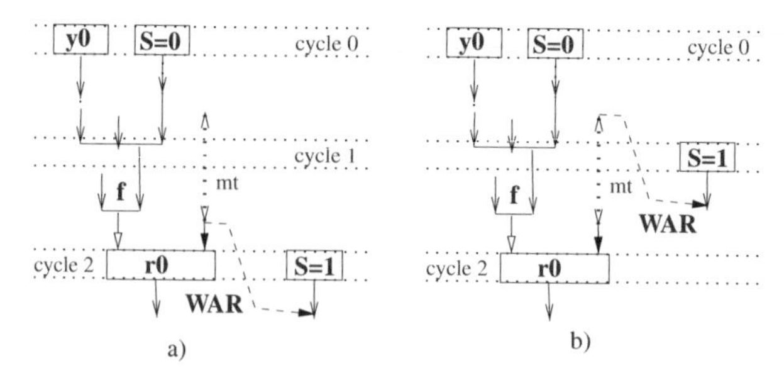

Fig. 10.13 Optimized scheduling of transfer graph

This allows when *mt* is large enough to anticipate the scheduling of the COP ($S = 1$) as shown on the Fig. 10.13b and so to get a better scheduling.

Furthermore, this feature allows to automatically schedule wave pipeline [12] provided that minimum and maximum probation times are close.

10.5.4 Scheduling Quality

The list order is used to set the *WaR* and *WaW* relations in the characterized transfer graph.

Unfortunately, the list of transfers only gives the data dependence relations (*RaW*) and thus defines only a partial order on the transfers. This fact induces that for a given list of transfers, there are in general several characterized transfer graphs, and as many different valid schedules.

To taxonomize this, we introduce three relations between the transfers.

$T_i \xrightarrow{DD} T_j$ (data dependent): SOP_i belongs to V^j_{source}. It means that T_j uses the result of T_i. It is the classical RAW relation.

$T_i \xleftrightarrow{SD} T_j$ (sequential dependent): SOP_i and SOP_j are the same and there is no direct or transitive $T_i \xrightarrow{DD} T_j$ relation nor $T_j \xrightarrow{DD} T_i$ relation. It means that the same resource is used to store two results of potentially concurrent transfers.

$T_i \xleftrightarrow{CD} T_j$ (concurrent dependent): there is a COP element of both V^i_{source} and V^j_{source} which selects a different value and there is no direct or transitive $T_i \xrightarrow{DD} T_j$ relation nor $T_j \xrightarrow{DD} T_i$ relation. It means that the same functional operator is used in both transfers but performs different functions in each transfer.

These relations allow to define three transfer graph classes.

Sequential-ordered transfer graph: This is the initial data, with all the $\xrightarrow{DD}$, $\xleftrightarrow{SD}$ and $\xleftrightarrow{CD}$ relations,

Concurrent-ordered transfer graph: It is a sequential-ordered transfer graph where all the $\xleftrightarrow{SD}$ relations have been resolved. Let X_k the transfers verifying $T_i \xrightarrow{DD} X_k$ and Y_k the one verifying $T_j \xrightarrow{DD} Y_k$. Resolving a $T_i \xleftrightarrow{SD} T_j$ relation means replacing it by either the pseudo relations $X_k \xrightarrow{DD} T_j$ or the pseudo relations $Y_k \xrightarrow{DD} T_i$ as shown Fig. 10.14a2. Note that resolving a $\xleftrightarrow{SD}$ adds $\xrightarrow{DD}$ relations and may suppress others $\xleftrightarrow{SD}$ or $\xleftrightarrow{CD}$.

Resolved-ordered transfer graph: It is a concurrent-ordered transfer graph in the which all the $\xleftrightarrow{CD}$ relations have been resolved. Resolving a $\xleftrightarrow{CD}$ means replacing it by either the pseudo relations $T_i \xrightarrow{DD} T_j$ or $T_j \xrightarrow{DD} T_i$. Fig. 10.14a3 shows the two possible resolutions of the sequential ordered graph of Fig. 10.14a1.

Resolving a relation only adds $\xrightarrow{DD}$, thus the algorithm does not create new relations to be solved, avoiding cycles. So a sequential-ordered transfer graph gives a set of concurrent-ordered transfer graphs and each of those gives a set of resolved-ordered transfer graphs (Fig. 10.14b).

The FGS algorithm is optimum at the level of resolved-ordered transfer graphs. It will give the same result for all the transfer lists extracted from the resolved-ordered transfer graph respecting the partial order of $\xrightarrow{DD}$ relation. Of course, other resolved-ordered transfer graphs can be obtained from the initial sequential-ordered transfer graph. Their schedulings may be better or worse.

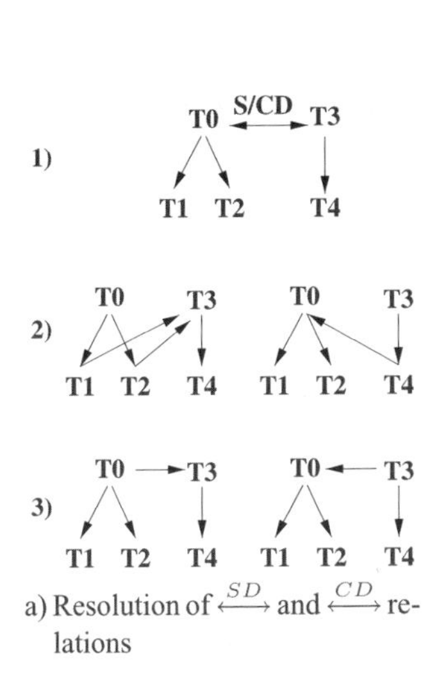

a) Resolution of $\xleftrightarrow{SD}$ and $\xleftrightarrow{CD}$ relations

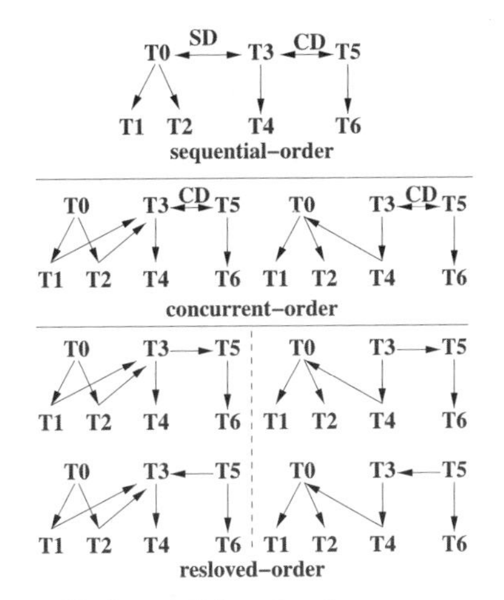

b) Sequential ordered, concurrent ordered and resolved ordered transfer graphs.

Fig. 10.14 $\xleftrightarrow{SD}$, $\xleftrightarrow{CD}$ relations and transfer graph orders

10.5.5 Scheduling of an Entire FSM

The previous sections have dealt with the FGS scheduling of a simple basic block. The FSM of an integrated circuit is however composed of a graph of basic blocks as shown on Fig. 10.15a. We call this graph $G(V,A)$, where V is the set of basic blocks and A is the set of transitions. A global approach is needed to optimize the scheduling.

Transition Function The first problem is to introduce the transition function (the arcs $a \in A$) in the transfer graph. Actually, we must compute the conditions of the transition arcs at the end of a basic block. For instance, the basic block BB2 in the Fig. 10.15a can branch to BB3, to BB4 or to BB5 only once the conditions $X+Y<0$ and $R \neq 0$ have been evaluated. This problem is solved by adding a transition transfer that loads the state register. The transition transfer of BB2 is shown on Fig. 10.15b. The TF operator corresponds to the transition function of the FSM. Once the basic block is FGS scheduled, the minimal number of cycles of the basic block is given by the cycle in which the state register is set.

The electrical characteristics (propagation delays) of the TF operator are unknown in the FGS scheduling. We must set them to arbitrary values, these values becoming constraints of the FSM synthesis tools. In practice, we set this value to the half of the cycle time.

Historic Given an integer N, we define $historic_N$ as a scheduled transfer graph of N cycles containing SOPs and COPs that cover all the bits of MIR. In the following, we build $historic_N$ in two ways.

The first way is the *worst-historic*$_N$ presented Fig. 10.16. We use the *worst* SOP for the sequential resources, for the COPs we use the value *unknown*. This *worst-historic*$_N$ is independent of the basic blocks. The worst SOP of a sequential resource is the operation that produces the data the latest. For a register that supports the load

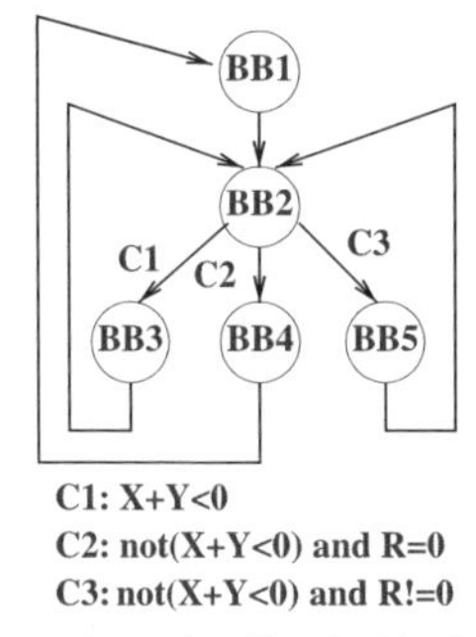

a) Control graph of basic block

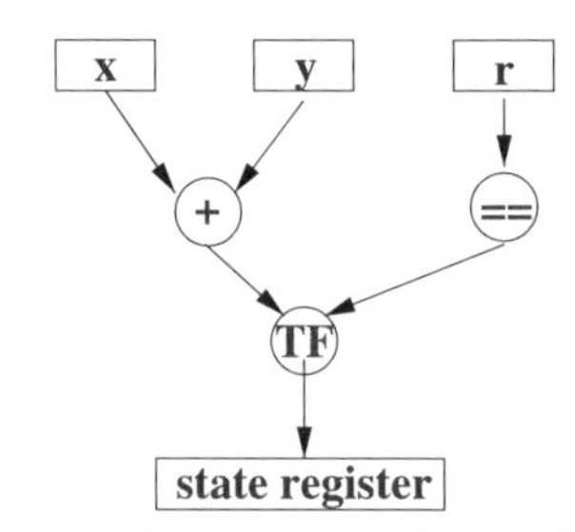

b) Transition transfer of the BB2 basic block

Fig. 10.15 Handling control

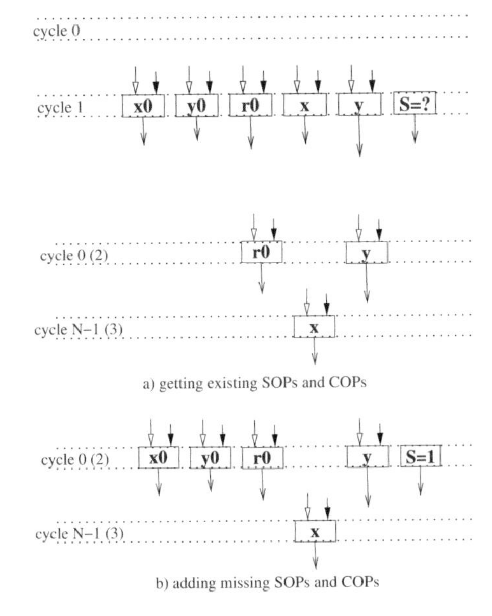

Fig. 10.16 *Worst-historic*$_2$ for the circuit of Fig. 10.9a

Fig. 10.17 $S_{2,x}$ of scheduled transfer graph of Fig. 10.12b

and the clear operations, it is the operation that has the greatest propagation time. Normally a COP can be shared by several transfers. The unknown COP value ($S = ?$ in the figure) indicates that this COP cannot be used by a transfer.

The second way is the *current-historic*$_{N,b}$ of the basic block b. Let $P : \{p \in V | (p,b) \in A\}$ the set of direct predecessors of b and $S_{N,x}$ the *historic*$_N$ summarizing the x basic block. The steps for building $S_{N,x}$ are given below and illustrated by the Fig. 10.17 for the scheduled basic block of the Fig. 10.12b and for $N = 2$. In this figure, the numbering of the cycles in parenthesis refers to the cycles of the Fig. 10.12b.

1. Perform the FGS scheduling of x basic block to get the scheduled transfer graph.
2. Take the COPs and SOPs of the last N cycles of the scheduled transfer graph, as shown in Fig. 10.17a.
3. Place the latest SOPs and COPs of the scheduled transfer graph except the formers on the first cycle of $S_{N,x}$ (Fig. 10.17b).

The *current-historic*$_{N,b}$ consists of merging the $S_{N,p}$. Merging means choosing the latest *worst* SOP for a sequential resource, and the latest COP for a concurrent resource. If there are two latest COPs with different values, the value of the COP is set to unknown.

FGS Scheduling with an Historic The scheduling of a basic block using an historic is similar to the algorithm presented in Sect. 10.5.2. When the transfer graph is build, the transfers are attached to the historic COPs and SOPs, and the scheduling must respect the following rules: The SOPs and COPs of the historic are already on the grid and must not be changed. The SOPs of the basic block must not be scheduled in the cycle of the historic. So the COPs may be scheduled in the historic, allowing to start the transfers in its predecessor basic blocks. The resulting scheduled transfer graphs do not directly correspond to the circuit FSM. Actually, the historic cycles must be suppressed and the COPs of these cycles must be transferred in the cycles of the preceding basic blocks.

Global Scheduling The algorithmic principles are presented in Algorithm 2. The main idea is to schedule each basic block taking in account the scheduling of its predecessors to start the scheduling of its transfers as soon as possible. Let p a predecessor of two basic blocks b_1 and b_2, the scheduling of b_1 can alter the historic of p, and so does the scheduling of b_2. We must ensure that after scheduling, the historic of p does not have different values for the same COP. This is done by the point noted $\{\dagger\}$ in the algorithm.

This algorithm may not converge if a cycle is present in G. To avoid endless iterations, it is necessary to break the loop after a predefined number of iterations. In our implementation, we break the loop by forcing the scheduling of one of the unscheduled basic blocks (new and old historics are different) with the *worst-historic*, suppressing it from G and then restarting the loop. The *pseudo-topological-sort* used in the algorithm labels the nodes so that the number of return arcs is minimal. It allows to schedule the maximum of basic blocks with their actual *current-historic*$_N$ at the first iteration.

Algorithm 2 Global scheduling algorithm

Record couple: {basic_block b, historic h}
Require: G the graph of basic blocks
Ensure: S the set of couples
  $S \leftarrow \cup_{b \in G}\{(b, \textit{worst-historic})\}$
  $S \leftarrow \textit{pseudo-topological-sort}(S)$
  for all $c \in S$ **do**
    $c.b \leftarrow$ schedule $c.b$ with $c.h$
  end for
  $end \leftarrow false$
  while not *end* **do**
    $end \leftarrow true$
    for $c \in S$ **do**
      $c.b \leftarrow$ schedule $c.b$ with $c.h$
      $h \leftarrow$ *current-historic* of $c.b$
      if $h \neq c.h$ **then**
        $c.h \leftarrow h$
        $c.b \leftarrow$ schedule $c.b$ with $c.h$
        transfer the COPs of $[0, N-1]$ cycles of $c.b$ into the $\{p \in V | (p, c.b) \in A\}$ $\{\dagger\}$
        $end \leftarrow false$
      end if
    end for
  end while

10.6 Experimentation

The UGH approach has been applied to the synthesis of several examples from various sources. Some are from the multimedia application field: the MPEG Variable Length Decoder [9], and a Motion-JPEG decoder [1]. Some are from the communication area: DMA controller for the PCI bus. And others are synthetic dataflow benchmarks.

The Table 10.2 summarizes the synthesis results and runtimes. The 4 first columns characterize the complexity of the design, in number of lines of the input *C* code, in circuit size in terms of inverters, and in FSM state number. The *ugh* column gives the runtime of both CGS and FGS, the *mapping* column gives the time required to generate the data-path including the UGH mapping (see Sect. 10.4) and the RTL synthesis. These results show that the tools are able to handle large descriptions and can produce circuits of more than 100,000 gates. They also show that the approach is usable for pure data-flow (i.e., IDCT), control oriented (i.e., VLD) or mixed (i.e., LIBU) type of algorithms. The CGS and FGS tools run very fast even for large designs, making the flow suitable for exploring several architectural solutions. The mapping is quite long and 95–99% of its time is due to RTL synthesis. However, this stage can be skipped during design space exploration and debug by using default delays.

The Table 10.3 details two implementations of the IDCT based on Loffler algorithm [16]. The first implementation is area optimized. The algorithm is straightforward sequential *C* code. Regarding the second implementation, it is optimized for speed. The parallelism has been implicitly exposed by unrolling the loops and introducing variables to make pipelining and parallelism possible. The design work to obtain this second implementation is not trivial. These two implementations are extremes, but all intermediates implementations can be described. This shows that UGH allows to cover the whole design space.

Table 10.2 Results and runtimes for a complete synthesis

Module		Complexity (in lines)	Size (in inverters)	FSM (in states)	Data flow	Time	
						ugh (s)	Mapping
MPEG	VLD	704	10.060	109	No	11	0h40
M-JPEG	VLD	513	182.102	167	No	18	1h57
	IQ	93	12.520	38	Yes	3	1h00
	ZZ	44	49.645	14	Yes	7	1h31
	IDCT	186	73.776	113	Yes	47	1h08
	LIBU	170	47.331	43	Mix	22	0h22
FD	FD 1	144	47.644	32	Yes	30	0h55
	FD 2	144	51.826	32	Yes	35	0h50
	FD 3	144	9.371	32	Yes	1	0h13
DMA	TX	346	48.536	442	No	1	0h15
	RX	287	40.730	111	No	1	0h21
	WD	212	16.394	43	No	1	0h18

Table 10.3 Illustration of UGH synthesis tuning capabilities

	Clock period (ns)	FSM states	Execution cycles	Execution time (μs)	Area (mm^2)
Area	17	90	1,466	24.92	10.9
Speed	17	460	460	7.82	18.4

Table 10.4 Impact of the binding constraints

Links	Clock period (ns)	FSM states	Execution cycles	Execution time (μs)	Area	
					(mm^2)	Inverter number
All	5	109	6.530.912	32.6	1.13	10.060
Some	5	112	6.905.663	34.5	1.14	10.168
None	5	115	6.936.683	34.6	1.14	10.134

In our approach, the data-path is fixed, so we fundamentally perform FSM retiming. Using the usual HLS approaches means that the logic synthesis tool has to perform data-path retiming for a given finite state machine. This is fine when the data-path is not too complex, however when logic synthesis enters procrastination and gate duplications techniques [23], the number of gates increases drastically and leads to unacceptable circuits.

We have experimented UGH with various levels of constraints in the DDP on several examples. The DDP is fully given (registers, operators and links between the resources), minimally given (registers and operators, no links at all), or partially given (registers, operators and links expected to be critical). Most of the time, their impact is weak, as illustrated by the Motion-JPEG VLD example whose synthesis results are given in Table 10.4. So given a sequential behavior, the functional operators and the registers with the allocation of the variables of the behavior, we conjecture that a unique optimal data-path exists.

10.7 Conclusion and Perspectives

UGH produces circuits better or similar in quality compared to other recent high level synthesis tools (see Chap. 7 of [8] for these comparisons), without using classic constraint or unconstrained scheduling algorithms such as list scheduling [11], force directed scheduling [22] or path scheduling [3] but by introducing the draft data-path and the retiming of the finite state machine.

The introduction of the *DDP* allows the circuit designer to target directly the circuit implementation he wants. This is to compare to the other high level synthesis tools that usually need a lot of lengthy iterations to achieve acceptable solutions. So UGH is dedicated to circuit designers. The most important point is that UGH does not disturb the designer working habits, as opposed to all other HLS tools. Indeed, the *DDP* is more or less the back of the envelope draft that any designer does before starting the description of the design. This part of the designer work is the more creative and the most interesting one. UGH leaves this to the designer, and handles the unrewarding ones automatically.

The introduction of the retiming of the finite state machine guarantees that the generated circuits run at the required frequency, as opposed to the vast majority of HLS tools for which frequency is a constraint given to the data-path synthesis tools. More often, data-path synthesis tools enter into procrastination algorithms to obey the frequency constraint and lead to unacceptable circuits. The retiming of the finite state machine just adds a few states which do not change significantly the circuit size. The only disadvantage is that the generated circuit requires asynchronous inputs and outputs.

UGH gives very good results for control dominated circuits. It does not implement, as dedicated data-flow synthesis tools do, neither the usual techniques such as loop folding and unrolling and unnesting nor the usual scheduling algorithms for pipelining data-flow blocks. Dedicated data-flow synthesis tools such as [13, 15, 17, 25] implement these techniques and algorithms but have difficulties to handle control dominated circuits. This is an handicap for the usage of data-flow oriented tools, because most circuits mix control and data flow parts.

For a circuit mixing control and data-flow parts, one can apply the specific data-flow techniques and algorithms by hand on the data-flow parts and make a UGH description (*C* program + *DDP*) of the circuit. So UGH inputs are at an adequate level for the outputs of a HLS compiler mixing both data and control parts. Such a compiler taking as input a *C* description could make the data-flow specific treatments on the data-flow parts and generate a UGH description as a *C* program and a *DDP*.

Finally, to make a parallel with a software compiler (compilation, assembly and link), for us UGH is at the assembly and link level. Indeed, it treats the electrical and timing aspects, and links with the back-end tools.

References

1. Augé, I., Pétrot, F., Donnet, F., and Gomez, P. (2005). Platform-based design from parallel C specifications. *IEEE Transactions on Computer-Aided Design of Integrated Circuits and Systems*, 24(12):1811–1826.
2. Bryant, R. E. (1986). Graph-based algorithms for boolean function manipulation. *IEEE Transactions on Computer*, C-35(8):677–691.
3. Camposano, R. (1991). Path-based scheduling for synthesis. *IEEE Transactions on Computer-Aided Design of Integrated Circuits and Systems*, 10(1):85–93.
4. Chang, E.-S. and Gajski, D. D. (1996). A connection-oriented binding model for binding algorithms. Technical Report ICS-TR-96-49, UC Irvine.
5. Chen, D. and Cong, J. (2004). Register binding and port assignment for multiplexer optimization. In *Proc. of the Asia and South Pacific Design Automation Conf.*, pages 68–73, Yokohama, Japan. IEEE.
6. Coussy, P., Corre, G., Bomel, P., Senn, E., and Martin, E. (2005). High-level synthesis under I/O timing and memory constraints. In *Proc. of the Int. Symp. on Circuits and Systems,*, volume 1, pages 680–683, Kobe, Japan. IEEE.

7. Darte, A. and Quinson, C. (2007). Scheduling register-allocated codes in user-guided high-level synthesis. In *Proc. of the 18th Int. Conf. on Application-specific Systems, Architectures and Processors*, pages 219–224, Montreal, Canada. IEEE.
8. Donnet, F. (2004). *Synthése de haut niveau contrôlée par l'utilisateur*. PhD, Université Pierre et Marie Curie (Paris VI).
9. Dwivedi, B. K., Hoogerbrugge, J., Stravers, P., and Balakrishnan, M. (2001). Exploring design space of parallel realizations: Mpeg-2 decoder case study. In *Proc. of the 9th Int. Symp. on Hardware/Software Codesign*, pages 92–97, Copenhagen, Denmark. IEEE.
10. Gajski, D. D., Dutt, N. D., Wu, Allen C.-H., and Lin, S. Y.-L. (1992). *High-Level Synthesis: Introduction to Chip and System Design*. Berlin Heidelberg New York. Springer.
11. Graham, R. L. (1969). Bounds on multiprocessing timing anomalies. *Journal of Applied Mathematics*, 17:416–429.
12. Gray, C. T., Liu, W., and Cavin, R. K. III (1994). Timing constraints for wave-pipelined systems. *IEEE Transactions on Computer-Aided Design of Integrated Circuits and Systems*, 13(8):987–1004.
13. Guillou, A.-C., Quinton, P., and Risset, T. (2003). Hardware synthesis for multi-dimensional time. In *Proc. of the Int. Conf. on Application-Specific Systems, Architectures, and Processors*, pages 40–50. IEEE.
14. Huang, S.-H., Cheng, C.-H., Nieh, Y.-T., and Yu, W.-C. (2006). Register binding for clock period minimization. In *Proc. of the Design Automation Conf.*, pages 439–444, San Francisco, CA. IEEE.
15. Ko, M.-Y., Zissulescu, C., Puthenpurayil, S., Bhattacharyya, S. S., Kienhuis, B., and Deprettere, E. F. (2007). Parameterized loop schedules for compact representation of execution sequences in dsp hardware and software implementation. *IEEE Transactions on Signal Processing*, 55(6):3126–3138.
16. Loeffler, C., Ligtenberg, A., and Moschytz, G. S. (1989). Practical fast 1-D DCT algorithms with 11 multiplications. In *Proc. of the Int. Conf. on Acoustics, Speech and Signal Processing*, volume 2, pages 988–991, Glasgow, UK.
17. Martin, E., Sentieys, O., Dubois, H., and Philippe, J.-L. (1993). An architectural synthesis tool for dedicated signal processors. In *Proc. of the European Design Automation Conf.*, pages 14–19.
18. Michel, P., Lauter, U., and Duzy, P. (1992). *The synthesis approach to digital system design*, chapter 6, pages 151–154. Dordrecht. Kluwer Academic.
19. Micheli, De G. (1994). *Synthesis and Optimization of Digital Circuits*, chapter 9, page 441. New York. McGraw-Hill.
20. Pangrle, B. M. and Gajski, D. D. (1987). Design tools for intelligent silicon compilation. *IEEE Trans. on Computer-Aided Design of Integrated Circuits and Systems*, 6(6):1098–1112.
21. Parameswaran, S., Jha, P., and Dutt, N. (1994). Resynthesizing controllers for minimum execution time. In *Proc. of the 2nd Conf. on Computer Hardware Description Languages and Their Applications*, pages 111–117. IFIP.
22. Paulin, P. G. and Knight, J. P. (1989). Force-directed scheduling for the behavioral synthesis of asics. *IEEE Trans. on Computer-Aided Design of Integrated Circuits and Systems*, 8(6):661–679.
23. Srivastava, A., Kastner, R., Chen, C., and Sarrafzadeh, M. (2004). Timing driven gate duplication. *IEEE Trans. on Very Large Scale Integration Systems*, 12(1):42–51.
24. Toi, T., Nakamura, N., Kato, Y., Awashima, T., Wakabayashi, K., and Jing, L. (2006). High-level synthesis challenges and solutions for a dynamically reconfigurable processor. In *Proc. of the Int. Conf. on Computer Aided Design*, pages 702–708, San José, CA. ACM.
25. van Meerbergen, J. L., Lippens, P. E. R., Verhaegh, W. F. J., and van der Werf, A. (1995). Phideo: High-level synthesis for high throughput applications. *Journal of VLSI Signal Processing*, 9(1–2):89–104.
26. Zhu, J. and Gajski, D. D. (1999). Soft scheduling in high level synthesis. In *Proc. of the 36th Design Automation Conf.*, pages 219–224, New Orleans, LA.

Chapter 11
Synthesis of DSP Algorithms from Infinite Precision Specifications

Christos-Savvas Bouganis and George A. Constantinides

Abstract Digital signal processing (DSP) technology is the core of many modern application areas. Computer vision, data compression, speech recognition and synthesis, digital audio and cameras, are a few of the many fields where DSP technology is essential.

Although Moore's law continues to hold in the semiconductor industry, the computational demands of modern DSP algorithms outstrip the available computational power of modern microprocessors. This necessitates the use of custom hardware implementations for DSP algorithms. Design of these implementations is a time consuming and complex process. This chapter focuses on techniques that aim to partially automate this task.

The main thesis of this chapter is that domain-specific knowledge for DSP allows the specification of behaviour at infinite precision, adding an additional 'axis' of arithmetic accuracy to the typical design space of power consumption, area, and speed. We focus on two techniques, one general and one specific, for optimizing DSP designs.

Keywords: DSP, Synthesis, Infinite precision, 2D filters.

11.1 Introduction

The aim of this chapter is to provide some insight into the process of synthesising digital signal processing circuits from high-level specifications. As a result, the material in this chapter relies on some fundamental concepts both from signal processing and from hardware design. Before delving into the details of design automation for DSP systems, we provide the reader with a brief summary of the necessary prerequisites. Much further detail can be found in the books by Mitra [14] and Wakerly [18], respectively.

P. Coussy and A. Morawiec (eds.) *High-Level Synthesis.*

Digital Signal Processing refers to the processing of signals using digital electronics, for example to extract, suppress, or highlight certain signal properties. A signal can be thought of as a 'wire' or variable, through which information is passed or streamed. A signal can have one or many dimensions; a signal that represents audio information is a one-dimensional signal, whereas a signal that represents video information is a two dimensional signal.

A discrete-time signal x is usually represented by using the notation $x[n]$. The value $x[n]$ of the signal x refers to the value of the corresponding continues-time signal at sampling time nT, where T denotes the sampling period.

The z transform is one of the main tools that is used for the analysis and processing of digital signals. For a signal $x[n]$, its z transform is given by (11.1).

$$X(z) = \sum_{n=-\infty}^{\infty} x[n]z^{-n} \tag{11.1}$$

The chapter will mainly focus on linear time invariant (LTI) systems, thus it is worthwhile to see how the z transform is useful for such systems. The output signal $y[n]$ of an LTI system with impulse response $h[n]$ and input signal $x[n]$ is given by the convolution of the input signal and the impulse response (11.2).

$$y[n] = \sum_{k=-\infty}^{\infty} h[k]x[n-k] \tag{11.2}$$

Using the z transform, (11.2) can be written as (11.3), where $Y(z)$, $H(z)$, and $X(z)$ are the z transforms of the $y[n]$, $h[n]$ and $x[n]$ signals, respectively.

$$Y(z) = H(z)X(z) \tag{11.3}$$

Thus convolution in the time domain is equivalent to multiplication in the z domain, a basic result used throughout this chapter.

11.1.1 Fixed Point Representation and Computational Graphs

In this chapter, the representation of DSP algorithm is the *computational graph*, a specialization of a data flow graph graph of Lee et al. [12]. In a computational graph each element in the set V corresponds to an atomic computation or input/output port, and $S \subseteq V \times V$ is the set of directed edges representing the data flow. An element of S is referred as a *signal*.

In the case of an LTI system, the computations in a computational graph can only be one of several types: input port, output port, gain (constant coefficient multiplier), addition, unit-sample delay and a fork (branching of data). These computations should satisfy the constrains of indegree and outdegree given in Table 11.1. A visualization of the different node types is shown in Fig. 11.1. An example of a computational graph is shown in Fig. 11.2.

Table 11.1 Degrees of a node in a computational graph

Type	Indegree	Outdegree
Inport	0	1
Outport	1	0
Add	2	1
Delay	1	1
Gain	1	1
Fork	1	≥ 2

Fig. 11.1 Type of nodes

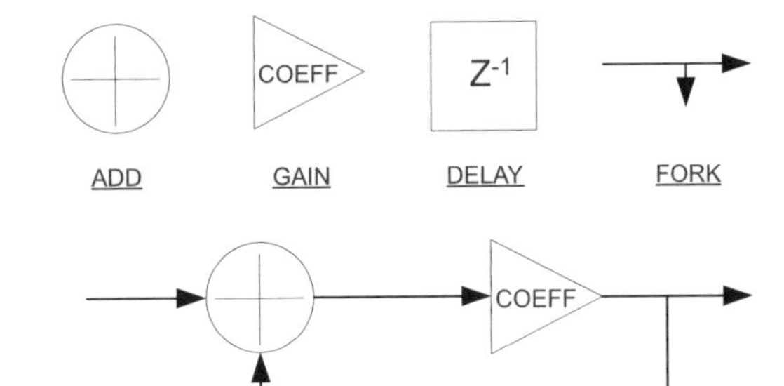

Fig. 11.2 An example of a computation graph

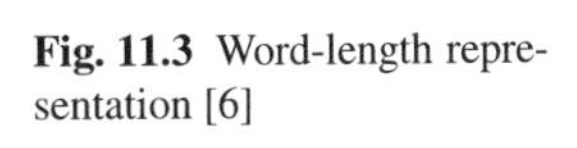

Fig. 11.3 Word-length representation [6]

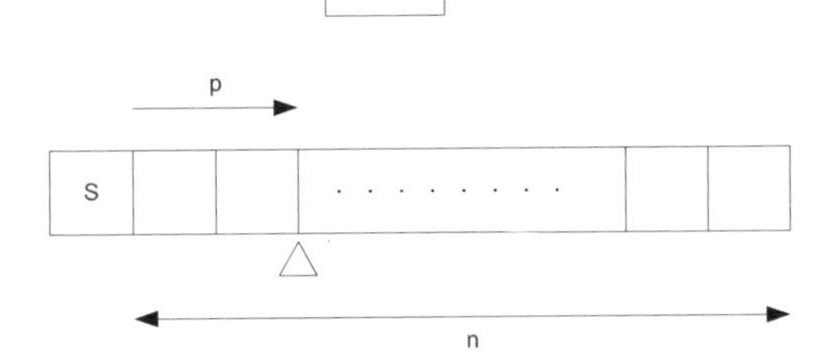

In digital hardware, the traditional number representations are floating-point and fixed-point. The former representation is typically used in the general purpose processing units, whereas the latter is commonly adopted by the DSP processors due to the low area requirements and high throughput that can be achieved. The representation, introduced in [6], that is used in this is the multiple word-length representation, an extension of fixed-point for parallel hardware design.

According to this scheme, each signal $j \in S$ in a computation graph $G(V,S)$ has two parameters n_j and p_j. The first parameter, n_j, specifies the number of bits in the representation of the signal (excluding the sign bit), while the second parameter, p_j, represents the displacement of the binary point from the sign bit. Figure 11.3 shows an example of a signal j.

Figure 11.4 illustrates the same system using (a) a fixed-point representation and (b) a multiple word-length representation. In the first case all the signals use the same number of bits ($\forall i, j \in S.n_i = n_j$) and scale ($\forall i, j \in S.p_i = p_j$). In the multiple word-length example, each signal can use a different representation.

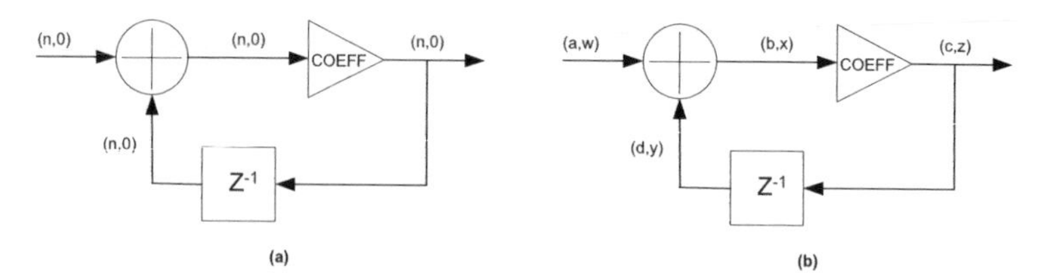

Fig. 11.4 An example of a system using (**a**) fixed-point representation and (**b**) multiple word-length representation

11.1.2 Peak Value Estimation

In order to make efficient use of the available resources, the scaling of each signal should be selected appropriately. The chosen representation should not be over-wasteful, by allowing the representation of values that are impossible to ever occur, but at the same time should not allow overflow errors to regularly occur.

If the absolute maximum value P of a signal is known, a scaling of $p = \lfloor \log_2 P \rfloor + 1$ should be used since a power of two multiplication is cost free in bit-parallel hardware.

There are three main approaches that can be applied in order to estimate the maximum value of the signals in a system. These are analytic peak estimation, range propagation, and simulation based approaches. We shall briefly elaborate on each of these schemes, below.

11.1.2.1 Analytic Peak Estimation

In the case of an LTI system, the peak value of each signal in the system can be estimated analytically. This is achieved by calculating the transfer function from each primary input to each signal. In the case where the system in non-recursive (i.e. the computational graph does not contain cycles), the calculation of the transfer function is a simple task, leading to polynomials in z^{-1}.

In the case of a recursive system, the calculation of the transfer function is more complex. The set of nodes whose outputs correspond to a system *state* should be identified. In this context, this set of nodes is the one which if removed from the computation graph it will break all the cycles. After breaking up the feedback loops, the transfer function matrix $S(z)$ from each input signal to the output of each of these state nodes is expressed as a function of the transfer function matrix $A(z)$ between state nodes and state nodes, and the transfer function matrix $B(z)$ between the primary inputs and state nodes as in (11.4).

$$S(z) = A(z)S(z) + B(z) \tag{11.4}$$

$$H(z) = C(z)S(z) + D(z) \tag{11.5}$$

The transfer function matrix $H(z)$ is calculated as in (11.5), where $C(z)$ is the transfer function matrix between the state-nodes outputs and the outputs of all nodes, where $D(z)$ is the transfer function matrix between the primary inputs and the outputs of all nodes. For a more detailed description for the calculation of $H(z)$, the reader is referred to [7].

Given the transfer function matrix $H(z)$, with associated impulse response $h[t] = Z^{-1}\{H(z)\}$, the worst-case peak value P_j of any signal j can be found by maximizing the convolution sum (11.6) [14], where $x_i[t]$ is the value of input i at time index t. Solving this maximization problem leads to (11.7), where M_i is the maximum absolute value of the signal $x_i[t]$.

$$P_j = \pm \sum_{i \in V_I} \max_{x_i[t']} \left(\sum_{t=0}^{N_{ij}-1} x_i[t'-t] h_{ij}[t] \right) \tag{11.6}$$

$$P_j = \sum M_i \sum_{t=0}^{\infty} |h_{ij}[t]| \tag{11.7}$$

11.1.2.2 Data Range Propagation

In the case where the system under consideration is not linear, or is not time-invariant, one mechanism to estimate the maximum value of the signals is by considering the propagation of data ranges. It should be noted that this approach is only applicable in the case of non-recursive systems.

Under this mechanism, the data ranges are propagated through the operations of the system. This approach has been formally stated in terms of interval analysis [2]. It should be noted that this mechanism can lead to pessimistic results in the case where the system includes data branches which later reconverge.

11.1.2.3 Simulation Driven Analysis

Another approach is to use simulation to estimate the peak value of each signal in a system. In this case, the system is simulated using as representative input data as possible and the maximum values for each signal are recorded. After the end of the simulation, the recorded peak values are multiplied by a user-supplied ‘safety-factor’ $k > 1$, in order to accommodate for values of the signal that did not occur during the simulation, but may occur in practice leading to overflow. More complex forms of the safety-factor have also been considered by researchers in the field [10].

This approach is more suitable for non-linear or time-varying systems where the propagation range methodology provides overly pessimistic results (such as recursive systems). The dependence of the final result on the input data is the main drawback of this approach.

Summarizing, there are three methodologies that can be applied for determining the peak value of the signals in the system. In the case where the system is LTI, the analytic method provides a tight bound on the peak value estimation of the signals. In the case where the system is nonlinear or time-variant, and non-recursive the propagation range method can be applied to provide an upper bound on the peak value of the signals. In the general case, simulation methods can always be applied, which provide a lower bound on the estimation of the peak value of the signals.

11.2 Word-Length Optimization

The previous section has focused on the scale determination of each signal in the system. This section concentrates on the estimation of the remaining parameter: the word-length of the signals.

In order to optimize the word-length of each signal in the system, a model that determines the error at the outputs of the system for a given set of word-length and scaling parameters is required. We call this problem error estimation. Given an error estimation model, the problem of word-length optimization reduces to a problem of utilizing the available resources, the area of the design in our case, satisfying a set of constraints for the outputs of the system.

11.2.1 Error Estimation Model

The quality of a fixed-point algorithm implementation is usually measured using the signal-to-noise ratio (SNR). The *fixed-point error* is calculated by subtracting the output sequence of the system under a fixed-point implementation from the output sequence of the same system under an infinite precision implementation. The ratio of the output power resulting from an infinite precision implementation to the fixed-point error power defines the signal-to-noise ratio. In this chapter we assume that the *signal* powers at the outputs are fixed, since they are determined only by the input signal statistics and the computation graph. Thus, it is sufficient to concentrate on noise power estimation.

11.2.2 Word-Length Propagation

In order to predict the quantization effects in the system, we need to propagate the word-length and scaling parameters from the inputs of each atomic operation to its outputs. Table 11.2 summarizes the word-length and scaling propagation rules for the different atomic operations. The superscript q denotes the signal before the quantization take place, i.e. without loss of information.

In a multiple word-length implementation, it is important to ensure that sub-optimal implementations are avoided. For example, consider the case of a GAIN

Table 11.2 Word-length and scaling propagation

Type	Propagation rules
GAIN	For input (n_a, p_a) and coefficient (n_b, p_b): $p_j = p_a + p_b$ $n_j^q = n_a + n_b$
ADD	For inputs (n_a, p_a) and (n_b, p_b): $p_j = \max(p_a, p_b) + 1$ $n_j^q = \max(n_a, n_b + p_a - p_b) - \min(0, p_a - p_b) + 1$ (for $n_a > p_a - p_b$ or $n_b > p_b - p_a$)
DELAY or FORK	For input (n_a, p_a): $p_j = p_a$ $n_j^q = n_a$

node where the input signal j_1 is (n_{j_1}, p_{j_1}), and the coefficient has format (n, p). If the output signal j_2 has $n_{j_2} > n_{j_1} + n$, then this choice is suboptimal since at most $n_{j_1} + n$ bits are required for representing the results in full precision. Ensuring that these cases do not arise, is referred to as 'conditioning' of the annotated computation graph [7]. During the optimization process, ill-conditioned computation graphs may rise, which should be transformed to well-conditioned ones [7].

11.2.3 Linear Time Invariant Systems

We will first address the error estimation model for linear time-invariant systems. In this case, we can derive analytic models of how the noise due to truncation of a signal is propagated through the computation graph to its outputs.

11.2.3.1 Noise Model

A common approach in DSP is that a truncation or roundoff operation will be performed after a multiplication or a multiplication-accumulation operation. This corresponds to the case of a processor where the result of an n-bit signal multiplied by an n-bit signal, which is a $2n$-bit signal, should be truncated to n-bits in order to fit in an n-bit register. Thus, for a two's complement representation, the error that is introduced to the system assuming $p = 0$ ranges between 0 and $2^{-2n} - 2^n \approx -2^n$. As long as the $2n$-bit result has sufficient dynamic range, it has been observed that the values in that range are equally likely to happen [13, 15]. This leads to the formulation of a uniform distribution model of the noise with variance $\sigma^2 = \frac{1}{12}2^{-2n}$, when $p = 0$ [13]. Moreover, it has been observed that the spectrum of the noise tends to be white, due to the fact the the truncation occurs in the low significant bits of the signals, and that roundoff errors that occur at different parts of the system are uncorrelated.

However, in our case the above noise model cannot be applied. Consider the truncation of a signal (n_1, p) to a signal (n_2, p). In the case where $n_1 \approx n_2$, the model will suffer in accuracy due to the discretization of the error probability density function. Also, the lower bound of the error can not be simplified as before since $2^{-n_2} - 2^{-n_1} \approx -2^{-n_1}$ no longer holds. Moreover, in the case of a branching node where the output signals can be truncated to different lengths, the preceding model does not consider the different error signals.

The solution to the above problems comes by considering a discrete probability distribution for the injected signals [5]. In the case of a truncation of a signal (n_1, p) to a signal (n_2, p), the error that is inserted to the system is bounded by (11.8).

$$-2^p(2^{-n_2} - 2^{-n_1}) \le e[t] \le 0 \tag{11.8}$$

Assuming, as before, that $e[t]$ takes values from the above range with equal probability, the expectation of $e[t]$, $E\{e[t]\}$ and its variance σ_e^2 are given by (11.9) and (11.10) respectively.

$$\begin{aligned} E\{e[t]\} &= -\frac{1}{2^{n_1-n_2}} \sum_{i=0}^{2^{n_1-n_2}-1} i \cdot 2^{p-n_1} \\ &= -2^{p-1}(2^{-n_2} - 2^{-n_1}) \end{aligned} \tag{11.9}$$

$$\begin{aligned} \sigma_e^2 &= \frac{1}{2^{n_1-n_2}} \sum_{i=0}^{2^{n_1-n_2}-1} (i \cdot 2^{p-n_1})^2 - E\{e[t]\} \\ &= \frac{1}{12} 2^{2p}(2^{2n_2} - 2^{2n_1}) \end{aligned} \tag{11.10}$$

11.2.3.2 Noise Propagation

By considering a computation graph, the truncation of a signal j from (n_j, p_j) to (n_j^q, p_j) in the graph injects a noise in the system according to (11.10). The application of this model is straight forward apart from the case of a fork. Figure 11.5 shows two different approaches for modelling the truncation of the signals. In the first approach, noise is injected at each output of the fork, leading to correlated injected noise. In the second approach, there is cascaded noise injection, leading to a less correlated noise injection, which is in line with the assumption about the noise propagation model.

Given an annotated graph, a set $F = \{(\sigma_p^2, R_p)\}$ of injected input variances, σ_p^2, and their transfer functions to the primary outputs, $R_p(z)$, can be constructed. From this set, and under the assumption that the noise sources have white spectrum and are uncorrelated, L_2 scaling [14] can be used to estimate the power of the injected noise at each output k of the system according to (11.11). The L_2 scaling of a transfer function is given in (11.12), where $Z^{-1}\{\cdot\}$ denotes the inverse z-transform.

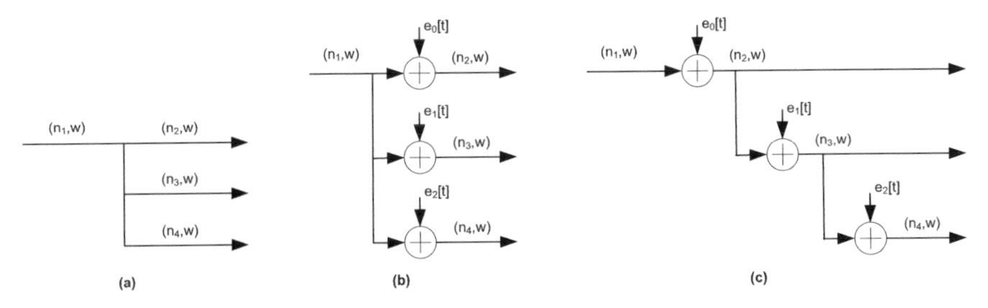

Fig. 11.5 Approaches for modelling post-FORK truncation

$$E_k = \sum_{(\sigma_p^2, R_p) \in F} \sigma_p^2 L_2^2\{R_{pk}\} \tag{11.11}$$

$$L_2\{H(z)\} = \left(\sum_{n=0}^{\infty} |Z^{-1}\{H(z)\}[n]|^2 \right)^{\frac{1}{2}} \tag{11.12}$$

11.2.4 Extension to Non-Linear Systems

The same methodology can be applied in an approximate way for non-linear systems, by linearizing the system around some operating point. In the DSP domain, the most common occurrence of non-linearities is from the introduction of general multipliers, which can be found, for example, in adaptive filter systems. A way to approach the problem is by approximating these non-linearities by using the first terms of a Taylor expansion, an idea that is derived from small-signal analysis usually found in analog electronics [17].

Let us consider an n-input function $Y[t] = f(X_1[t], X_2[t], \ldots, X_n[t])$, where t is the time index. If we consider a small perturbation x_i in variable X_i, then the perturbation $y[t]$ on variable $Y[t]$ can be approximated as $y[t] \approx x_1[t]\frac{\partial f}{\partial X_1} + x_2[t]\frac{\partial f}{\partial X_2} + \ldots + x_n[t]\frac{\partial f}{\partial X_n}$.

This approximation is linear in each x_i, but the coefficients may vary with time since $\frac{\partial f}{\partial X_i}$ is a function of $X_1, X_2, \ldots, X_n$. Using the above approximation we have managed to transform a non-linear time-invariant system into a linear time-varying system. This linearity allows us to predict the error at the output of the system due to any scaling of a small perturbation of a signal s analytically, given the simulation-obtained error by a single such perturbation at s.

For the case of a general multiplier, $f(X_1, X_2) = X_1 X_2$, $\frac{\partial f}{\partial X_1} = X_2$ and $\frac{\partial f}{\partial X_2} = X_1$.

Within a synthesis tool, such Taylor coefficients can be recorded during a simulation run through the modification of the computational graph to include so-called *monitors* [5]. These data can then be used later for the error calculation step. Figure 11.6 shows a multiplier node and its transformation prior to simulation where the appropriate signals for monitoring the derivatives have been inserted.

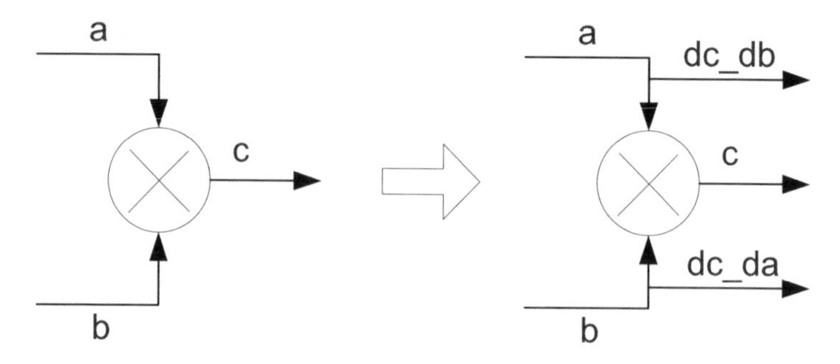

Fig. 11.6 Transformation of a multiplier node to insert derivative monitors [5]

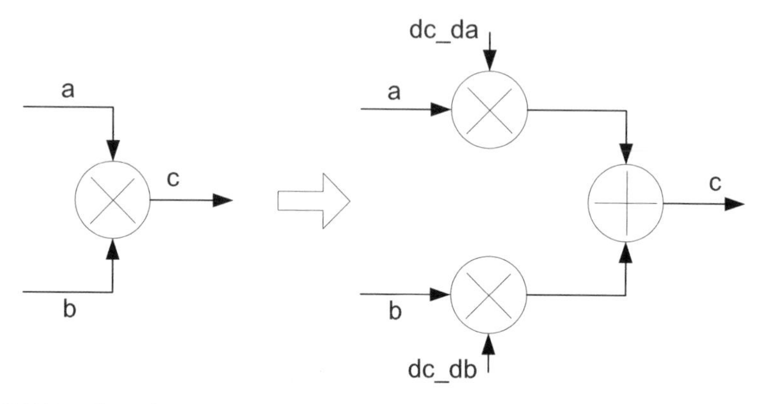

Fig. 11.7 Transformation of a multiplier node to produce a linear model [5]

11.2.4.1 Linearization

The linearization of the general multiplier is performed by transforming the general multiplier component in the computational graph into its Taylor approximation component as it is shown in Fig. 11.7. Note that the model still has a general multiplier node, however one input is external to the model ensuring linearity. These new added external signals to the system read data from the derivative monitor files created by the above large-scale simulation.

11.2.4.2 Noise Injection

In the case of a linear time-invariant system, the L_2 scaling was used to analytically estimate the variance of the noise at the outputs of the system. In this section, an extension of this approach is proposed for the case of non-linear systems.

The advantage of transforming the non-linear components of the system to linear components in the small-signal model is that if the variance of an output signal is V when it is excited by an error with variance σ^2 that it is injected to a signal of the system, then the same output signal will have a variance aV when the injected error has variance $a\sigma^2$. This implies that if we can calculate the variance of an output signal for a given known variance of the injected error only once through simulation, then we can analytically calculate the variance of the output signal for any variance of the injected error.

In order to achieve that, an additional adder node is added in the system, which injects the error to the signal under investigation, and a simulation is performed for a known error variance. In the simulation, the error that is injected to the system due to truncation of two's complement signal is independent and identically distributed over the range $[-2\sqrt{3},0]$. The selection of unit variance for the injected noise allows us to make the measured output response an unscaled 'sensitivity' measure. To finalize the small-signal model, zeros are propagated through the original inputs of the system during the simulation leading to faster simulation results [1].

11.2.5 Word-Length Optimization Algorithm

Given a computation graph $G(V,S)$ of a system, Sect. 11.1.2 has already described how a scaling vector $\mathbf{p}$ can be derived. The total required area of the system can be expressed by a single metric $A_G(\mathbf{n},\mathbf{p})$, which combines the area models of the components of the system. Finally, the error variances of the output of the system can be combined in a single vector $E_G(\mathbf{n},\mathbf{p})$.

The Word-length optimization problem can be formulated as follows: Given a computation graph $G(V,S)$, select $\mathbf{n}$ such that $A_G(\mathbf{n},\mathbf{p})$ is minimized subject to $\mathbf{n} \in N^{|S|}$ and $E_G(\mathbf{n},\mathbf{p}) < \mathcal{E}$ where $\mathcal{E}$ is a vector that defines the maximum acceptable error variance for each output of the system.

It can be demonstrated that the error variance at the outputs of the system may not be a monotonically decreasing function in each internal word-length. Moreover, it can be shown that error non-convexity may occur, causing the constraint space to be non-convex in $\mathbf{n}$. As it is demonstrated in [7], as long as the system remains well-conditioned, increasing the word-length of the output of the node types GAIN, ADD or DELAY can not lead to an increase of the observed error at the outputs of a system. However, a computation graph containing a 2-way FORK can exhibit such behavior that is not monotonic in the word-length vector. Moreover, in the case of systems that incorporate a 3-way FORK, non-convexity may arise. This non-convexity makes the word-length optimization problem a harder problem to find solutions [9].

11.2.5.1 A Heuristic Approach

It has been shown in [4] that the word-length optimization problem is NP-hard. Thus, a heuristic approach has been developed to find the word-length vector that minimizes the area of a system given under the set of constraints on the error variance at the outputs of the system.

The method starts from determining the scaling vector **p** of the system. After that, the algorithm estimates the minimum uniform word-length that can be used for all the signals in the system such that the error constraints are not violated. Each word-length is scaled up by a factor $k > 1$ which defined the upper bound that the signal can reach in the final design. A conditioning step is performed to transform an ill-conditioned graph that may has arise to a well-conditioned one.

In each iteration of the algorithm, each signal is visited in turn and its word-length is reduced until the maximum reduction in its word-length is found that does not violate the error constraints. The signal with the largest reduction in the area is chosen. Each signal's word-length is explored using binary search.

For completeness, in the case where the DSP system under investigation is an LTI system, optimum solutions can be found using a Mixed Integer Linear Programming formulation (MILP). However, it should be noted that the solution time of MILP formulations render the synthesis of large systems intractable. The interested reader is referred to [7].

11.2.6 Some Results

Figure 11.8 shows the place-and-routed resource usage versus the specified error variance at the output of an IIR biquadratic filter. The target device is an Altera Flex10k. This plot is a representative plot of the plots obtained by many designs. The plot shows the results obtained when uniform word-length is used in the system and when the multiple word-length scheme is applied. It can be seen that the multiple word-length approach results in designs that use between 2 and 15% less area for the same error specification at the output.

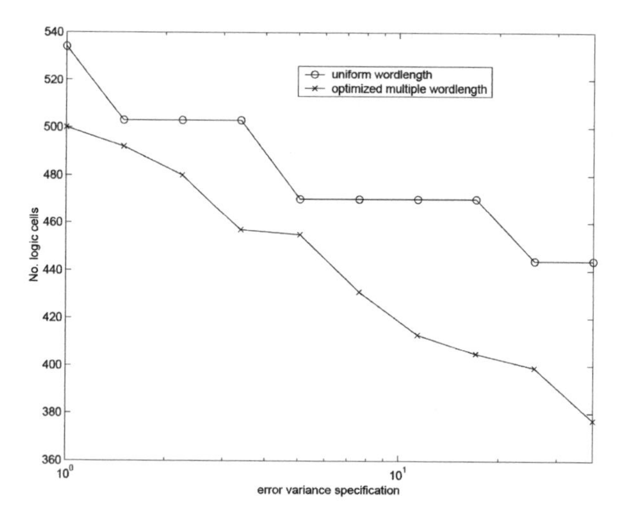

Fig. 11.8 Area resources versus specified error variance for an IIR biquadratic filter [7] (published with kind permission of Springer Science and Business Media)

11.3 Synthesis and Optimization of 2D FIR Filter Designs

The previous section discusses the optimization of general DSP designs, focusing on peak value estimation and word-length optimization of the signals. This section focuses on the problem of resource optimization in Field Programmable Gate Array (FPGA) devices for a specific class of DSP designs. The class under consideration is the class of designs performing two-dimensional convolution, i.e. 2D FIR filters.

The two-dimensional convolution is a widely used operator in image processing field. Moreover, in applications that require real-time performance, in many cases engineers select as a target hardware platform an FPGA device due to its fine grain parallelism and reconfigurability properties. Contrary to the firstly introduced FPGA devices consisting of reconfigurable logic only, modern FPGA devices contain a variety of hardware components like embedded multipliers and memories.

This section focuses on the optimization of a pipelined 2D convolution filter implementation in a heterogeneous device, given a set of constraints regarding the number of embedded multipliers and reconfigurable logic (4-LUTs). As before, we are interested in a "lossy synthesis" framework, where an approximation of the original 2D filter is targeted which minimizes the error at the output of the system and at the same time meets the user's constraints on resource usage. Contrary to the previous section, we are not interested in the quantization/truncation of the signals, but to alter the impulse response of the system optimizing the resource utilization of the design. The exploration of the design space is performed at a higher level than the word-length optimization methods or methods that use common subexpressions [8,16] to reduce the area, since they do not consider altering the computational structure of the filter. Thus, the proposed technique is complementary to these previous approaches.

11.3.1 Objective

We are interested to find a mapping of the 2D convolution kernel into hardware that given a bound on the available resources, it achieves a minimum error at the output of the system. As before, the metric that is employed to measure the accuracy of the result is the variance of the noise at the output of the system.

From [14] the variance of a signal at the output of a LTI system, and in our specific case of a 2D convolution, when the input signal is a white random process is given by (11.13), where σ_y^2 is the variance of the signal at the output of the system, σ_x^2 is the variance of the signal at the input, and $h[n]$ is the impulse response of the system.

$$\sigma_y^2 = \sigma_x^2 \sum_{n=-\infty}^{\infty} |h[n]|^2 \tag{11.13}$$

Under the proposed framework, the impulse response of the new system $\hat{h}[n]$ can be expressed as the sum of the impulse response of the original system $h[n]$ and an

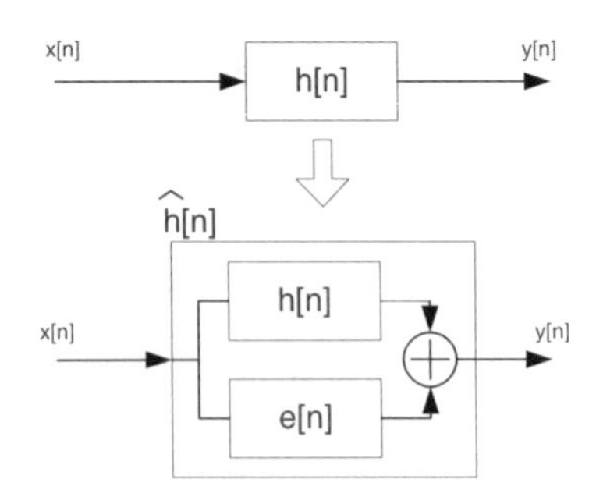

Fig. 11.9 The *top graph* shows the original system, where the *second graph* shows the approximated system and its decomposition to the original impulse response and to the error impulse response

error impulse response $e[n]$ as in (11.14).

$$\hat{h}[n] = h[n] + e[n] \tag{11.14}$$

The new system can be decomposed into two parts as shown in Fig. 11.9. The first part has the original impulse response $h[n]$, where the second part has the error impulse response $e[n]$. Thus, the variance of the noise at the output of the system due to the approximation of the original impulse response is given by (11.15), where *SSE* denotes the sum of square errors in the filter's impulse response approximation.

$$\sigma_{noise}^2 = \sigma_x^2 \sum_{n=-\infty}^{\infty} |e[n]|^2 = \sigma_x^2 \cdot SSE \tag{11.15}$$

It can be concluded that the uncertainty at the output of the system is proportional to the sum of square error of the impulse response approximation, which is used as a measure to access the system's accuracy.

11.3.2 2D Filter Optimization

The main idea is to decompose the original filter into a set of separable filters, and to one non-separable filter which encodes the trailing error of the decomposition.

A 2D filter is called separable if its impulse response $h[n_1,n_2]$ is a separable sequence, i.e.

$$h[n_1,n_2] = h_1[n_1]h_2[n_2].$$

The important property is that a 2D convolution with a separable filter can be decomposed into two one-dimensional convolutions as $y[n_1,n_2] = h_1[n_1] \otimes (h_2[n_2] \otimes x[n_1,n_2])$. The symbol $\otimes$ denotes the convolution operation.

The separable filters can potentially reduce the number of required multiplications from $m \times n$ to $m+n$ for a filter with size $m \times n$ pixels. The non-separable part encodes the trailing error of the approximation and still requires $m \times n$ multiplications. However, the coefficients are intended to need fewer bits for representation and therefore their multiplications are of low complexity. Moreover, we want a decomposition that that enforces a ranking on the separable levels according to their impact on the accuracy of the original filter's approximation.

The above can be achieved by employing the Singular Value Decomposition (SVD) algorithm, which decomposes the original filter into a linear combination of the fewest possible separable matrices [3].

By applying the SVD algorithm, the original filter **F** can be decomposed into a set of separable filters $\mathbf{A}_j$ and into a non-separable filter **E** as follows:

$$\mathbf{F} = \sum_{j=1}^{r} \mathbf{A}_j + \mathbf{E} \tag{11.16}$$

where r notes the levels of decompositions. The initial decomposition levels capture most of the information of the original filter **F**.

11.3.3 Optimization Algorithm

This section describes the optimization algorithm which has two stages. In the first stage the allocation of reconfigurable logic is performed, where in the second stage the constant coefficient multipliers that require the most resources are identified and mapped to embedded multipliers.

11.3.3.1 Reconfigurable Logic Allocation Stage

In this stage the algorithm decomposes the original filter using the SVD algorithm and manifests the constant coefficient multiplications using only reconfigurable logic. However, due to the coefficient quantization in a hardware implementation, quantization error is inserted at each level of the decomposition. The algorithm reduces the effect of the quantization error by propagating the error inserted in each decomposition level to the next one during the sequential calculation of the separable levels [3].

Given that the variance of the noise at the output of the system due the quantization of each coefficient is proportional to the variance of the signal at the input of the coefficient multiplier, which is the same for the coefficients that belong to the same 1D filter, the algorithm keeps the coefficients of the same 1D filter to the same accuracy. It should be noted that only one coefficient for each 1D FIR filter is considered for optimization at each iteration, leading to solutions that are computational efficient.

11.3.3.2 Embedded Multipliers Allocation

In the second stage, the algorithm determines the coefficients that will be placed into embedded multipliers. The coefficients that have the largest cost in terms of reconfigurable logic in the current design and reduce the filter's approximation error when are allocated to embedded multipliers, are selected. The second condition is necessary due to the limited precision of the embedded multipliers (e.g. 18 bits in Xilinx devices), which in some cases may restrict the approximation of the multiplication and consequently to violate the user's specifications.

11.3.4 Some Results

The performance of the proposed algorithm is compared to a direct pipelined implementation of a 2D convolution using Canonic Signed Digit recoding [11] for the constant coefficient multipliers. Filters that are common in the computer vision field are used to evaluate the performance of the algorithm (see Table 11.3). The first filter is a Gabor filter which yields images which are locally normalized in intensity and decomposed in terms of spatial frequency and orientation. The second filter is a Laplacian of Gaussian filter which is mainly used for edge detection.

Figure 11.10a shows the achieved variance of the error at the output of the filter as a function of the area, for the described and the reference algorithms. In all

Table 11.3 Filters tests

Test number	Description
1	9 × 9 Gabor filter $F(x,y) = \alpha \sin\theta e^{-\rho^2(\frac{\alpha}{\sigma})^2}, \rho^2 = x^2+y^2, \theta = \alpha x,$ $\alpha = 4, \sigma = 6$
2	9 × 9 Laplacian of Gaussian filter $LoG(x,y) = -\frac{1}{\pi\sigma^4}[1 - \frac{x^2+y^2}{2\sigma^2}]e^{-\frac{x^2+y^2}{2\sigma^2}},$ $\sigma = 1.4$

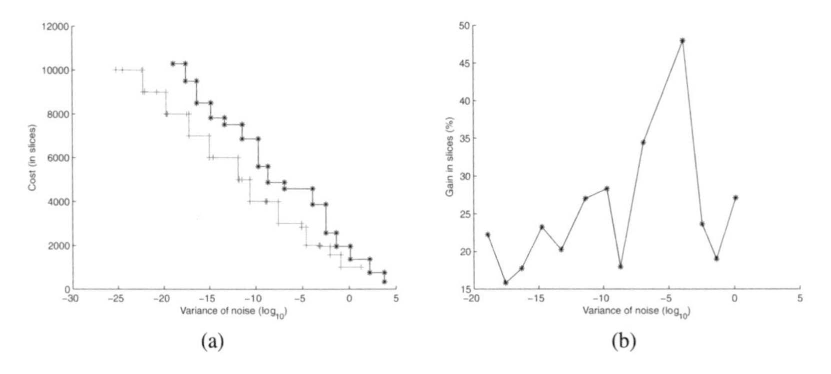

Fig. 11.10 (**a**) Achieved variance of the noise at the output of the design versus the area usage of the proposed design (*plus*) and the reference design (*asterisks*) for Test case 1. (**b**) illustrates the percentage gain in slices of the proposed framework for different values of the variance of the noise. A *slice* is a resource unit used in Xilinx devices

cases, the described algorithm leads to designs that use less area than the reference algorithm, for the same error variance at the output. Figure 11.10b illustrates the relative reduction in area achieved. An average reduction of 24.95 and 12.28% is achieved for the Test case 1 and 2 respectively. Alternative, the proposed methodology produces designs with up to 50 dB improvement in the signal to noise ratio requiring the same area in the device with designs that are derived from the reference algorithm. Moreover, Test filter 1 was used for evaluation of the performance of the algorithm when embedded multipliers are available. Thirty embedded multipliers of 18×18 bits are made available in the algorithm. The relative percentage reduction achieved by the algorithm between designs that use the embedded multipliers and designs that realized without any embedded multiplier is around 10%.

11.4 Summary

This chapter focused on the optimization of the synthesis of DSP algorithms into hardware. The first part of the chapter described techniques that produce area-efficient designs from general block-based high level specifications. These techniques can be applied to LTI systems as well as to non-linear systems. Examples of these systems vary from finite impulse response (FIR) filters and infinite impulse response (IIR) filters to polyphase filter banks and adaptive least mean square (LMS) filters. The chapter focused on peak value estimation, using analytic and simulation based techniques, and on word-length optimization.

The second part of the chapter focused on a specific DSP synthesis problem, which is the efficient mapping into hardware of 2D FIR filter designs, a widely-used class of designs in the image processing community. The chapter described a methodology that explores the space of possible implementation architectures of 2D FIR filters targeting the minimization of the required area and optimizes the usage of the different components in a heterogeneous device.

References

1. Aho, A. V., Sethi, R., and Ullman, J. D. (1986). *Compilers: Principles, Techniques and Tools*. Addison-Wesley, Reading, MA.
2. Benedetti, K. and Prasanna, V. K. (2000). Bit-width optimization for configurable dsps by multi-interval analysis. In *34th Asilomar Conference on Signals, Systems and Computers*.
3. Bouganis, C.-S., Constantinides, G. A., and Cheung, P. Y. K. (2005). A novel 2d filter design methodology for heterogeneous devices. In *IEEE Symposium on Field-Programmable Custom Computing Machines*, pages 13–22.
4. Constantinides, G. A. and Woeginger, G. J. (2002). The complexity of multilple wordlength assignment. *Applied Mathematics Letters*, 15(2):137–140.
5. Constantinides, George A. (2003). Perturbation analysis for word-length optimization. In *11th Annual IEEE Symposium on Field-Programmable Custom Computing Machines*.

6. Constantinides, George A., Cheung, Peter Y. K., and Luk, Wayne (2002). Optimum wordlength allocation. In *10th Annual IEEE Symposium on Field-Programmable Custom Computing Machines*, pages 219–228.
7. Constantinides, George A., Cheung, Peter Y. K., and Luk, Wayne (2004). *Synthesis and Optimization of DSP Algorithms*. Kluwer, Norwell, MA, 1st edition.
8. Dempster, A. and Macleod, M. D. (1995). Use of minimum-adder multiplier blocks in FIR digital filters. *IEEE Trans. Circuits Systems II*, 42:569–577.
9. Fletcher, R. (1981). *Practical Methods of Optimization, Vol. 2: Constraint Optimization*. Wiley, New York.
10. Kim, S., Kum, K., and Sung, W. (1998). Fixed-point optimization utility for C and C++ based digital signal processing programs. *IEEE Transactions on Circuits and Systems II*, 45(11):1455–1464.
11. Koren, Israel (2002). *Computer Arithmetic Algorithms*. Prentice-Hall, New Jersey, 2nd edition.
12. Lee, E. A. and Messerschmitt, D. G. (1987). Synchronous data flow. *IEEE Proceedings*, 75(9).
13. Liu, B. (1971). Effect of finite word length on the accuracy of digital filters – a review. *IEEE Transactions on Circuit Theory*, 18(6):670–677.
14. Mitra, Sanjit K. (2006). *Digital Signal Processing: A Computer-Based Approach*. McGraw-Hill, Boston, MA, 3rd edition.
15. Oppenheim, A. V. and Schafer, R. W. (1972). Effects of finite register length in digital filtering and the fast fourier transform. *IEEE Proceedings*, 60(8):957–976.
16. Pasko, R., Schaumont, P., Derudder, V., Vernalde, S., and Durackova, D. (1999). A new algorithm for elimination of common subexpressions. *IEEE Transactions on Computer-Aided Design of Integrated Circuit and Systems*, 18(1):58–68.
17. Sedra, A. S. and Smith, K. C. (1991). *Microelectronic Circuits*. Saunders, New York.
18. Wakerly, John F. (2006). *Digital Design Principles and Practices*. Pearson Education, Upper Saddle River, NJ, 4th edition.

Chapter 12
High-Level Synthesis of Loops Using the Polyhedral Model

The MMAlpha Software

Steven Derrien, Sanjay Rajopadhye, Patrice Quinton, and Tanguy Risset

Abstract High-level synthesis (HLS) of loops allows efficient handling of intensive computations of an application, e.g. in signal processing. Unrolling loops, the classical technique used in most HLS tools, cannot produce regular parallel architectures which are often needed. In this Chapter, we present, through the example of the MMAlpha testbed, basic techniques which are at the heart of loop analysis and parallelization. We present here the point of view of the *polyhedral model* of loops, where iterative calculations are represented as recurrence equations on integral polyhedra. Illustrated from an example of string alignment, we describe the various transformations allowing HLS and we explain how these transformations can be merged in a synthesis flow.

Keywords: Polyhedral model, Recurrence equations, Regular parallel arrays, Loop transformations, Space–time mapping, Partitioning.

12.1 Introduction

One of the main problems that High Level Synthesis (HLS) tools have not solved yet is the efficient handling of nested loops. Highly computational programs occurring for example in signal processing and multimedia applications make extensive use of deeply nested loops. The vast majority of HLS tools either provide loop unrolling to take advantage of parallelism, or treat loops as sequential when unrolling is not possible. Because of the increasing complexity of embedded code, complete unrolling of loops is often impossible. Partial unrolling coupled with software pipelining techniques has been successfully used, in the Pico tool [29] for instance, but a lot of other loop transformations, such as loop tiling, loop fusion or loop interchange, can be used to optimize the hardware implementation of nested loops. A tool able to propose such loop transformations in the source code before performing HLS should necessarily have an internal representation in which the loop nest structure

P. Coussy and A. Morawiec (eds.) *High-Level Synthesis*.

is kept. This is a serious problem and this is why, for instance, source level loop transformations are still not available is commercial compilers, whereas the loop transformation theory is quite mature.

The work presented in this chapter proposes to perform HLS from the source language ALPHA. The ALPHA language is based on the so-called *polyhedral model* and is dedicated to the manipulation of *recurrence equations* rather than loops. The MMAlpha programming environment allows a user to transform ALPHA programs in order to refine the ALPHA initial description until it can be translated down to VHDL. The target architecture of MMAlpha is currently limited to regular parallel architectures described in a register transfer level (RTL) formalism. This paradigm, as opposed to the control+datapath formalism, is useful for describing highly pipelined architectures where computations of several successive samples are overlapped.

This chapter gives an overview of the possibilities of the MMAlpha design environment focusing on its use for HLS. The concepts presented in this chapter are not limited to the context were a specification is described using an applicative language such as ALPHA: they can also be used in a compiler environment as it has been done for example in the WraPit project [3].

The chapter is organized as follows. In Sect. 12.2, we present an overview of this system by describing the ALPHA language, its relationship with loop nests, and the design-flow of the MMAlpha tool. Section 12.3 is devoted to the front-end which transforms an ALPHA software specification into a virtual parallel architecture. Section 12.4 shows how synthesizable VHDL code can be generated. All these first sections are illustrated on a simple example of string alignment, so that the main concepts are apparent. In Sect. 12.5, we explain how the virtual architecture can be further transformed in order to be adapted to resource constraints. Implementations of the string alignment application are shown and discussed in Sect. 12.6. Section 12.7 is a short review of other works in the field of hardware generation for loop nests. Finally, Sect. 12.8 concludes the chapter.

12.2 An Overview of the MMAlpha Project

Throughout this chapter, we shall consider the running example of a string matching algorithm for genetic sequence comparison, as shown in Fig. 12.1. This algorithm is expressed using the single-assignment language ALPHA. Such a program is called a `system`. Its name is `sequence`, and it makes use of integral parameters `X` and `Y`. These parameters are constrained (line 1) to satisfy the linear inequalities $3 \leq X$ and $X \leq Y-1$. This system has two inputs: a sequence `QS` (for *Query Sequence*) of size `X` and a sequence `DB` (for *Data Base* sequence) of size `Y`. It returns a sequence `res` of integers. The calculation described by this system is expressed by *equations* defining local variables `M` and `MatchQ` as well as result `res`. Each ALPHA variable is defined on the set of integral points of a convex polyhedron called its *domain*. For example, `M` is defined on the set $\{i,j|0 \leq i \leq X \wedge 0 \leq j \leq Y\}$. The definition of `M`

```
1  system sequence :{X,Y | 3<=X<=Y-1}
2                        (QS : {i | 1<=i<=X} of integer;
3                         DB : {j | 1<=j<=Y} of integer)
4            returns  (res : {j | 1<=j<=Y} of integer);
5  var
6    M : {i,j | 0<=i<=X; 0<=j<=Y} of integer;
7    MatchQ : {i,j | 1<=i<=X; 1<=j<=Y} of integer;
8  let
9    M[i,j] =
10         case
11           {| i=0} | {| 1<=i; j=0} : 0;
12           {| 1<=i; 1<=j} : Max4(0, M[i,j-1] - 8,
13                     M[i-1,j] - 8, M[i-1,j-1] + MatchQ[i,j]);
14         esac;
15   MatchQ[i,j] = if (QS[i] = DB[j]) then 15 else -12;
16   res[j] = M[X,j];
17 tel;
```

Fig. 12.1 ALPHA program for the string alignment algorithm

is given by a case statement, each branch of which covers a subset of its domain. If $i = 0$ or if $j = 0$, then its value is 0. Otherwise, it is the maximum of four quantities: 0, `M[i,j-1]` − 8, `M[i-1,j]` − 8, and `M[i-1,j-1]`+`MatchQ[i,j]`. This definition represents a recurrence equation. Its last term depends on whether the query character `QS[i]` is equal to the data base sequence character `DB[j]`. Such a set of recurrences is often represented as a dependence graph as shown in Fig. 12.2. It should be noted, however that the ALPHA language allows one to represent arbitrary linear recurrences, which in general, cannot be represented graphically as easily. ALPHA allows structured systems to be described: a given system can be instantiated inside another one, by using a use statement which operated as a higher order *map* operator. For example

```
use {k | 1<=k<=10} sequence[X,Y] (a, b) returns (res)
```

would allow ten instances of the above sequence program to be instantiated. For the sake of conciseness, we do not detail in this chapter structured systems and refer the reader to [12].

Figure 12.3 shows the typical design flow of MMAlpha. MMAlpha allows ALPHA programs to be transformed, under some conditions, into a VHDL synthesizable program. The input is nested loops which, in the current tools, are described as an ALPHA program, but could be generated from loop nests in an imperative language (see [16] for example). After parsing, we get an internal representation of the program as a set of recurrence equations. Scheduling, localization and space–time mapping are then performed to obtain the description of a virtual architecture also described using ALPHA: all these transformations form the *front-end* of MMAlpha. Several steps allow the virtual architecture to be transformed to synthesizable VHDL

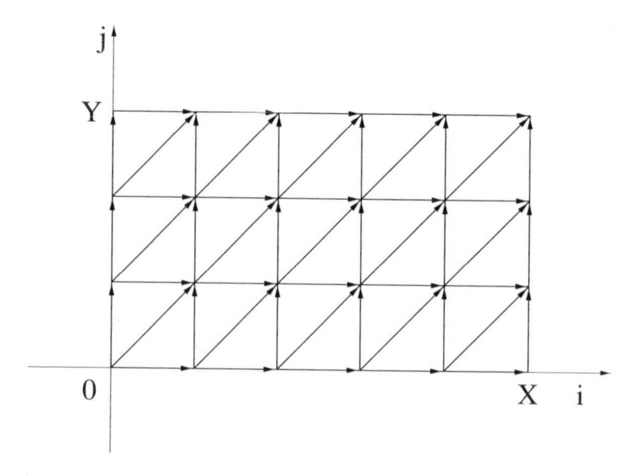

Fig. 12.2 Graphical representation of the string alignment. Each point in the graph represents a calculation `M[i,j]` and the arcs show dependences between the calculations

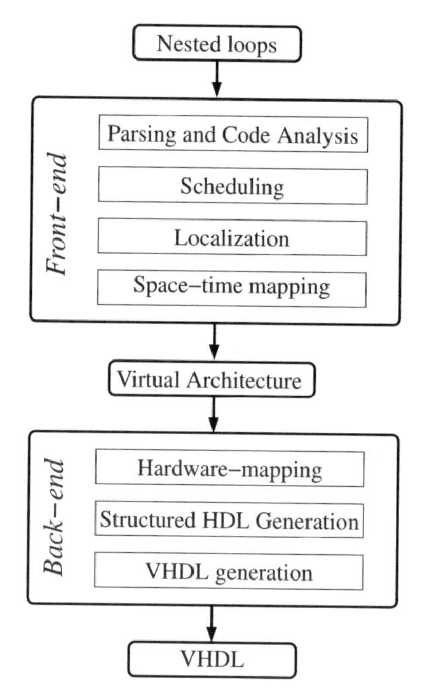

Fig. 12.3 Design flow of MMAlpha

code: hardware-mapping identifies ALPHA constructs with basic hardware elements such as registers, multiplexers, and generates boolean signal control instead of linear inequalities constraints. Then a structured HDL description incorporating a controller and data-path cells is produced. Finally, VHDL is generated.

In Sect. 12.3, we shall survey the front-end transformations whereas back-end will be presented in Sect. 12.4.

12.3 The MMAlpha Front-End: From Initial Specifications to a Virtual Architecture

The front-end of MMAlpha contains several tools to perform code analysis and transformations.

Code analysis and verification: The initial specification of the program, called here a *loop nest*, is translated into an internal representation in form of recurrence equations. Thanks to the polyhedral model, some properties of the loop nest can be checked by analysis: one can check for example that all elements of an *array* (represented by an ALPHA variable) are defined and used in a system, by means of calculations on domains. More complex properties of code can also be checked using verification techniques [8].

Scheduling: This is the central step of MMAlpha. It consists in analyzing the dependences between the variables, and deriving for each variable, say `V[i,j]` a timing-function $t_{\mathrm{V}}(i,j)$ which gives the time instant at which this variable can be computed. Timing-functions are usually *affine*, of the form $t_{\mathrm{V}}(i,j) = \alpha_{\mathrm{V}} i + \beta_{\mathrm{V}} j + \gamma_{\mathrm{V}}$ with coefficients depending on variable `V`. Finding out a schedule is performed by solving an integer linear problem using parameterized integer programming and is described in [17]. More complex schedules can be found: multi-dimensional timing functions, for example, allow some forms of loop tiling to be represented, but code generation is still not available for such functions.

Localization: It is an optional transformation (also sometimes referred to as uniformization or pipelining) that helps removing long interconnections [28]. It is inherited from the theory of systolic arrays where data which are re-used in a calculation should be read only once from memory, thus saving input–outputs. MMAlpha performs automatically many such localization transformations described in the literature.

Space–time mapping: Once a schedule is found, the system of recurrence equations is rewritten by transforming indexes of each variable, say `V[i,j]`, in a new reference index set `V[t,p]` where t is the schedule of the variable instance and p is the processor where it can be executed. The space–time mapping amounts formally to a *change of basis* of the domain of each variable. Finding out the basis is done by algebraic methods described in the literature (unimodular completion). Simple heuristics are incorporated in MMAlpha to discover quickly reasonable, if not always optimal, changes of basis.

After front-end processing, the initial ALPHA specification becomes a *virtual architecture* where each equation can be interpreted in term of hardware. To illustrate this, consider a sketch of the virtual architecture produced by the front-end from the string alignment specification as shown in Fig. 12.4. In this program, only

```
system sequence :{X,Y | 3<=X<=Y-1}
                 (QS : {i | 1<=i<=X} of integer;
                  DB : {j | 1<=j<=Y} of integer)
        returns  (res : {j | 1<=j<=Y} of integer);
var
  QQS_In : {t,p | 2p-X+1<=t<=p+1; 1<=p} of integer;
  ...
  M : {t,p | p<=t<=p+Y; 0<=p<=X} of integer;
  ...
let
...
  M[t,p] =
      case
        { | p=0} : 0;
        { | t=p; 1<=p} : 0;
        { | p+1<=t; 1<=p} :
            Max4( 0[],
                  M[t-1,p] - 8,
                  M[t-1,p-1] - 8,
                  M[t-2,p-1] + MatchQ[t,p] );
      esac;
  QQS[t,p] =
      case
        { | t=p+1} : QQS_In;
        { | p+2<=t} : QQS[t-1,p];
      esac;

...
tel;
```

Fig. 12.4 Sketch of the virtual parallel architecture produced by the front-end of MMAlpha. Only variables `M` and `QQS` are represented. Variable `QQS` was produced by localization to propagate the query sequence to each cell of this array

the declaration and the definition of variable `M` (present in the initial program) and of a new `QQS` variable are kept. In the declaration of `M`, we can see that the domain of this variable in now indexed by t and p. The constraints on these indexes let us infer that the calculation of this variable is going to be done on a linear array of $X+1$ processors. The definition of `M` reveals several informations. It shows that the calculation of `M[t,p]` is the maximum of four quantities: the constant 0, the previous value `M[t-1,p]` which can be interpreted as a register in processor p, the previous value `M[t-1,p-1]` which was held in neighboring processor $p-1$, and value `M[t-2,p-1]`, also held in processor $p-1$. All these informations can be directly interpreted in term of hardware elements. However, the linear inequalities guarding the branches of this definition are much less straightforward to translate into hardware. Moreover, the number of processors of this architecture is directly linked to the size parameter `X`, which may not be appropriate for the requirements of a practical application: this is the rôle of the back-end of MMAlpha to transform this virtual architecture into a real one. The `QQS` variable requires some more explanations, as it is not present in the initial specification. It is produced by the localization transformation, in order to propagate the query value `QS` from processor to processor. A careful examination of its declaration and its definition reveals that this variable is present only in processors 1 to `X` and initialized by reading the value of another variable `QQS_In` when $t = p+1$, otherwise, it is kept in a register of processor p. As for `M`, the guards of this equation must be translated into simpler hardware elements.

12.4 The Back-End Process: Generating VHDL

The back-end of MMAlpha comprises a set of transformations allowing a virtual parallel architecture to be transformed into a synthesizable VHDL description. These transformations can be regrouped into three parts (see Fig. 12.3): hardware-mapping, structured HDL Generation, and VHDL generation.

In this section, we review these back-end transformations as they are implemented in MMAlpha by highlighting the concepts underlying them rather than the implementation details.

12.4.1 Hardware-Mapping

The virtual architecture is essentially an *operational parallel* description of the initial specification: each computation occurs at a particular date on a particular processor. The two main transformations needed to obtain an architectural description are: *control signal generation* and *simple expression generation*. They are implemented in the hardware-mapping component which produces a subset of ALPHA traditionally referred to as ALPHA0.

12.4.1.1 Control Signal Generation

It consists in replacing complex, linear inequalities by the propagation of simple control signals and is better explained here on an example. Consider for instance the definition of the `QQS` variable in the program of Fig. 12.4. It can be interpreted as a multiplexer controlled by a signal which is true at step `t=p` in processor number `p` (Fig. 12.5a). It is easy to see intuitively that this control can be implemented by a signal initialized in the first processor (i.e., value 1 at step 0 in processor 0) and then transmitted to the neighboring processor with a one cycle delay (i.e., value 1 at step 1 in processor 1, and so on). This is illustrated on Fig. 12.5b: the control signal `QQS_ctl` is inferred and is pipelined through the array. This is what the control signal generation achieves: to produce a particular cell (the *controller*) at the boundary of the regular array and to pipeline (or broadcast) this control signal through the array.

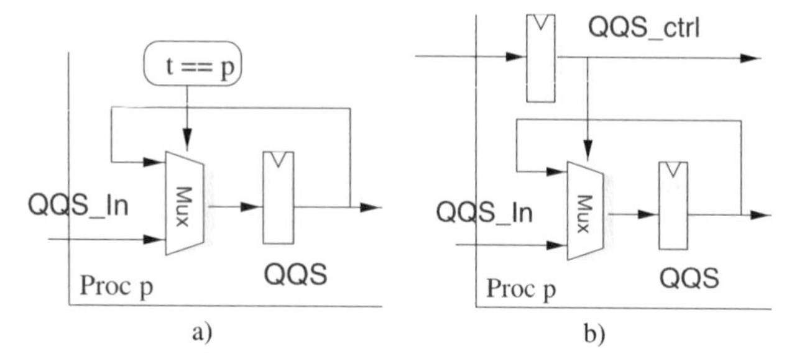

Fig. 12.5 Control signal inference for QQS updating

```
1  QQSReg6[t,p] = QQS[t-1,p];
2  QQS_In[t,p] = QQSReg6[t,p-1];
3  QQS[t,p] =
4      case
5        { |  1<=p<=X;} : if (QQSXctl1) then
6                case
7                  { | t=p+1;} : QQS_In;
8                  { | p+2<=t<=p+Y;} : 0[];
9                esac else
10           case
11             { | t=p+1; } : 0[];
12             { | p+2<=t<=p+Y; } : QQSReg6;
13           esac;
14     esac;
```

Fig. 12.6 Description in ALPHA0 of the hardware of Fig. 12.5b

12.4.1.2 Generation of Simple Expressions

This transformation deals with splitting complex equations in several simpler equations so that each one corresponds to a single hardware component: a register, an operator or a simple wire.

In the ALPHA0 subset of ALPHA, the RTL architecture can be very easily deduced from the code. For instance Fig. 12.6 shows three equations which represent: a register (line 1), a connexion between two processors (line 2) and a multiplexer (lines 3–14). They are interconnected to produce the hardware shown in Fig. 12.5b.

12.4.2 Structured HDL *Generation*

The second step of the back-end deals with generating a structured hardware description from the ALPHA0 format so that the re-use of identical cells explicitly appears in the structuration of the program and provision is made to include other components in the description. The subset of ALPHA which is used at this level is called ALPHARD and is illustrated in Fig. 12.7. Here, we have a module including

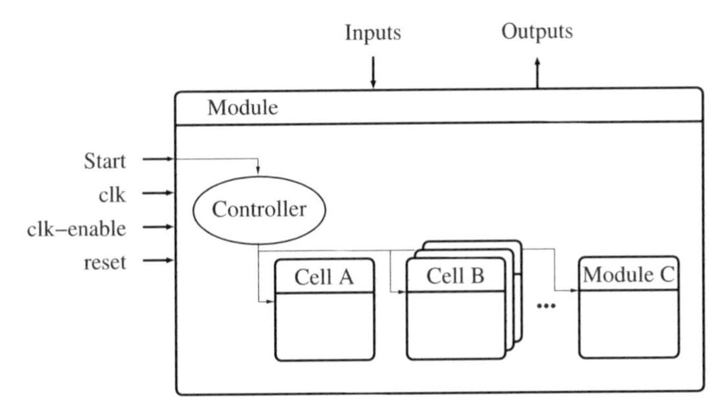

Fig. 12.7 An ALPHARD program is a complex module containing a controller and various instantiations of cells or modules

a local controller, a single instance of a `A` cell, several instances of a `B` cell and an instance of another module. Cells are simple data-paths whereas modules include controllers and can instantiate other cells and modules. Thanks to the hierarchical structure of ALPHA, it is easy to represent such a system in our language while keeping its semantics.

In the case of the string alignment application, the hardware structure contains, in addition to the controller, an instance of a particular cell representing processor $p = 0$, and $X - 1$ instances of another cell representing processors 1 to X. It is depicted in Fig. 12.8. (for the sake of clarity the controller and the control signal are not represented).

The main difficulty of this step is to uncover, in the set of recurrence equations of ALPHA0, the least number of common cells. To this end, the polyhedral domains of all equations are projected on the space indexes and combined to form space maximal regions sharing the same behavior. Each such region defines a cell of the architecture. This operation is made possible thanks to the polyhedral model which allows projection, intersection, unions, etc. of domains to be computed easily.

12.4.3 *Generating* VHDL

The VHDL generation is basically a syntax-directed translation of the ALPHARD program as each ALPHA construct corresponds to a VHDL construct. For instance, the VHDL code that corresponds to the ALPHA0 code shown in Fig. 12.6 is given in Fig. 12.9. Line 1 is a simple connexion, line 3 represents a multiplexer and lines 5–8 model a register. One can notice that the time index `t` disappears (except in the controller) as it is implemented by the `clk` and a clock enable signal.

If the variable sizes are not specified in the ALPHA program, the translator assumes 16-bit fixed-point arithmetics (using `std_logic_vector` VHDL type) but other signal types can be specified. VHDL test benches are also generated to ease the testing of the resulting VHDL.

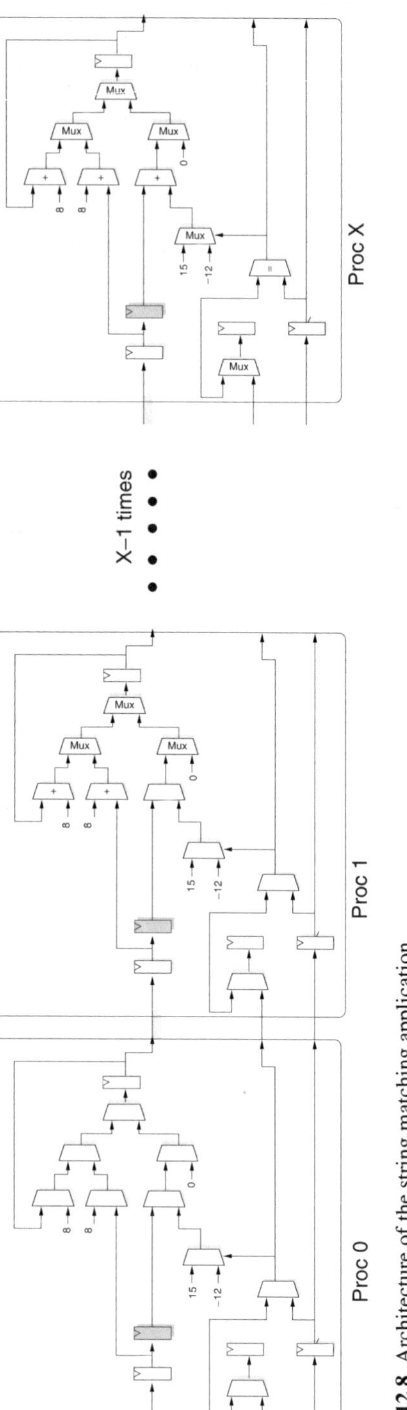

Fig. 12.8 Architecture of the string matching application

```
1  QQS_In <= QQSReg6_In;
2
3  QQS <= QQS_In WHEN QQSXctl1 = '1' ELSE QQSReg6;
4
5  PROCESS(clk) BEGIN IF (clk = '1' AND clk'EVENT) THEN
6    IF CE='1' THEN QQSReg6 <= QQS; END IF;
7                END IF;
8  END PROCESS;
```

Fig. 12.9 VHDL code corresponding to the ALPHA0 code shown in Fig. 12.6

12.5 Partitioning for Resource Management

In MMAlpha, the choice of the various scheduling and/or space–time mappings can be seen as a design space exploration step. However there are practical situations in which none of the *virtual architectures* obtained through the flow matches the user requirements. This is often the case when iteration domains involved in the loop nests are very wide: in such situations, the mapping may result in an architecture with a very large number of processing elements, which often exceeds the allowed silicon budget. As an example, assuming a string alignment program with a query size $X = 10^3$, the architecture corresponding to the mapping proposed in Sect. 12.3 and shown in Fig. 12.4 would result in 10^3 processing elements, which represents a huge cost in term of hardware resources.

Many methods can be used to overcome such a difficulty. In the context of regular parallel architectures, *partitioning* transformations are the method of choice. Here, we consider a processor array partitioning transformation, which can be applied directly on the virtual architecture (i.e., at the RTL level).

Partitioning is a well studied problem [14, 25] and it is essentially based on the combination of two techniques. *Locally Sequential Globally Parallel* (LSGP) partitioning consists in merging several virtual PE into a single PE with modified time-sliced schedule. *Locally Parallel Globally Sequential* (LPGS) partitioning consists in *tiling* the virtual processor array into a set of virtual sub-arrays, and in executing the whole computations as a sequence of passes on the sub-array.

In the following, we present an LSGP technique based on *serialization* [13]: serialization merges σ virtual processors along a given processor axis into a single physical processor. One can show that a complete LSGP partitioning can be obtained through the use of successive serializations along the processor space axis.

To explain the principles of serialization, consider the processor datapath of the string alignment architecture shown in Fig. 12.10. We distinguish *temporal* registers (shown in grey) which have both their source and sink in the same processor, and *spatial* registers, the source and sink of which are in distinct processors. (We assume that registers have always a single sink, which is easy to ensure by transformation if needed.) Besides we assume that the communications between processing elements are unidirectional and pipelined.

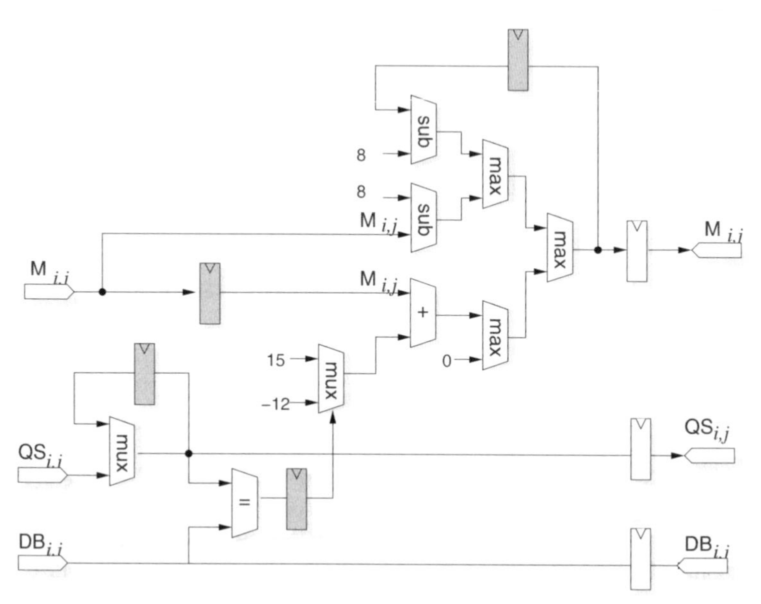

Fig. 12.10 Original datapath of the string alignment processor

Under these assumptions, serialization can be done in two steps:

- Any temporal register is transformed into a shift register line of depth σ.
- A one cycle delay feedback loop is associated to each *spatial register*; this feedback loop is controlled (through an additional multiplexer) by a signal activated every σ cycles.

Obviously, a serialization by a factor σ replaces an array of X processors by a *partitioned* array containing $\lceil X/\sigma \rceil$ processors. Figure 12.11 shows the effect of a serialization with $\sigma = 3$. This kind of transformation can be used to adjust the number of processors to the needs of the application. It can also be combined with various other transformations to cover a large set of potential hardware configurations. An example of hardware resource exploration for a bioinformatics application is presented in [11].

12.6 Implementation and Performance

To illustrate the typical performance of a parallel implementation of an application, we implemented on a Xilinx Virtex-4 device several configurations of string alignment with or without partitioning. The results are shown in Table 12.1. For each configuration, the number X of processors, the total resources of the device, – look-up tables, flip-flops and number of slices – the clock frequency and the performance, in Giga Cell Update per second (GCUps) are given. The last four lines present partitioned versions. As a reference, we show the typical performance of a software implementation of the string aligment on a desktop computer which

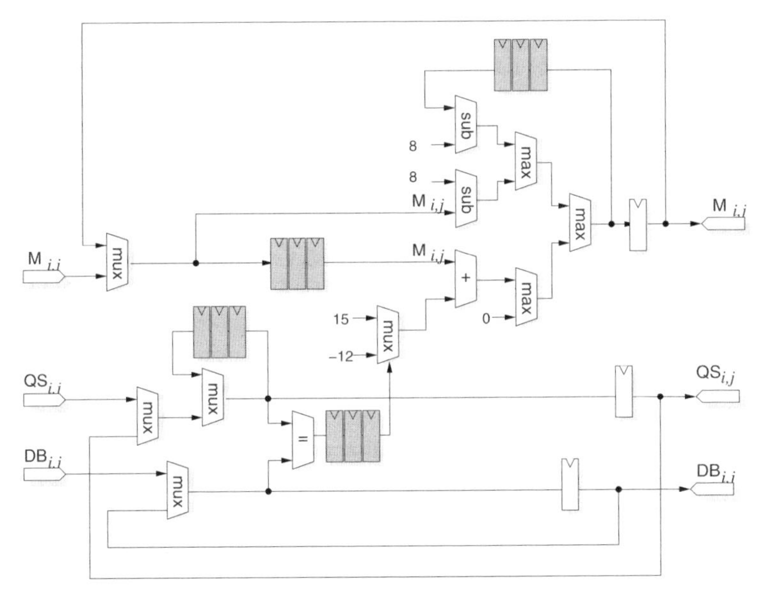

Fig. 12.11 The string alignment processor datapath after serialization by $\sigma = 3$

Table 12.1 Performance of various string alignment hardware configurations measured in Giga Cell Updates per seconds

Description	LUT/DFF/Slices	Clock (MHz)	Perf. (GCUps)
Software	–	–	0,1
$X = 10$	1,047/1,619/1,047	110	1.1
$X = 50$	8,088/4,130/4,771	110	5.5
$X = 100$	16,300/8,233/9,542	110	11
$X = 100, \sigma = 2$	10.8 K/4,308/6,628	95	5.5
$X = 100, \sigma = 10$	2 K/1 K/1,543	102	≈1.0
$X = 100, \sigma = 20$	1.2 K/550/931	93	≈0.45
$X = 100, \sigma = 50$	5.6 K/231/517	98	≈0.2

LUT is the number of look-up tables, *DFF* is the number of data flip-flops, and *Slices* is the number of Virtex-4 FPGA slices used by the designs

achieves 100 MCUps. The speed-up factor reaches up to two orders of magnitude depending on the number of processors. It is also noteworthy that the derived architecture is scalable: the achievable clock period does not suffer from an increase in the number of processing elements, and the hardware resource cost grows linearly with that number.

12.7 Other Works: The Polyhedral Model

The polyhedral model has been used for memory modeling [9, 15], communication modeling [33], cache misses [24], but its most important use was done in parallelizing compilers and HLS tools.

There is an important trend in commercial high-level synthesis tools to perform hardware synthesis from C programs: CatapultC (Mentor Graphics), Pico (Synfora) [30], Cynthesizer (Forte Design System) [18], and Cascade (Critical Blue) [4]. However all these tools suffer from inefficient handling of arbitrary nested loops algorithms.

Academic HLS tools are numerous and reflect the focus of recent researches on efficient synthesis of application-specific algorithms. Among the most important tools: Spark [19], Compaan/Laura [32], ESPAM [27], MMAlpha [26], Paro [6], Gaut [31], UGH [2], Streamroller [22], xPilot [7]. Compaan, Paro and MMAlpha have focused of the efficient compilation of loops, and they use the polyhedral model to perform loop analysis and/or transformations. Another formalism, called *Array-OL*, has been used for multidimensional signal processing [10] and revisited recently [5].

Parallelizing compiler prototypes have also provided a lot of research results on loop transformations [23]: Tiny [34], LooPo [20], Suif [1] or Pips [21]. Recently, WraPit [3], integrated in the `Open64` compiler, proposed an explicit polyhedral internal representation for loop nest, very close to the representation used by MMAlpha.

12.8 Conclusion

We have shown the main principles of high-level synthesis for loops targeting parallel architectures. Our presentation has used the MMAlpha tools as an example to explain the polyhedral model, the basic loops transformations, and the way these transformations may be arranged in order to produce parallel hardware. MMAlpha uses the ALPHA single-assignment language to represent the architecture, from its initial specification to its practical, synthesizable hardware implementation.

The polyhedral model, which underlies the representation and transformation of loops, is a very powerful vehicle to express the variety of transformations that can be used to extract parallelism et take benefit of it for hardware implementations. Future SoC architectures will increasingly need such techniques to exploit available multi-core architectures. We therefore believe that it is a good basis for carrying research on HLS whenever parallelism is considered.

References

1. S. Amarasinghe et al. Suif: An Infrastructure for Research on Parallelizing and Optimizing Compilers. Technical report, Stanford University, May 1994.
2. I. Augé, F. Pétrot, F. Donnet, and P. Gomez. Platform-Based Design From Parallel C Specifications. *IEEE Transactions on Computer-Aided Design of Integrated Circuits and Systems*, 24(12):1811–1826, 2005.
3. C. Bastoul, A. Cohen, S. Girbal, S. Sharma, and O. Temam. Putting Polyhedral Loop Transformations to Work. In *LCPC*, pages 209–225, 2003.

4. Critical Blue. Boosting Software Processing Performance With Coprocessor Synthesis, 2005. http://www.criticalblue.com.
5. P. Boulet. Array-OL Revisited, Multidimensional Intensive Signal Processing Specification. Research Report 6113, INRIA, February 2007.
6. M. Bednara and J. Teich. Automatic Synthesis of FPGA Processor Arrays from Loop Algorithms. *Journal of Supercomputer*, 26(2):149–165, 2003.
7. J. Cong, Y. Fan, G. Han, W. Jiang, and Z. Zhang. Platform-Based Behavior-Level and System-Level Synthesis. In *International SOC Conference*, pages 199–202. IEEE, 2006.
8. D. Cachera and K. Morin-Allory. Verification of Safety Properties for Parameterized Regular Systems. *Transaction on Embedded Computing Systems*, 4(2):228–266, May 2005.
9. F. Catthoor, S. Wuytack, E. De Greef, F. Balasa, L. Nachtergaele, and A. Vandecappelle. *Custom Memory Management Methodology*. Kluwer Academic Publishers, 1998.
10. A. Demeure and Y. Del Gallo. An Array Approach for Signal Processing Design. In *SAME 98*, October 1998.
11. S. Derrien and P. Quinton. Parallezing HMMER for Hardware Acceleration on FPGAs. In *ASAP07*, pages 10–17, Montreal, Quebec, July 2007.
12. F. Dupont de Dinechin, P. Quinton, and T. Risset. Structuration of the Alpha Language. In *Int. Conf. on Massively Parallel Programming Models*, Berlin, Germany, October 1995.
13. S. Derrien, S. V. Rajopadhye, and S. Sur-Kolay. Combined Instruction and Loop Parallelism in Array Synthesis for FPGAs. In *ISSS'01 : Proceedings of the International Symposium on System Synthesis*, pages 165–170, 2001.
14. A. Darte, R. Schreiber, B. R. Rau, and F. Vivien. Constructing and Exploiting Linear Schedules with Prescribed Parallelism. *ACM Trans. Des. Autom. Electron. Syst.*, 7(1):159–172, 2002.
15. A. Darte, R. Schreiber, and G. Villard. Lattice-Based Memory Allocation. *IEEE Transactions on Computers*, 54(10):1242–1257, 2005.
16. P. Feautrier. Dataflow Analysis of Array and Scalar References. *Int. J. Parallel Programming*, 20(1):23–53, February 1991.
17. P. Feautrier. Some Efficient Solutions to the Affine Scheduling Problem, Part I, One Dimensional Time. *Int. J. of Parallel Programming*, 21(5), October 1992.
18. Forte Design Systems. Cynthesizer Closes the ESL-to-Silicon Gap. http://www.forteds.com/products/cynthesizer.asp.
19. S. Gupta, R. Gupta, N. Dutt, and A. Nicolau. *SPARK: A Parallelizing Approach to the High-Level Synthesis of Digital Circuits*. Kluwer Academic, 2004.
20. M. Griebl and C. Lengauer. The Loop Parallelizer LooPo. In M. Gerndt, editor, *Proceedings of Sixth Workshop on Compilers for Parallel Computers*, volume 21 of *Konferenzen des Forschungszentrums Jülich*, pages 311–320. Forschungszentrum Jülich, 1996.
21. F. Irigoin, P. Jouvelot, and R. Triolet. Semantical Interprocedural Parallelization: An Overview of the PIPS Project. In *ACM International Conference on Supercomputing, ICS'91, Cologne*, June 1991.
22. M. Kudlur, K. Fan, and S. Mahlke. Streamroller: Automatic Synthesis of Prescribed Throughput Accelerator Pipelines. In *CODES+ISSS '06: Proceedings of the 4th International Conference on Hardware/Software Codesign and System Synthesis*, pages 270–275, New York, NY, USA, 2006. ACM Press, New York.
23. C. Lengauer. Loop Parallelization in the Polytope Model. In E. Best, editor, *CONCUR'93*, Lecture Notes in Computer Science 715, pages 398–416. Springer, Berlin Heidelberg New York, 1993.
24. V. Loechner, B. Meister, and P. Clauss. Precise Data Locality Optimization of Nested Loops. *The Journal of Supercomputing*, 21(1):37–76, 2002.
25. D. I. Moldovan and J. A. B. Fortes. Partitioning and Mapping Algorithms into Fixed Size Systolic Arrays. *IEEE Transactons on Computers*, 35(1):1–12, 1986.
26. A. Mozipo, D. Massicotte, P. Quinton, and T. Risset. Automatic Synthesis of a Parallel Architecture for Kalman Filtering using MMAlpha. In *International Conference on Parallel Computing in Electrical Engineering (PARELEC 98)*, pages 201–206, Bialystok, Poland, September 1998.

27. H. Nikolov, T. Stefanov, and E. Deprettere. Efficient Automated Synthesis, Programming, and Implementation of Multi-Processor Platforms on FPGA Chips. In *16th International Conference on Field Programmable Logic and Applications (FPL'06)*, pages 323–328, Madrid, Spain, August 2006.
28. P. Quinton and V. Van Dongen. The Mapping of Linear Recurrence Equations on Regular Arrays. *The Journal of VLSI Signal Processing*, 1:95–113, 1989.
29. R. Schreiber et al. PICO-NPA: High-Level Synthesis of Nonprogrammable Hardware Accelerators (HPL-2001-249), October 2001.
30. R. Schreiber, S. Aditya, B. R. Rau, V. Kathail, S. Mahlke, S. Abraham, and G. Snider. High-Level Synthesis of Nonprogrammable Hardware Accelerators. In *ASAP'00: Proceedings of the IEEE International Conference on Application-Specific Systems, Architectures, and Processors*, page 113, Washington, DC, USA, 2000. IEEE Computer Society, Washington, DC.
31. O. Sentieys, J. P. Diguet, and J. L. Philippe. GAUT: A High Level Synthesis Tool Dedicated to Real Time Signal Processing Application. In *European Design Automation Conference*, September 2000. University booth stand.
32. T. Stefanov, C. Zissulescu, A. Turjan, B. Kienhuis, and E. Deprettere. System Design Using Kahn Process Networks: The Compaan/Laura Approach. In *DATE '04: Proceedings of the Conference on Design, Automation and Test in Europe*, page 10340, Washington, DC, USA, 2004. IEEE Computer Society, Washington, DC.
33. A. Turjan, B. Kienhuis, and E. F. Deprettere. Translating Affine Nested-Loop Programs to Process Networks. In *International Conference on Compilers, Architecture, and Synthesis for Embedded Systems*, pages 220–229, 2004.
34. M. Wolfe. A Loop Restructuring Research Tool. Technical Report CSE 90-014, Oregon Graduate Institute, August 1990.

Chapter 13
Operation Scheduling: Algorithms and Applications

Gang Wang, Wenrui Gong, and Ryan Kastner

Abstract Operation scheduling (OS) is an important task in the high-level synthesis process. An inappropriate scheduling of the operations can fail to exploit the full potential of the system. In this chapter, we try to give a comprehensive coverage on the heuristic algorithms currently available for solving both timing and resource constrained scheduling problems. Besides providing a broad survey on this topic, we focus on some of the most popularly used algorithms, such as List Scheduling, Force-Directed Scheduling and Simulated Annealing, as well as the newly introduced approach based on the Ant Colony Optimization meta-heuristics. We discuss in details on their applicability and performance by comparing them on solution quality, performance stability, scalability, extensibility, and computation cost. Moreover, as an application of operation scheduling, we introduce a novel uniformed design space exploration method that exploits the duality of the time and resource constrained scheduling problems, which automatically constructs a high quality time/area tradeoff curve in a fast, effective manner.

Keywords: Design space exploration, Ant colony optimization, Instruction scheduling, MAX–MIN ant system

13.1 Introduction

As fabrication technology advances and transistors become more plentiful, modern computing systems can achieve better system performance by increasing the amount of computation units. It is estimated that we will be able to integrate more than a half billion transistors on a 468 mm^2 chip by the year of 2009 [38]. This yields tremendous potential for future computing systems, however, it imposes big challenges on how to effectively use and design such complicated systems.

As computing systems become more complex, so do the applications that can run on them. Designers will increasingly rely on automated design tools in order

P. Coussy and A. Morawiec (eds.) *High-Level Synthesis.*

to map applications onto these systems. One fundamental process of these tools is mapping a behavioral application specification to the computing system. For example, the tool may take a C function and create the code to program a microprocessor. This is viewed as software compilation. Or the tool may take a transaction level behavior and create a register transfer level (RTL) circuit description. This is called hardware or behavioral synthesis [31]. Both software and hardware synthesis flows are essential for the use and design of future computing systems.

Operation scheduling (OS) is an important problem in software compilation and hardware synthesis. An inappropriate scheduling of the operations can fail to exploit the full potential of the system. Operation scheduling appears in a number of different problems, e.g., compiler design for superscalar and VLIW microprocessors [23], distributed clustering computation architectures [4] and behavioral synthesis of ASICs and FPGAs [31].

Operation scheduling is performed on a behavioral description of the application. This description is typically decomposed into several blocks (e.g., basic blocks), and each of the blocks is represented by a data flow graph (DFG).

Operation scheduling can be classified as *resource constrained* or *timing constrained*. Given a DFG, clock cycle time, resource count and resource delays, a resource constrained scheduling finds the minimum number of clock cycles needed to execute the DFG. On the other hand, a timing constrained scheduling tries to determine the minimum number of resources needed for a given deadline.

In the timing constrained scheduling problem (also called fixed control step scheduling), the target is to find the minimum computing resource cost under a set of given types of computing units and a predefined latency deadline. For example, in many digital signal processing (DSP) systems, the sampling rate of the input data stream dictates the maximum time allowed for computation on the present data sample before the next sample arrives. Since the sampling rate is fixed, the main objective is to minimize the cost of the hardware. Given the clock cycle time, the sampling rate can be expressed in terms of the number of cycles that are required to execute the algorithm.

Resource constrained scheduling is also found frequently in practice. This is because in a lot of the cases, the number of resources are known a priori. For instance, in software compilation for microprocessors, the computing resources are fixed. In hardware compilation, DFGs are often constructed and scheduled almost independently. Furthermore, if we want to maximize resource sharing, each block should use same or similar resources, which is hardly ensured by time constrained schedulers. The time constraint of each block is not easy to define since blocks are typically serialized and budgeting global performance constraint for each block is not trivial [30].

Operation scheduling methods can be further classified as *static scheduling* and *dynamic scheduling* [40]. Static operation scheduling is performed during the compilation of the application. Once an acceptable scheduling solution is found, it is deployed as part of the application image. In dynamic scheduling, a dedicated system component makes scheduling decisions on-the-fly. Dynamic scheduling methods must minimize the program's completion time while considering the overhead paid for running the scheduler.

13.2 Operation Scheduling Formulation

Given a set of operations and a collection of computational units, the resource constrained scheduling (RCS) problem schedules the operations onto the computing units such that the execution time of these operations are minimized, while respecting the capacity limits imposed by the number of computational resources. The operations can be modeled as a data flow graph (DFG) $G(V,E)$, where each node $v_i \in V(i=1,\ldots,n)$ represents an operation op_i, and the edge e_{ij} denotes a dependency between operations v_j and v_i. A DFG is a directed acyclic graph where the dependencies define a partially ordered relationship (denoted by the symbol $\preccurlyeq$) among the nodes. Without affecting the problem, we add two virtual nodes *root* and *end*, which are associated with no operation (NOP). We assume that *root* is the only starting node in the DFG, i.e., it has no predecessors, and node *end* is the only exit node, i.e., it has no successors.

Additionally, we have a collection of computing resources, e.g., ALUs, adders, and multipliers. There are R different types and $r_j > 0$ gives the number of units for resource type j $(1 \leqslant j \leqslant R)$. Furthermore, each operation defined in the DFG must be executable on at least one type of the resources. When each of the operations is uniquely associated with one resource type, we call it *homogenous* scheduling. If an operation can be performed by more than one resource types, we call it *heterogeneous* scheduling [44]. Moreover, we assume the cycle delays for each operation on different type resources are known as $d(i,j)$. Of course, *root* and *end* have zero delays. Finally, we assume the execution of the operations is non-preemptive, that is, once an operation starts execution, it must finish without being interrupted.

A resource constrained schedule is given by the vector

$$\{(s_{root}, f_{root}), (s_1, f_1), \ldots, (s_{end}, f_{end})\}$$

where s_i and f_i indicate the starting and finishing time of the operation op_i. The resource-constrained scheduling problem is formally defined as $min(s_{end})$ with respect to the following conditions:

1. An operation can only start when all its predecessors have finished, i.e., $s_i \geqslant f_j$ if $op_j \preccurlyeq op_i$
2. At any given cycle t, the number of resources needed is constrained by r_j, for all $1 \leqslant j \leqslant R$

The timing constrained scheduling (TCS) is a dual problem of the resource constrained version and can be defined using the same terminology presented above. Here the target is to minimize total resources $\sum_j r_j$ or the total cost of the resources (e.g., the hardware area needed) subject to the same dependencies between operations imposed by the DFG and a given deadline D, i.e., $s_{end} < D$.

13.3 Operation Scheduling Algorithms

13.3.1 ASAP, ALAP and Bounding Properties

The simplest scheduling task occurs when we have unlimited computing resources for the given application while trying to minimize its latency. For this task, we can simply solve it by schedule an operation as soon as all of its predecessors in the DFG have completed, which gives it the name *As Soon As Possible*. Because of its ASAP, nature, it is closely related with finding the longest path between an operation and the starting of the application op_{root}. Furthermore, it can be viewed as a special case of resource constrained scheduling where there is no limit on the number computing unit. The result of ASAP provides the lower bound for the starting time of each operation, together with the lower bound of the overall application latency.

Correspondingly, with a given latency, we have the so called *As Late As Possible* (ALAP) scheduling, where each operation is scheduled to the latest opportunity. This can be done by computing the longest path between the operation node and the end of the application op_{end}. The result scheduling provides a upper bound for the starting time of each operation given the latency constraint on the application. However, different from ASAP, it typically does not have any significance regarding to how efficient the resources are used. On the contrary, it often yields a bad solution in the sense of timing constrained scheduling since the operations tends to cluster towards the end.

Though not directly useful in typical practice, ASAP and ALAP are often critical components for more advanced scheduling methods. This is because their combined results provide the possible scheduling choices for each operation. Such range is often referred as the *mobility* of an operation.

Finally, the upper bound of the application latency (under a given technology mapping) can be obtained by serializing the DFG, that is to perform the operations sequentially based on a topologically sorted sequence of the operations. This is equivalent to have only one unit for each type of operation.

13.3.2 Exact Methods

Though scheduling problems are $\mathcal{NP}$-hard [8], both time and resource constrained problems can be formulated using integer linear programming (ILP) method [27], which tries to find an optimal schedule using a branch-and-bound search algorithm. It also involves some amount of backtracking, i.e., decisions made earlier are changed later on. A simplified formulation of the ILP method for the time constrained problem is given below:

First it calculates the mobility range for each operation $M = \{S_j | E_k \leqslant j \leqslant L_k\}$, where E_k and L_k are the ASAP and ALAP values respectively. The scheduling problem in ILP is defined by the following equations:

$$Min(\sum_{k=1}^{n}(C_k \cdot N_k)) \text{ while } \sum_{E_i \leqslant j \leqslant L_i} x_{ij} = 1$$

where $1 \leqslant i \leqslant n$ and n is the number of operations. There are $1 \leqslant k \leqslant m$ operation types available, and N_k is the number of computing units for operation type k, and C_k is the cost of each unit. Each x_{ij} is 1 if the operation i is assigned in control step j and 0 otherwise. Two more equations that enforce the resource and data dependency constraints are: $\sum_{i=1}^{n} x_{ij} \leqslant N_i$ and $((q * x_{j,q}) - (p * x_{i,p})) \leqslant -1, p \leqslant q$, where p and q are the control steps assigned to the operations x_i and x_j respectively.

We can see that the ILP formulation increases rapidly with the number of control steps. For one unit increase in the number of control steps we will have n additional x variables. Therefore the time of execution of the algorithm also increases rapidly. In practice the ILP approach is applicable only to very small problems.

Another exact method is Hu's algorithm [22], which provides an optimal solution for a limited set of applications. Though can be modified to address generic acyclic DFG scheduling problem, the optimality only applies when the DFG is composed of a set of trees and each unit has single delay with uniformed computing units. Essentially, Hu's method is a special list scheduling algorithm with a priority based on longest paths [31].

13.3.3 Force Directed Scheduling

Because of the limitations of the exact approaches, a range of heuristic methods with polynomial runtime complexity have been proposed. Many timing constrained scheduling algorithms used in high level synthesis are derivatives of the force-directed scheduling (FDS) algorithm presented by Paulin and Knight [34, 35]. Verhaegh et al. [45, 46] provide a theoretical treatment on the original FDS algorithm and report better results by applying gradual time-frame reduction and the use of global spring constants in the force calculation.

The goal of the FDS algorithm is to reduce the number of functional units used in the implementation of the design. This objective is achieved by attempting to uniformly distribute the operations onto the available resource units. The distribution ensures that resource units allocated to perform operations in one control step are used efficiently in all other control steps, which leads to a high utilization rate.

The FDS algorithm relies on both the ASAP and the ALAP scheduling algorithms to determine the feasible control steps for every operation op_i, or the *time frame* of op_i (denoted as $[t_i^S, t_i^L]$ where t_i^S and t_i^L are the ASAP and ALAP times respectively). It also assumes that each operation op_i has a uniform probability of being scheduled into any of the control steps in the range, and zero probability of being scheduled elsewhere. Thus, for a given time step j and an operation op_i which needs $\triangle_i \geqslant 1$ time steps to execute, this probability is given as:

$$\mathsf{p}_j(op_i) = \begin{cases} (\sum_{l=0}^{\triangle_i} h_i(j-l))/(t_i^L - t_i^S + 1) & \text{if } t_i^S \leqslant j \leqslant t_i^L \\ 0 & \text{otherwise} \end{cases} \tag{13.1}$$

where $h_i(\cdot)$ is a unit window function defined on $[t_i^S, t_i^L]$.

Based on this probability, a set of *distribution graphs* can be created, one for each specific type of operation, denoted as q_k. More specifically, for type k at time step j,

$$q_k(j) = \sum_{op_i} \mathsf{p}_j(op_i) \qquad \text{if type of } op_i \text{ is } k \tag{13.2}$$

We can see that $q_k(j)$ is an estimation on the number of type k resources that are needed at control step j.

The FDS algorithm tries to minimize the overall concurrency under a fixed latency by scheduling operations one by one. At every time step, the effect of scheduling each unscheduled operation on every possible time step in its frame range is calculated, and the operation and the corresponding time step with the smallest negative effect is selected. This effect is equated as the force for an unscheduled operation op_i at control step j, and is comprised of two components: the self-force, SF_{ij}, and the predecessor–successor forces, PSF_{ij}.

The self-force SF_{ij} represents the direct effect of this scheduling on the overall concurrency. It is given by:

$$SF_{ij} = \sum_{l=t_i^S}^{t_i^L+\triangle_i} q_k(l)(H_i(l) - \mathsf{p}_i(l)) \tag{13.3}$$

where, $j \in [t_i^S, t_i^L]$, k is the type of operation op_i, and $H_i(\cdot)$ is the unit window function defined on $[j, j+\triangle_i]$.

We also need to consider the predecessor and successor forces since assigning operation op_i to time step j might cause the time frame of a predecessor or successor operation op_l to change from $[t_l^S, t_l^L]$ to $[\widetilde{t_l}^S, \widetilde{t_l}^S]$. The force exerted by a predecessor or successor is given by:

$$PSF_{ij}(l) = \sum_{m=\widetilde{t_i}^S}^{\widetilde{t_i}^L+\triangle_l} (q_k(m) \cdot \widetilde{\mathsf{p}}_m(op_l)) - \sum_{m=t_i^S}^{t_i^L+\triangle_l} (q_k(m) \cdot \mathsf{p}_m(op_l)) \tag{13.4}$$

where $\widetilde{\mathsf{p}}_m(op_l)$ is computed in the same way as (13.1) except the updated mobility information $[\widetilde{t_l^S}, \widetilde{t_l^S}]$ is used. Notice that the above computation has to be carried for all the predecessor and successor operations of op_i. The total force of the hypothetical assignment of scheduling op_i on time step j is the addition of the self-force and all the predecessor–successor forces, i.e.,

$$\text{total force}_{ij} = SF_{ij} + \sum_l PSF_{ij}(l) \tag{13.5}$$

where op_l is a predecessor or successor of op_i. Finally, the total forces obtained for all the unscheduled operations at every possible time step are compared. The operation and time step with the best force reduction is chosen and the partial scheduling result is incremented until all the operations have been scheduled.

The FDS method is "constructive" because the solution is computed without performing any backtracking. Every decision is made in a greedy manner. If there are two possible assignments sharing the same cost, the above algorithm cannot accurately estimate the best choice. Based on our experience, this happens fairly often as the DFG becomes larger and more complex. Moreover, FDS does not take into account future assignments of operators to the same control step. Consequently, it is likely that the resulting solution will not be optimal, due to the lack of a look ahead scheme and the lack of compromises between early and late decisions.

Our experiments show that a baseline FDS implementation based on [34] fails to find the optimal solution even on small testing cases. To ease this problem, a look-ahead factor was introduced in the same paper. A second order term of the displacement weighted by a constant η is included in force computation, and the value η is experimentally decided to be $1/3$. In our experiments, this look-ahead factor has a positive impact on some testing cases but does not always work well. More details regarding FDS performance can be found in Sect. 13.4.

13.3.4 List Scheduling

List scheduling is a commonly used heuristic for solving a variety of RCS problems [36, 37]. It is a generalization of the ASAP algorithm with the inclusion of resource constraints [25]. A list scheduler takes a data flow graph and a priority list of all the nodes in the DFG as input. The list is sorted with decreasing magnitude of priority assigned to each of the operation. The list scheduler maintains a ready list, i.e., nodes whose predecessors have already been scheduled. In each iteration, the scheduler scans the priority list and operations with higher priority are scheduled first. Scheduling an operator to a control step makes its successor operations ready, which will be added to the ready list. This process is carried until all of the operations have been scheduled. When there exist more than one ready nodes sharing the same priority, ties are broken randomly.

It is easy to see that list scheduler always generates feasible schedule. Furthermore, it has been shown that a list scheduler is always capable of producing the optimal schedule for resource-constrained instruction scheduling problem if we enumerate the topological permutations of the DFG nodes with the input priority list [25].

The success of the list scheduler is highly dependent on the priority function and the structure of the input application (DFG) [25,31,43]. One commonly used priority function assigns the priority inversely proportional to the mobility. This ensures that the scheduling of operations with large mobilities are deferred because they have more flexibility as to where they can be scheduled. Many other priority functions have been proposed [2, 5, 18, 25]. However, it is commonly agreed that there is no single good heuristic for prioritizing the DFG nodes across a range of applications using list scheduling. Our results in Sect. 13.4 confirm this.

13.3.5 Iterative Heuristic Methods

Both FDS and List Scheduling are greedy constructive methods. Due to the lack of a look ahead scheme, they are likely to produce a sub-optimal solution. One way to address this issue is the iterative method proposed by Park and Kyung [33] based on Kernighan and Lin's heuristic [24] method used for solving the graph-bisection problem. In their approach, each operation is scheduled into an earlier or later step using the move that produces the maximum gain. Then all the operations are unlocked and the whole procedure is repeated with this new schedule. The quality of the result produced by this algorithm is highly dependent upon the initial solution. There have been two enhancements made to this algorithm: (1) Since the algorithm is computationally efficient it can be run many times with different initial solution and the best solution can be picked. (2) A better look-ahead scheme that uses a more sophisticated strategy of move selection as in [kris84] can be used. More recently, Heijligers et al. [20] and InSyn [39] use evolutionary techniques like genetic algorithms and simulated evolution.

There are a number of iterative algorithms for the resource constrained problem, including genetic algorithm [7, 18], tabu search [6, 44], simulated annealing [43], graph theoretic and computational geometry approaches [4, 10, 30].

13.3.6 Ant Colony Optimization (ACO)

ACO is a cooperative heuristic searching algorithm inspired by ethological studies on the behavior of ants [15]. It was observed [13] that ants – who lack sophisticated vision – manage to establish the optimal path between their colony and a food source within a very short period of time. This is done through indirect communication known as *stigmergy* via the chemical substance, or *pheromone*, left by the ants on the paths. Each individual ant makes a decision on its direction biased on the "strength" of the pheromone trails that lie before it, where a higher amount of pheromone hints a better path. As an ant traverses a path, it reinforces that path with its own pheromone. A collective autocatalytic behavior emerges as more ants will choose the shortest trails, which in turn creates an even larger amount of pheromone on the short trails, making such short trails more attractive to the future ants. The ACO algorithm is inspired by this observation. It is a population based approach where a collection of agents cooperate together to explore the search space. They communicate via a mechanism imitating the pheromone trails.

One of the first problems to which ACO was successfully applied was the Traveling Salesman Problem (TSP) [15], for which it gave competitive results comparing with traditional methods. Researchers have since formulated ACO methods for a variety of traditional $\mathcal{NP}$-hard problems. These problems include the maximum clique problem, the quadratic assignment problem, the graph coloring problem, the shortest common super-sequence problem, and the multiple knapsack problem. ACO also has been applied to practical problems such as the vehicle routing problem, data mining, network routing problem and the system level task partitioning problem [12,48,49].

It was shown [19] that ACO converges to an optimal solution with probability of exactly one; however there is no constructive way to guarantee this. Balancing exploration to achieve close-to-optimal results within manageable time remains an active research topic for ACO algorithms. MAX–MIN Ant System (MMAS) [42] is a popularly used method to address this problem. MMAS is built upon the original ACO algorithm, which improves it by providing dynamically evolving bounds on the pheromone trails so that the heuristic never strays too far away from the best encountered solution. As a result, all possible paths will have a non-trivial probability of being selected; thus it encourages broader exploration of the search space while maintaining a good differential between alternative solutions. It was reported that MMAS was the best performing ACO approach on a number of classic combinatory optimization tasks.

Both time constrained and resource constrained scheduling problems can be effectively solved by using ACO. Unfortunately, in the consideration of space, we can only give a general introduction on the ACO formulation for the TCS problem. For a complete treatment of the algorithms, including detailed discussion on the algorithms' implementation, applicability, complexity, extensibility, parameter selection and performance, please refer to [47,50].

In its ACO-based formulation, the TCS problem is solved with an iterative searching process. the algorithms employ a collection of agents that collaboratively explore the search space. A stochastic decision making strategy is applied in order to combine global and local heuristics to effectively conduct this exploration. As the algorithm proceeds in finding better quality solutions, dynamically computed local heuristics are utilized to better guide the searching process. Each iteration consists of two stages. First, the ACO algorithm is applied where a collection of ants traverse the DFG to construct individual operation schedules with respect to the specified deadline using global and local heuristics. Secondly, these scheduling results are evaluated using their resource costs. The associated heuristics are then adjusted based on the solutions found in the current iteration. The hope is that future iterations will benefit from this adjustment and come up with better schedules.

Each operation or DFG node op_i is associated with D pheromone trails τ_{ij}, where $j = 1, \ldots, D$ and D is the specified deadline. These pheromone trails indicate the global favorableness of assigning the ith operation at the jth control step in order to minimize the resource cost with respect to the time constraint. Initially, based on ASAP and ALAP results, τ_{ij} is set with some fixed value τ_0 if j is a valid control step for op_i; otherwise, it is set to be 0.

For each iteration, m ants are released and each ant individually starts to construct a schedule by picking an unscheduled instruction and determining its desired control step. However, unlike the deterministic approach used in the FDS method, each ant picks up the next instruction for scheduling decision probabilistically. Once an instruction op_h is selected, the ant needs to make decision on which control step it should be assigned. This decision is also made probabilistically as illustrated in (13.6).

$$\mathsf{p}_{hj} = \begin{cases} \frac{\tau_{hj}(t)^{\alpha} \cdot \eta_{hj}^{\beta}}{\sum_l (\tau_{hl}^{\alpha}(t) \cdot \eta_{hl}^{\beta})} & \text{if } op_h \text{ can be scheduled at } l \text{ and } j \\ 0 & \text{otherwise} \end{cases} \tag{13.6}$$

Here j is the time step under consideration. The item η_{hj} is the local heuristic for scheduling operation op_h at control step j, and α and β are parameters to control the relative influence of the distributed global heuristic τ_{hj} and local heuristic η_{hj}. Assuming op_h is of type k, η_{hj} to simply set to be the inverse of the distribution graph value [34], which is computed based on partial scheduling result and is an indication on the number of computing units of type k needed at control step j. In other words, an ant is more likely to make a decision that is globally considered "good" and also uses the fewest number of resources under the current partially scheduled result. We do not recursively compute the forces on the successor nodes and predecessor nodes. Thus, selection is much faster. Furthermore, the time frames are updated to reflect the changed partial schedule. This guarantees that each ant will always construct a valid schedule.

In the second stage of our algorithm, the ant's solutions are evaluated. The quality of the solution from ant h is judged by the total number of resources, i.e., $Q_h = \sum_k r_k$. At the end of the iteration, the pheromone trail is updated according to the quality of individual schedules. Additionally, a certain amount of pheromone evaporates. More specifically, we have:

$$\tau_{ij}(t) = \rho \cdot \tau_{ij}(t) + \sum_{h=1}^{m} \Delta\tau_{ij}^{h}(t) \qquad \text{where } 0 < \rho < 1. \tag{13.7}$$

Here ρ is the evaporation ratio, and

$$\Delta\tau_{ij}^{h} = \begin{cases} Q/Q_h & \text{if } op_i \text{ is scheduled at } j \text{ by ant } h \\ 0 & \text{otherwise} \end{cases} \tag{13.8}$$

Q is a fixed constant to control the delivery rate of the pheromone. Two important operations are performed in the pheromone trail updating process. Evaporation is necessary for ACO to effectively explore the solution space, while reinforcement ensures that the favorable operation orderings receive a higher volume of pheromone and will have a better chance of being selected in the future iterations. The above process is repeated multiple times until an ending condition is reached. The best result found by the algorithm is reported.

Comparing with the FDS method, the ACO algorithm differs in several aspects. First, rather than using a one-time constructive approach based on greedy local decisions, the ACO method solves the problem in an evolutionary manner. By using simple local heuristics, it allows individual scheduling result to be generated in a faster manner. With a collection of such individual results and by embedding and adjusting global heuristics associated with partial solutions, it tries to learn during the searching process. By adopting a stochastic decision making strategy considering both global experience and local heuristics, it tries to balance the efforts of exploration and exploitation in this process. Furthermore, it applies positive feedback to strengthen the "good" partial solutions in order to speed up the convergence. Of course, the negative effect is that it may fall into local minima, thus requires compensation measures such as the one introduced in MMAS. In our experiments, we implemented both the basic ACO and the MMAS algorithms. The latter consistently achieves better scheduling results, especially for larger DFGs.

13.4 Performance Evaluation

13.4.1 Benchmarks and Setup

In order to test and evaluate our algorithms, we have constructed a comprehensive set of benchmarks named *ExpressDFG*. These benchmarks are taken from one of two sources: (1) popular benchmarks used in previous literature; (2) real-life examples generated and selected from the MediaBench suite [26].

The benefit of having classic samples is that they provide a direct comparison between results generated by our algorithm and results from previously published methods. This is especially helpful when some of the benchmarks have known optimal solutions. In our final testing benchmark set, seven samples widely used in instruction scheduling studies are included. These samples focus mainly on frequently used numeric calculations performed by different applications. However, these samples are typically small to medium in size, and are considered somewhat old. To be representative, it is necessary to create a more comprehensive set with benchmarks of different sizes and complexities. Such benchmarks shall aim to:

- Provide real-life testing cases from real-life applications
- Provide more up-to-date testing cases from modern applications
- Provide challenging samples for instruction scheduling algorithms with regards to larger number of operations, higher level of parallelism and data dependency
- Provide a wide range of synthesis problems to test the algorithms' scalability

For this purpose, we investigated the MediaBench suite, which contains a wide range of complete applications for image processing, communications and DSP applications. We analyzed these applications using the SUIF [3] and Machine SUIF [41] tools, and over 14,000 DFGs were extracted as preliminary candidates for our

Table 13.1 ExpressDFG benchmark suite

Benchmark name	No. nodes	No. edges	ID
HAL	11	8	4
horner_bezier†	18	16	8
ARF	28	30	8
motion_vectors†	32	29	6
EWF	34	47	14
FIR2	40	39	11
FIR1	44	43	11
h2v2_smooth_downsample†	51	52	16
feedback_points†	53	50	7
collapse_pyr†	56	73	7
COSINE1	66	76	8
COSINE2	82	91	8
write_bmp_header†	106	88	7
interpolate_aux†	108	104	8
matmul†	109	116	9
idctcol	114	164	16
jpeg_idct_ifast†	122	162	14
jpeg_fdct_islow†	134	169	13
smooth_color_z_triangle†	197	196	11
invert_matrix_general†	333	354	11

Benchmarks with † are extracted from MediaBench

benchmark set. After careful study, thirteen DFG samples were selected from four MediaBench applications: JPEG, MPEG2, EPIC and MESA.

Table 13.1 lists all 20 benchmarks that were included in our final benchmark set. Together with the names of the various functions where the basic blocks originated are the number of nodes, number of edges and instruction depth (assuming unit delay for every instruction) of the DFG. The data, including related statistics, DFG graphs and source code for the all testing benchmarks, is available online [17].

For all testing benchmarks, operations are allocated on two types of computing resources, namely MUL and ALU, where MUL is capable of handling multiplication and division, and ALU is used for other operations such as addition and subtraction. Furthermore, we define the operations running on MUL to take two clock cycles and the ALU operations take one. This definitely is a simplified case from reality. However, it is a close enough approximation and does not change the generality of the results. Other choices can easily be implemented within our framework.

13.4.2 Time Constrained Scheduling: ACO vs. FDS

With the assigned resource/operation mapping, ASAP is first performed to find the critical path delay L_c. We then set our predefined deadline range to be $[L_c, 2L_c]$, i.e., from the critical path delay to two times of this delay. This results 263 testing cases in total. For each delay, we run FDS first to obtain its scheduling result. Following this, the ACO algorithm is executed five times to obtain enough data for performance evaluation. We report the FDS result quality, the average and best result quality for the ACO algorithm and the standard deviation for these results. The execution time information for both algorithms is also reported.

We have implemented the ACO formulation in C for the TCS problem. The evaporation rate ρ is configured to be 0.98. The scaling parameters for global and local heuristics are set to be $\alpha = \beta = 1$ and delivery rate $Q = 1$. These parameters are not changed over the tests. We also experimented with different ant number m and the allowed iteration count N. For example, set m to be proportional to the average branching factor of the DFG under study and N to be proportional to the total operation number. However, it is found that there seems to exist a fixed value pair for m and N which works well across the wide range of testing samples in our benchmark. In our final settings, we set m to be 10, and N to be 150 for all the timing constrained scheduling experiments.

Based on our experiments, the ACO based operation scheduling achieves better or much better results. Our approach seems to have much stronger capability in robustly finding better results for different testing cases. Furthermore, it scales very well over different DFG sizes and complexities. Another aspect of scalability is the pre-defined deadline. The average result quality generated by the ACO algorithm is better than or equal to the FDS results in 258 out of 263 cases. Among them, for 192 testing cases (or 73% of the cases) the ACO method outperforms the FDS method. There are only five cases where the ACO approach has worse average quality results. They all happened on the *invert_matrix_general* benchmark. On average, we can expect a 16.4% performance improvement over FDS. If only considering the best results among the five runs for each testing case, we achieve a 19.5% resource reduction averaged over all tested samples. The most outstanding results provided by the ACO method achieve a 75% resource reduction compared with FDS. These results are obtained on a few deadlines for the *jpeg_idct_ifast* benchmark.

Besides absolute quality of the results, one difference between FDS and the ACO method is that ACO method is relatively more stable. In our experiments, it is observed that the FDS approach can provide worse quality results as the deadline is relaxed. Using the *idctcol* as an example, FDS provides drastically worse results for deadlines ranging from 25 to 30 though it is able to reach decent scheduling qualities for deadline from 19 to 24. The same problem occurs for deadlines between 36 and 38. One possible reason is that as the deadline is extended, the time frame of each operation is also extended, which makes the force computation more likely to clash with similar values. Due to the lack of backtracking and good look-ahead capability, an early mistake would lead to inferior results. On the other hand, the ACO algorithm robustly generates monotonically non-increasing results with fewer resource requirements as the deadline increases.

13.4.3 Resource Constrained Scheduling: ACO vs. List Scheduling and ILP

We have implemented the ACO-based resource-constrained scheduling algorithm and compared its performance with the popularly used list scheduling and force-directed scheduling algorithms.

For each of the benchmark samples, we run the ACO algorithm with different choices of local heuristics. For each choice, we also perform five runs to obtain enough statistics for evaluating the stability of the algorithm. Again we fixed the number of ants per iteration 10 and in each run we allow 100 iterations. Other parameters are also the same as those used in the timing constrained problem. The best schedule latency is reported at the end of each run and then the average value is reported as the performance for the corresponding setting. Two different experiments are conducted for resource constrained scheduling – the homogenous case and the heterogenous case.

For the homogenous case, resource allocation is performed before the operation scheduling. Each operation is mapped to a unique resource type. In other words, there is no ambiguity on which resource the operation shall be handled during the scheduling step. In this experiment, similar to the timing constrained case, two types of resources (MUL/ALU) are allowed. The number of each resource type is predefined after making sure they do not make the experiment trivial (for example, if we are too generous, then the problem simplifies to an ASAP problem).

Comparing with a variety of list scheduling approaches and the force-directed scheduling method, the ACO algorithm generates better results consistently over all testing cases, which is demonstrated by the number of times that it provides the best results for the tested cases. This is especially true for the case when operation depth (OD) is used as the local heuristic, where we find the best results in 14 cases amongst 20 tested benchmarks. For other traditional methods, FDS generates the most hits (ten times) for best results, which is still less than the worst case of ACO (11 times). For some of the testing samples, our method provides significant improvement on the schedule latency. The biggest saving achieved is 22%. This is obtained for the COSINE2 benchmark when operation mobility (OM) is used as the local heuristic for our algorithm and also as the heuristic for constructing the priority list for the traditional list scheduler. For cases that our algorithm fails to provide the best solution, the quality of its results is also much closer to the best than other methods.

ACO also demonstrates much stronger stability over different input applications. As indicated in Sect. 13.3.4, the performance of traditional list scheduler heavily depends on the input application, while the ACO algorithm is much less sensitive to the choice of different local heuristics and input applications. This is evidenced by the fact that the standard deviation of the results achieved by the new algorithm is much smaller than that of the traditional list scheduler. The average standard deviation for list scheduling over all the benchmarks and different heuristic choices is 1.2, while for the ACO algorithm it is only 0.19. In other words, we can expect to achieve high quality scheduling results much more stably on different application DFGs regardless of the choice of local heuristic. This is a great attribute desired in practice.

One possible explanation for the above advantage is the different ways how the scheduling heuristics are used by list scheduler and the ACO algorithm. In list scheduling, the heuristics are used in a greedy manner to determine the order of the operations. Furthermore, the schedule of the operations is done all at once. Differently, in the ACO algorithm, local heuristics are used stochastically and combined with the pheromone values to determine the operations' order. This makes the solution exploration more balanced. Another fundamental difference is that the ACO algorithm is an iterative process. In this process, the pheromone value acts as an indirect feedback and tries to reflect the quality of a potential component based on the evaluations of historical solutions that contain this component. It introduces a way to integrate global assessments into the scheduling process, which is missing in the traditional list or force-directed scheduling.

In the second experiment, heterogeneous computing units are allowed, i.e., one type of operation can be performed by different types of resources. For example, multiplication can be performed by either a faster multiplier or a regular one. Furthermore, multiple same type units are also allowed. For example, we may have three faster multipliers and two regular ones.

We conduct the heterogenous experiments with the same configuration as for the homogenous case. Moreover, to better assess the quality of our algorithm, the same heterogenous RCS tasks are also formulated as integer linear programming problems and then optimally solved using CPLEX. Since the ILP solution is time consuming to obtain, our heterogenous tests are only done for the classic samples.

Compared with a variety of list scheduling approaches and the force-directed scheduling method, the ACO algorithm generates better results consistently over all testing cases. The biggest saving achieved is 23%. This is obtained for the FIR2 benchmark when the latency weighted operation depth (LWOD) is used as the local heuristic. Similar to the homogenous case, our algorithm outperforms other methods in regards to consistently generating high-quality results. The average standard deviation for list scheduler over all the benchmarks and different heuristic choices is 0.8128, while that for the ACO algorithm is only 0.1673.

Though the results of force-directed scheduler generally outperform the list scheduler, our algorithm achieves even better results. On average, comparing with the force-directed approach, our algorithm provides a 6.2% performance enhancement for the testing cases, while performance improvement for individual test sample can be as much as 14.7%.

Finally, compared to the optimal scheduling results computed by using the integer linear programming model, the results generated by the ACO algorithm are much closer to the optimal than those provided by the list scheduling heuristics and the force-directed approach. For all the benchmarks with known optima, our algorithm improves the average schedule latency by 44% comparing with the list scheduling heuristics. For the larger size DFGs such as COSINE1 and COSINE2, CPLEX fails to generate optimal results after more than 10 h of execution on a SPARC workstation with a 440 MHz CPU and 384 MB memory. In fact, CPLEX crashes for these two cases because of running out of memory. For COSINE1, CPLEX does provide a intermediate sub-optimal solution of 18 cycles before it crashes. This result is worse than the best result found by the ACO algorithm.

13.4.4 Further Assessment: ACO vs. Simulated Annealing

In order to further investigate the quality of the ACO-based algorithms, we compared them with a simulated annealing (SA) approach. Our SA implementation is similar to the algorithm presented in [43]. The basic idea is very similar to the ACO approach in which a meta-heuristic method (SA) is used to guide the searching process while a traditional list scheduler is used to evaluate the result quality. The scheduling result with the best resource usage is reported when the algorithm terminates.

The major challenge here is the construction of a *neighbor* selection in the SA process. With the knowledge of each operation's mobility range, it is trivial to see the search space for the TCS problem is covered by all the possible combinations of the operation/timestep pairs, where each operation can be scheduled into any time step in its mobility range. In our formulation, given a scheduling S where operation op_i is scheduled at t_i, we experimented with two different methods for generating a neighbor solution:

1. *Physical neighbor*: A neighbor of S is generated by selecting an operation op_i and rescheduling it to a physical neighbor of its current scheduled time step t_i, namely either $t_i + 1$ or $t_i - 1$ with even possibility. In case t_i is on the boundary of its mobility range, we treat the mobility range as a circular buffer;
2. *Random neighbor*: A neighbor of S is generated by selecting an operation and rescheduling it to any of the position in its mobility range excluding its currently scheduled position.

However, both of the above approaches suffer from the problem that a lot of these *neighbors* will be invalid because they may violate the data dependency posed by the DFG. For example, say, in S a single cycle operation op_1 is scheduled at time step 3, and another single cycle operation op_2 which is data dependent on op_1 is scheduled at time step 4. Changing the schedule of op_2 to step 3 will create an invalid scheduling result. To deal with this problem in our implementation, for each generated scheduling, we quickly check whether it is valid by verifying the operation's new schedule against those of its predecessor and successor operations defined in the DFG. Only valid schedules will be considered.

Furthermore, in order to give roughly equal chance to each operation to be selected in the above process, we try to generate multiple neighbors before any temperature update is taken. This can be considered as a local search effort, which is widely implemented in different variants of SA algorithm. We control this local search effort with a weight parameter θ. That is before any temperature update taking place, we attempt to generate θN valid scheduling candidates where N is the number of operations in the DFG. In our work, we set $\theta = 2$, which roughly gives each operation two chances to alter its currently scheduled position in each cooling step.

This local search mechanism is applied to both neighbor generation schemes discussed above. In our experiments, we found there is no noticeable difference between the two neighbor generation approaches with respect to the quality of the final scheduling results except that the *random neighbor* method tends to take significantly more computing time. This is because it is more likely to come up with an invalid scheduling which are simply ignored in our algorithm. In our final realization, we always use the *physical neighbor* method.

Another issue related to the SA implementation is how to set the initial seed solution. In our experiments, we experimented three different seed solutions: ASAP, ALAP and a randomly generated valid scheduling. We found that SA algorithm with a randomly generated seed constantly outperforms that using the ASAP or ALAP initialization. It is especially true when the *physical neighbor* approach is used. This is not surprising since the ASAP and ALAP solutions tend to cluster operations together which is bad for minimizing resource usage. In our final realization, we always use a randomly generated schedule as the seed solution.

The framework of our SA implementation for both timing constrained and resource constrained scheduling is similar to the one reported in [51]. The acceptance of a more costly neighboring solution is determined by applying the Boltzmann probability criteria [1], which depends on the cost difference and the annealing temperature. In our experiments, the most commonly known and used geometric cooling schedule [51] is applied and the temperature decrement factor is set to 0.9. When it reaches the pre-defined maximum iteration number or the stop temperature, the best solution found by SA is reported.

Similar to the ACO algorithm, we perform five runs for each benchmark sample and report the average savings, the best savings, and the standard deviation of the reported scheduling results. It is observed that the SA method provides much worse results compared with the ACO solutions. In fact, the ACO approach provides better results on every testing case. Though the SA method does have significant gains on select cases over FDS, its average performance is actually worse than FDS by 5%, while our method provides a 16.4% average savings. This is also true if we consider the best savings achieved amongst multiple runs where a modest 1% savings is observed in SA comparing with a 19.5% reduction obtained by the ACO method. Furthermore, the quality of the SA method seems to be very dependent on the input applications. This is evidenced by the large dynamic range of the scheduling quality and the larger standard deviation over the different runs. Finally, we also want to make it clear that to achieve this result, the SA approach takes substantially more computing time than the ACO method. A typical experiment over all 263 testing cases will run between 9 to 12 h which is 3–4 times longer than the ACO-based TCS algorithm.

As discussed above, our SA formulation for resource constrained scheduling is similar to that studied in [43]. It is relatively more straight forward since it will always provide valid scheduling using a list scheduler. To be fair, a randomly generated operation list is used as the seed solution for the SA algorithm. The neighbor solutions are constructed by swapping the positions of two neighboring operations in the current list. Since the algorithm always generates a valid scheduling, we can better control the runtime than in its TCS counterpart by adjusting the cooling scheme parameter. We carried experiments using execution limit ranging from 1 to 10 times of that of the ACO approach. It was observed that SA RCS algorithm provides poor performance when the time limit was too short. On the other hand, once we increase this time limit to over five times of the ACO execution time, there was no significant improvement on the results as the execution time increased. It is observed that the ACO-based algorithm consistently outperforms it while using much less computing time.

13.5 Design Space Exploration

As a direct application of the operation scheduling algorithms, we examine the Design Space Exploration problem in this section, which is not only of theoretical interest but also encountered frequently in real-life high level synthesis practice.

13.5.1 Problem Formulation and Related Work

When building a digital system, designers are faced with a countless number of decisions. Ideally, they must deliver the smallest, fastest, lowest power device that can implement the application at hand. More often than not, these design parameters are contradictory. Designers must be able to reason about the tradeoffs amongst a set of parameters. Such decisions are often made based on experience, i.e., this worked before, it should work again. Exploration tools that can quickly survey the design space and report a variety of options are invaluable.

From optimization point of view, design space exploration can be distilled to identifying a set of Pareto optimal design points according to some objective function. These design points form a curve that provides the best tradeoffs for the variables in the objective function. Once the curve is constructed, the designer can make design decisions based on the relative merits of the various system configurations. Timing performance and the hardware cost are two common objectives in such process.

Resource allocation and scheduling are two fundamental problems in constructing such Pareto optimal curves for time/cost tradeoffs. By applying resource constrained scheduling, we try to minimize the application latency without violating the resource constraints. Here allocation is performed before scheduling, and a different resource allocation will likely produce a vastly different scheduling result. On the other hand, we could perform scheduling before allocation; this is the time constrained scheduling problem. Here the inputs are a data flow graph and a time deadline (latency). The output is again a start time for each operation, such that the latency is not violated, while attempting to minimize the number of resources that are needed. It is not clear as to which solution is better. Nor is it clear on the order that we should perform scheduling and allocation.

Obviously, one possible method of design space exploration is to vary the constraints to probe for solutions in a point-by-point manner. For instance, you can use some time constrained algorithm iteratively, where each iteration has a different input latency. This will give you a number of solutions, and their various resource allocations over a set of time points. Or you can run some resource constrained algorithm iteratively. This will give you a latency for each of these area constraints.

Design space exploration problem has been the focus in numerous studies. Though it is possible to formulate the problems using Integer Linear Program (ILP), they quickly become intractable when the problem sizes get large. Much research has been done to cleverly use heuristic approaches to address these problems. Actually, one major motivation of the popularly used Force Directed Scheduling (FDS) algorithm [34] was to address the design space exploration task, i.e., by performing FDS to solve a series timing constrained scheduling problems. In the same paper, the authors also proposed a method called force-directed list scheduling (FDLS) to address the resource constrained scheduling problem. The FDS method is constructive since the solution is computed without backtracking. Every decision is made deterministically in a greedy manner. If there are two potential assignments with the same cost, the FDS algorithm cannot accurately estimate the best choice. Moreover, FDS does not take into account future assignments of operators to the same control step. Consequently, it is possible that the resulting solution will not be optimal due to it's greedy nature. FDS works well on small sized problems, however, it often results to inferior solutions for more complex problems. This phenomena is observed in our experiments reported in Sect. 13.4.

In [16], the authors concentrate on providing alternative "module bags" for design space exploration by heuristically solving the clique partitioning problems and using force directed scheduling. Their work focuses more on the situations where the operations in the design can be executed on alternative resources. In the Voyager system [11], scheduling problems are solved by carefully bounding the design space using ILP, and good results are reported on small sized benchmarks. Moreover, it reveals that clock selection can have an important impact on the final performance of the application. In [14, 21, 32], genetic algorithms are implemented for design space exploration. Simulated annealing [29] has also been applied in this domain. A survey on design space exploration methodologies can be found in [28] and [9].

In this chapter, we focus our attention on the basic design space exploration problem similar to the one treated in [34], where the designers are faced with the task of mapping a well defined application represented as a DFG onto a set of known resources where the compatibility between the operations and the resource types has been defined. Furthermore, the clock selection has been determined in the form of execution cycles for the operations. The goal is to find the a Pareto optimal tradeoff amongst the design implementations with regard to timing and resource costs. Our basic method can be extended to handle clock selection and the use of alternative resources. However, this is beyond the scope of this discussion.

13.5.2 Exploration Using Time and Resource Constrained Duality

As we have discussed, traditional approaches solve the design space exploration task solving a series of scheduling problems, using either a resource constrained method or a timing constrained method. Unfortunately, designers are left with individual tools for tackling either problem. They are faced with questions like: Where do we start the design space exploration? What is the best way to utilize the scheduling tools? When do we stop the exploration?

Moreover, due to the lack of connection amongst the traditional methods, there is very little information shared between time constrained and resource constrained solutions. This is unfortunate, as we are throwing away potential solutions since solving one problem can offer more insight into the other problem. Here we propose a novel design exploration method that exploits the duality of the time and resource constrained scheduling problems. Our exploration automatically constructs a time/area tradeoff curve in a fast, effective manner. It is a general approach and can be combined with any high quality scheduling algorithm.

We are concerned with the design problem of making tradeoffs between hardware cost and timing performance. This is still a commonly faced problem in practice, and other system metrics, such as power consumption, are closely related with them. Based on this, we have a 2-D design space as illustrated in Fig. 13.1a, where the x-axis is the execution deadline and the y-axis is the aggregated hardware cost. Each point represents a specific tradeoff of the two parameters.

For a given application, the designer is given R types of computing resources (e.g., multipliers and adders) to map the application to the target device. We define a specific design as a *configuration*, which is simply the number of each specific resource type. In order to keep the discussion simple, in the rest of the paper we

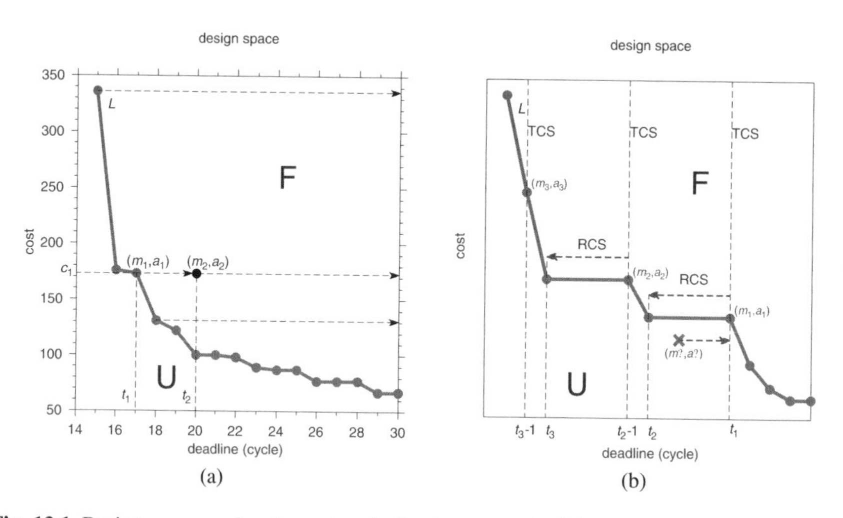

Fig. 13.1 Design space exploration using duality between schedule problems (curve L gives the optimal time/cost tradeoffs)

assume there are only two resource types M (multiply) and A (add), though our algorithm is not limited to this constraint. Thus, each configuration can be specified by (m,a) where m is the number of resource M and a is the number of A. For each specific configuration we have the following lemma about the portion of the design space that it maps to.

Lemma 1. *Let C be a feasible configuration with cost c for the target application. The configuration maps to a horizontal line in the design space starting at (t_{min}, c), where t_{min} is the resource constrained minimum scheduling time.*

The proof of the lemma is straightforward as each feasible configuration has a minimum execution time t_{min} for the application, and obviously it can handle every deadline longer than t_{min}. For example, in Fig. 13.1a, if the configuration (m_1, a_1) has a cost c_1 and a minimum scheduling time t_1, the portion of design space that it maps to is indicated by the arrow next to it. Of course, it is possible for another configuration (m_2, a_2) to have the same cost but a bigger minimum scheduling time t_2. In this case, their feasible space overlaps beyond (t_2, c_1).

As we discussed before, the goal of design space exploration is to help the designer find the optimal tradeoff between the time and area. Theoretically, this can be done by finding the minimum area c amongst all the configurations that are capable of producing $t \in [t_{asap}, t_{seq}]$, where t_{asap} is the ASAP time for the application while t_{seq} is the sequential execution time. In other words, we can find these points by performing time constrained scheduling (TCS) on all t in the interested range. These points form a curve in the design space, as illustrated by curve L in Fig. 13.1a. This curve divides the design space into two parts, labeled with F and U respectively in Fig. 13.1a, where all the points in F are feasible to the given application while U contains all the unfeasible time/area pairs. More interestingly, we have the following attribute for curve L:

Lemma 2. *Curve L is monotonically non-increasing as the deadline t increases.*[1]

Due to this lemma, we can use the dual solution of finding the tradeoff curve by identifying the minimum resource constrained scheduling (RCS) time t amongst all the configurations with cost c. Moreover, because the monotonically non-increasing property of curve L, there may exist horizontal segments along the curve. Based on our experience, horizontal segments appear frequently in practice. This motivates us to look into potential methods to exploit the duality between RCS and TCS to enhance the design space exploration process. First, we consider the following theorem:

Theorem 1. *If C is a configuration that provides the minimum cost at time t_1, then the resource constrained scheduling result t_2 of C satisfies $t_2 \leqslant t_1$. More importantly, there is no configuration C' with a smaller cost that can produce an execution time within $[t_2, t_1]$.*[2]

---

[1] Proof is omitted because of page limitation.

[2] Proof is omitted because of page limitation.

This theorem provides a key insight for the design space exploration problem. It says that if we can find a configuration with optimal cost c at time t_1, we can move along the horizontal segment from (t_1, c) to (t_2, c) without losing optimality. Here t_2 is the RCS solution for the found configuration. This enables us to efficiently construct the curve L by iteratively using TCS and RCS algorithms and leveraging the fact that such horizontal segments do frequently occur in practice. Based on the above discussion, we propose a new space exploration algorithm as shown in Algorithm 1 that exploits the duality between RCS and TCS solutions. Notice the *min* function in step 10 is necessary since a heuristic RCS algorithm may not return the true optimal that could be worse than t_{cur}.

By iteratively using the RCS and TCS algorithms, we can quickly explore the design space. Our algorithm provides benefits in runtime and solution quality compared with using RCS or TCS alone. Our algorithm performs exploration starting from the largest deadline t_{max}. Under this case, the TCS result will provide a configuration with a small number of resources. RCS algorithms have a better chance to find the optimal solution when the resource number is small, therefore it provides a better opportunity to make large horizontal jumps. On the other hand, TCS algorithms take more time and provide poor solutions when the deadline is unconstrained. We can gain significant runtime savings by trading off between the RCS and TCS formulations.

The proposed framework is general and can be combined with any scheduling algorithm. We found that in order for it to work in practice, the TCS and RCS algorithms used in the process require special characteristics. First, they must be fast, which is generally requested for any design space exploration tool. More importantly, they must provide close to optimal solutions, especially for the TCS problem. Otherwise, the conditions for Theorem 1 will not be satisfied and the generated curve L will suffer significantly in quality. Moreover, notice that we enjoy the biggest jumps when we take the minimum RCS result amongst all the configurations

Algorithm 1 Iterative design space exploration algorithm

```
procedure DSE
output: curve L
 1: interested time range [t_min, t_max], where t_min ⩾ t_asap and t_max ⩽ t_seq.
 2: L = ϕ
 3: t_cur = t_max
 4: while t_cur ⩾ t_min do
 5:     perform TCS on t_cur to obtain the optimal configurations C_i.
 6:     for configuration C_i do
 7:         perform RCS to obtain the minimum time t^i_rcs
 8:     end for
 9:     t_rcs = min_i (t^i_rcs) /* find the best rcs time */
10:     t_cur = min(t_cur, t_rcs) − 1
11:     extend L based on TCS and RCS results
12: end while
13: return L
```

that provide the minimum cost for the TCS problem. This is reflected in Steps 6–9 in Algorithm 1. For example, it is possible that both (m,a) and (m',a') provide the minimum cost at time t but they have different deadline limits. Therefore a good TCS algorithm used in the proposed approach should be able to provide multiple candidate solutions with the same minimum cost, if not all of them.

13.6 Conclusion

In this chapter, we provide a comprehensive survey on various operation scheduling algorithms, including List Scheduling, Force-Directed Scheduling, Simulated Annealing, Ant Colony Optimization (ACO) approach, together with others. We report our evaluation for the aforementioned algorithms against a comprehensive set of benchmarks, called ExpressDFG. We give the characteristics of these benchmarks and discuss suitability for evaluating scheduling algorithms. We present detailed performance evaluation results in regards of solution quality, stability of the algorithms, their scalability over different applications and their runtime efficiency. As a direct application, we present a uniformed design space exploration method that exploits duality between the timing and resource constrained scheduling problems.

Acknowledgment This work was partially supported by National Science Foundation Grant CNS-0524771.

References

1. Aarts, E. and Korst, J. (1989). *Simulated Annealing and Boltzmann Machines: A Stochastic Approach to Combinatorial Optimization and Neural Computing*. Wiley, New York, NY.
2. Adam, T. L., Chandy, K. M., and Dickson, J. R. (1974). A comparison of list schedules for parallel processing systems. *Communications of the ACM*, 17(12):685–690.
3. Aigner, G., Diwan, A., Heine, D. L., Moore, M. S. L. D. L., Murphy, B. R., and Sapuntzakis, C. (2000). *The Basic SUIF Programming Guide*. Computer Systems Laboratory, Stanford University.
4. Aletà, A., Codina, J. M., and and Antonio G., Jesús S. (2001). Graph-Partitioning Based Instruction Scheduling for ClusteredProcessors. In *Proceedings of the 34th Annual ACM/IEEE International Symposium on Microarchitecture*.
5. Auyeung, A., Gondra, I., and Dai, H. K. (2003). Integrating random ordering into multi-heuristic list scheduling genetic algorithm. *Advances in Soft Computing: Intelligent Systems Design and Applications*. Springer, Berlin Heidelberg New York.
6. Beaty, Steve J. (1993). Genetic algorithms versus tabu search for instruction scheduling. In *Proceedings of the International Conference on Artificial Neural Networks and Genetic Algorithms*.
7. Beaty, Steven J. (1991). Genetic algorithms and instruction scheduling. In *Proceedings of the 24th Annual International Symposium on Microarchitecture*.
8. Bernstein, D., Rodeh, M., and Gertner, I. (1989). On the Complexity of Scheduling Problems for Parallel/PipelinedMachines. *IEEE Transactions on Computers*, 38(9):1308–1313.

9. C. McFarland, M., Parker, A. C., and Camposano, R. (1990). The high-level synthesis of digital systems. In *Proceedings of the IEEE*, vol. 78, pp. 301–318.
10. Camposano, R. (1991). Path-based scheduling for synthesis. *IEEE Transaction on Computer-Aided Design*, 10(1):85–93.
11. Chaudhuri, S., Blythe, S. A., and Walker, R. A. (1997). A solution methodology for exact design space exploration in a three-dimensional design space. *IEEE Transactions on very Large Scale Integratioin Systems*, 5(1):69–81.
12. Corne, D., Dorigo, M., and Glover, F., editors (1999). *New Ideas in Optimization*. McGraw Hill, London.
13. Deneubourg, J. L. and Goss, S. (1989). Collective Patterns and Decision Making. *Ethology, Ecology and Evolution*, 1:295–311.
14. Dick, R. P. and Jha, N. K. (1997). MOGAC: A Multiobjective Genetic Algorithm for the Co-Synthesis of Hardware-Software Embedded Systems. In *IEEE/ACM Conference on Computer Aided Design*, pp. 522–529.
15. Dorigo, M., Maniezzo, V., and Colorni, A. (1996). Ant System: Optimization by a Colony of Cooperating Agents. *IEEE Transactions on Systems, Man and Cybernetics, Part-B*, 26(1):29–41.
16. Dutta, R., Roy, J., and Vemuri, R. (1992). Distributed design-space exploration for high-level synthesis systems. In *DAC '92*, pp. 644–650. IEEE Computer Society Press, Los Alamitos, CA.
17. ExpressDFG (2006). ExpressDFG benchmark web site. *http://express.ece.ucsb.edu/benchmark/*.
18. Grajcar, M. (1999). Genetic List Scheduling Algorithm for Scheduling and Allocationon a Loosely Coupled Heterogeneous Multiprocessor System. In *Proceedings of the 36th ACM/IEEE Conference on Design Automation Conference*.
19. Gutjahr, W. J. (2002). Aco algorithms with guaranteed convergence to the optimal solution. *Information Processing Letters*, 82(3):145–153.
20. Heijligers, M. and Jess, J. (1995). High-level synthesis scheduling and allocation using genetic algorithms based on constructive topological scheduling techniques. In *International Conference on Evolutionary Computation*, pp. 56–61, Perth, Australia.
21. Heijligers, M. J. M., Cluitmans, L. J. M., and Jess, J. A. G. (1995). High-level synthesis scheduling and allocation using genetic algorithms. p. 11.
22. Hu, T. C. (1961). Parallel sequencing and assembly line problems. *Operations Research*, 9(6):841–848.
23. Kennedy, K. and Allen, R. (2001). *Optimizing Compilers for Modern Architectures: A Dependence-basedApproach*. Morgan Kaufmann, San Francisco.
24. Kernighan, B. W. and Lin, S. (1970). An efficient heuristic procedure for partitioning graphs. *Bell System Technical Journal*, 49(2):291–307.
25. Kolisch, R. and Hartmann, S. (1999). Heuristic algorithms for solving the resource-constrained project scheduling problem: classification and computational analysis. *Project Scheduling: Recent Models, Algorithms and Applications*. Kluwer Academic, Dordrecht.
26. Lee, C., Potkonjak, M., and Mangione-Smith, W. H. (1997). Mediabench: A tool for evaluating and synthesizing multimedia and communications systems. In *Proceedings of the 30th Annual ACM/IEEE International Symposium on Microarchitecture*.
27. Lee, J.-H., Hsu, Y.-C., and Lin, Y.-L. (1989). A new integer linear programming formulation for the scheduling problem in data path synthesis. In *Proceedings of ICCAD-89*, pp. 20–23, Santa Clara, CA.
28. Lin, Y.-L. (1997). Recent developments in high-level synthesis. *ACM Transactions on Design of Automation of Electronic Systems*, 2(1):2–21.
29. Madsen, J., Grode, J., Knudsen, P. V., Petersen, M. E., and Haxthausen, A. (1997). LYCOS: The Lyngby Co-Synthesis System. *Design Automation for Embedded Systems*, 2(2):125–63.
30. Memik, S. O., Bozorgzadeh, E., Kastner, R., and MajidSarrafzadeh (2001). A super-scheduler for embedded reconfigurable systems. In *IEEE/ACM International Conference on Computer-Aided Design*.

31. Micheli, G. De (1994). *Synthesis and Optimization of Digital Circuits*. McGraw-Hill, New York.
32. Palesi, M. and Givargis, T. (2002). Multi-Objective Design Space Exploration Using GeneticAlgorithms. In *Proceedings of the Tenth International Symposium on Hardware/Software-Codesign*.
33. Park, I.-C. and Kyung, C.-M. (1991). Fast and near optimal scheduling in automatic data path synthesis. In *DAC '91: Proceedings of the 28th conference on ACM/IEEE design automation*, pp. 680–685. ACM Press, New York, NY.
34. Paulin, P. G. and Knight, J. P. (1987). Force-directed scheduling in automatic data path synthesis. In *24th ACM/IEEE Conference Proceedings on Design Automation Conference*.
35. Paulin, P. G. and Knight, J. P. (1989). Force-directed scheduling for the behavioral synthesis of asic's. *IEEE Transactions on Computer-Aided Design*, 8:661–679.
36. Poplavko, P., van Eijk, C. A. J., and Basten, T. (2000). Constraint analysis and heuristic scheduling methods. In *Proceedings of 11th Workshop on Circuits, Systems and Signal Processing(ProRISC2000)*, pp. 447–453.
37. Schutten, J. M. J. (1996). List scheduling revisited. *Operation Research Letter*, 18:167–170.
38. Semiconductor Industry Association (2003). National Technology Roadmap for Semiconductors.
39. Sharma, A. and Jain, R. (1993). Insyn: Integrated scheduling for dsp applications. In *DAC*, pp. 349–354.
40. Smith, J. E. (1989). Dynamic instruction scheduling and the astronautics ZS-1. *IEEE Computer*, 22(7):21–35.
41. Smith, M. D. and Holloway, G. (2002). *An Introduction to Machine SUIF and Its Portable Librariesfor Analysis and Optimization*. Division of Engineering and Applied Sciences, Harvard University.
42. Stützle, T. and Hoos, H. H. (2000). MAX–MIN Ant System. *Future Generation Computer Systems*, 16(9):889–914.
43. Sweany, P. H. and Beaty, S. J. (1998). Instruction scheduling using simulated annealing. In *Proceedings of 3rd International Conference on Massively Parallel Computing Systems*.
44. Topcuouglu, H., Hariri, S., and you Wu, M. (2002). Performance-effective and low-complexity task scheduling for heterogeneous computing. *IEEE Transactions on Parallel and Distributed Systems*, 13(3):260–274.
45. Verhaegh, W. F. J., Aarts, E. H. L., Korst, J. H. M., and Lippens, P. E. R. (1991). Improved force-directed scheduling. In *EURO-DAC '91: Proceedings of the Conference on European Design Automation*, pp. 430–435. IEEE Computer Society Press, Los Alamitos, CA.
46. Verhaegh, W. F. J., Lippens, P. E. R., Aarts, E. H. L., Korst, J. H. M., van der Werf, A., and van Meerbergen, J. L. (1992). Efficiency improvements for force-directed scheduling. In *ICCAD '92: Proceedings of the 1992 IEEE/ACM international Conference on Computer-Aided Design*, pp. 286–291. IEEE Computer Society Press, Los Alamitos, CA.
47. Wang, G., Gong, W., DeRenzi, B., and Kastner, R. (2006). Ant Scheduling Algorithms for Resource and Timing Constrained Operation Scheduling. *IEEE Transactions of Computer-Aided Design of Integrated Circuits and Systems (TCAD)*, 26(6):1010–1029.
48. Wang, G., Gong, W., and Kastner, R. (2003). A New Approach for Task Level Computational ResourceBi-partitioning. *15th International Conference on Parallel and Distributed Computing and Systems*, 1(1):439–444.
49. Wang, G., Gong, W., and Kastner, R. (2004). System level partitioning for programmable platforms using the ant colony optimization. *13th International Workshop on Logic and Synthesis, IWLS'04*.
50. Wang, G., Gong, W., and Kastner, R. (2005). Instruction scheduling using MAX–MIN ant optimization. In *15th ACM Great Lakes Symposium on VLSI, GLSVLSI'2005*.
51. Wiangtong, T., Cheung, P. Y. K., and Luk, W. (2002). Comparing Three Heuristic Search Methods for FunctionalPartitioning in Hardware-Software Codesign. *Design Automation for Embedded Systems*, 6(4):425–49.

Chapter 14
Exploiting Bit-Level Design Techniques in Behavioural Synthesis

María Carmen Molina, Rafael Ruiz-Sautua, José Manuel Mendías, and Román Hermida

Abstract Most conventional high-level synthesis algorithms and commercial tools handle specification operations in a very conservative way, as they assign operations to one or several consecutive clock cycles, and to one functional unit of equal or larger width. Independently of the parameter to be optimized, area, execution time, or power consumption, more efficient implementations could be derived from handling operations at the bit level. This way, one operation can be decomposed into several smaller ones that may be executed in several inconsecutive cycles and over several functional units. Furthermore, the execution of one operation fragment can begin once its input operands are available, even if the calculus of its predecessors finishes at a later cycle, and also arithmetic properties can be partially applied to specification operations. These design strategies may be either exploited within the high-level synthesis, or applied to optimize behavioural specifications or register-transfer-level implementations.

Keywords: Scheduling, Allocation, Binding, Bit-level design

14.1 Introduction

Conventional High-Level Synthesis (HLS) algorithms and commercial tools derive Register-Transfer-Level (RTL) implementations from behavioural specifications subject to some constraints in terms of area, execution time, or power consumption. Most algorithms handle specification operations in a very conservative way. In order to reduce one or more of the already mentioned parameters, they assign operations to one or several consecutive clock cycles, and to one functional unit (FU) of equal or larger width. These bindings represent quite a limited subset of all possible ones, whose consideration would surely lead to better designs.

If circuit area becomes the parameter to be minimized, then conventional HLS algorithms usually try to balance the number of operations executed per cycle, as

P. Coussy and A. Morawiec (eds.) *High-Level Synthesis.*

well as keep HW resources busy in most cycles. Since a perfect distribution of operations among cycles is nearly impossible to be reached, some *HW waste* (idle HW resources) appears in almost every cycle. This waste is mainly due to the following two factors: *operation mobility* (range of cycles in which every operation may start its execution, subject to data dependencies and given timing constraints), and *specification heterogeneity* (in function of the number of different types, widths, and data formats present in the behavioural specification).

O*peration mobility* influences the HW waste because a limited mobility makes perfect distributions of operations among cycles unreachable. Even in the hypothetical case of specifications without data dependencies, some waste appears as long as the latency is not a divisor of the number of operations.

Specification heterogeneity influences the HW waste because HLS algorithms usually treat separately operations with different types, data formats or widths, preventing them from sharing the same HW resource. In consequence, a particular *mobility dependent waste* emerges for every different (type, data format, width) triplet in the specification. This waste exists even if more efficient HLS algorithms are used. For example, many algorithms are able to allocate operations of different widths to the same FU. But if an operation is executed over a wider FU (extending its arguments) it is partially wasted. Besides, in most cases, the cycle length is longer than necessary because the most significant bits (MSB) of the calculated results are discarded. The HW waste also arises in implementations with arithmetic-logic units (ALU) able to execute different types of operations. In this case, part of the ALU always remains unused while it executes any operation.

On one hand, the mobility dependent waste could be reduced through the balance in the number of bits of every different operation type calculated per cycle, instead of the number of operations. In order to obtain homogeneous distributions, some specification operations should be transformed into a set of new ones, whose types and widths might be different from the original. On the other hand, the heterogeneity dependent waste could be reduced if all compatible operations were synthesized jointly (two operations are compatible if they can be executed over same type FUs and some glue logic). To do so, algorithms able to fragment compatible operations into their common operative kernel plus some glue logic are needed.

In both cases, each operation fragment inherits the mobility of the original operation and is scheduled separately. Therefore, one original operation may be executed across a set of cycles, not necessarily consecutive (saving the partial results and carry outs calculated in every cycle), and bound to a set of linked HW resources in each cycle. It might occur that the first operation fragment executed is not the one that uses the least significant bits (LSB) of the input operands, although this feature does not imply that the MSB of the result can be calculated before its LSB. In the datapaths designed following these strategies the number, type, and width of the HW resources are, in general, independent of the number, type, and width of the specification operations and variables.

The example in Fig. 14.1 illustrates the main features of these design strategies. It shows a fragment of a data flow graph obtained from a multiple precision specification with multiplications and additions, and two schedules proposed by

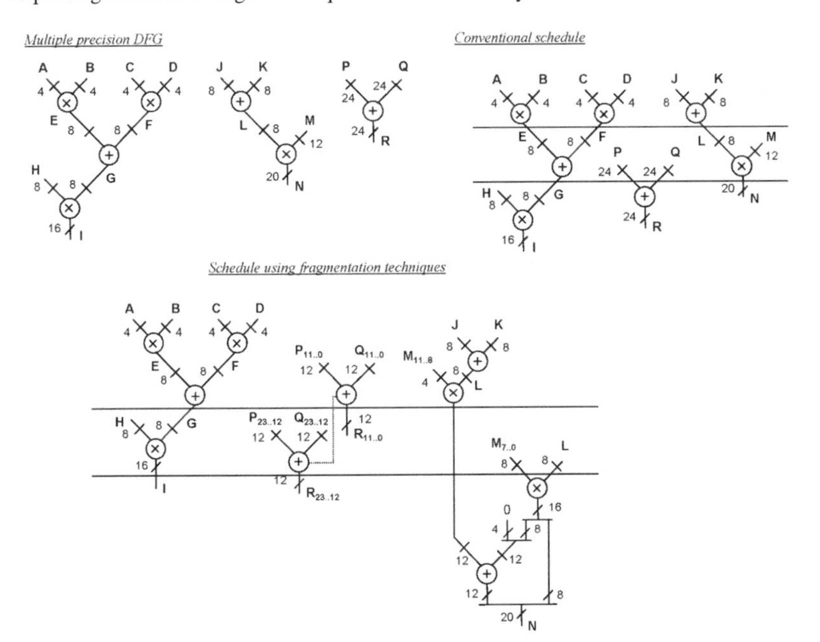

Fig. 14.1 DFG of a behavioural specification, conventional schedule, and more efficient schedule based on operation fragmentations

a conventional algorithm and by a HLS algorithm including some fragmentation techniques. While every operation is executed in a single cycle in the conventional schedule, some operations are executed during several cycles in the second one. However, this feature should not be confused with multi cycle operators. The main advantages of this new design method follow below: one operation can be scheduled in several non consecutive cycles, different FUs can be used to compute every operation fragment, the FUs needed to execute every fragment are narrower than the original operation, and the storage of all the input operand bits is not needed all the time the operation is being executed. In the example, the addition $R = P + Q$ is scheduled in the first and second cycles, and the multiplication $N = L \times M$ in the first and third cycles. In this case, the set of cycles selected to execute $N = L \times M$ are not consecutive. Note also that the multiplication fragment scheduled in the first cycle ($L \times M_{11..8}$) is not the one that calculates the LSB of the original operation.

Table 14.1 shows the set of cycles and FUs selected to execute every specification operation. Operations $I = H \times G$, $N = L \times M$, and $R = P + Q$ have been fragmented into several new operations as shown in the table. In order to balance the computational cost of the operations executed per cycle, $N = L \times M$ and $R = P + Q$ have been fragmented by the proposed scheduling algorithm into seven and two new operations, respectively. And in order to minimize the FUs area (increasing FUs reuse), $I = H \times G$ has been fragmented into five new operations, which are then executed over three multipliers and two adders. Figure 14.2 illustrates how the execution of $N = L \times M$, that begins in the first cycle, is completed in the third one over three multipliers and three adders.

Table 14.1 Schedule and binding based on operation fragmentations

Operation	Fragmentation	Cycle 1	Cycle 2	Cycle 3
$E = A \times B$		$\otimes 4 \times 4$		
$F = C \times D$		$\otimes 4 \times 4$		
$G = E + F$		$\oplus 8$		
$I = H \times G$	$I1 = H \times G_{3..0}$		$\otimes 8 \times 4$	
	$I2 = H_{3..0} \times G_{7..4}$		$\otimes 4 \times 4$	
	$I3 = H_{7..4} \times G_{7..4}$		$\otimes 4 \times 4$	
	$I4 = I1_{11..4} + I2$		$\oplus 8$	
	$I5 = (\text{'000'}\ \&\ I4_{8..4}) + I3$		$\oplus 8$	
$L = J + K$		$\oplus 8$		
$N = L \times M$	$N1 = L \times M_{11..8}$	$\otimes 8 \times 4$		
	$N2 = L \times M_{3..0}$			$\otimes 8 \times 4$
	$N3 = L_{3..0} \times M_{7..4}$			$\otimes 4 \times 4$
	$N4 = L_{7..4} \times M_{7..4}$			$\otimes 4 \times 4$
	$N5 = N3 + N2_{11..4}$			$\oplus 8$
	$N6 = N4 + (\text{'000'}\ \&\ N5_{8..4})$			$\oplus 8$
	$N7 = N1 + (\text{'000'}\ \&\ N6)$			$\oplus 12$
$R = P + Q$	$R1 = P_{11..0} + Q_{11..0}$	$\oplus 12$		
	$R2 = P_{23..12} + Q_{23..12}$		$\oplus 12$	

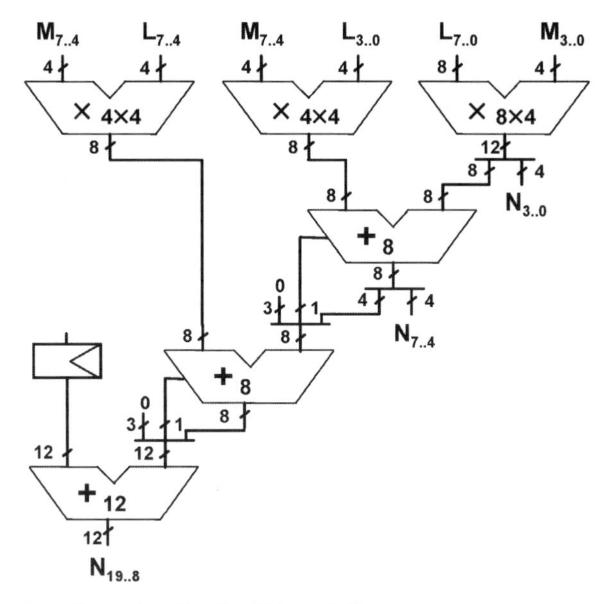

Fig. 14.2 Execution of operation $N = L \times M$ in cycle 3

Table 14.2 Main features of the synthesized implementations

	Conventional	Proposed	Saved (%)
Datapath FUs	$\otimes 12\times 8$, $\otimes 4\times 4$, $\oplus 24$	$\otimes 8\times 4$, $2\otimes 4\times 4$, $\oplus 12$, $2\oplus 8$	–
FU's area	1,401 inverters	911 inverters	35
Circuit area	3,324 inverters	2,298 inverters	30.86
Cycle length	23.1 ns	21.43 ns	7.22

In the second implementation of this example every datapath FU is used once per cycle to execute one operation of its same width, and therefore the HW waste present in the conventional implementation has been totally removed. The circuit area (including FUs, registers, multiplexers, and controller) and the cycle length have been reduced around 31 and 7%, respectively, as summarized in Table 14.2.

14.2 Design Techniques to Reduce the HW Waste

The first step to reduce the HW waste present in conventional implementations becomes the extraction of the common operative kernel of operations in the behavioural specification. This implies the fragmentation of *compatible operations* (that can be executed over FUs of a same type) into their common operative kernel plus some glue logic. These transformations can be applied not only to trivial cases like the compatibility between additions and subtractions, but to more complex ones like the compatibility between additions and multiplications. The common operative kernel extraction increments the number of operations that can be executed over the same HW resources, and in consequence, helps augment the FUs reuse.

In addition to the extraction of the common operative kernels, several other advanced design techniques can be applied during the scheduling or allocation phases to further reduce the HW waste. These new methods handle specification operations at the bit level, and produce datapaths where the number, type, and width of the HW resources do not depend on the number, type, and width of the specification operations and variables. This independence from the description style used in the specification brings the applicability of HLS closer to inexperienced designers, and offers expert ones more freedom to specify behaviours.

14.2.1 Kernel Extraction of Compatible Operations

Compatible operations can be executed over the same FUs in order to improve HW reuse, provided that their common operative kernel has been extracted in advance. This kernel extraction can be applied to the behavioural specifications as a preprocessing phase prior to the synthesis process. This way, additive operations such as subtractions, comparisons, maximums, minimums, absolute values, data format converters, or data limiters can be substituted for additions and logic operations.

And more complex operations like multiplication, division, MAC, factorial or power operations can be decomposed into several multiplications, additions, and some glue logic.

In order to increase the number of operations that may share one FU, it is also desirable the unification of the different representation formats used in the specification. Signed operations can be transformed into several unsigned ones, e.g. a two's complement signed multiplication of $m \times n$ bits (being $m \geq n$) is transformed using a variant of the Baugh and Wooley algorithm [1] into one multiplication of $(m-1) \times (n-1)$ bits, and two additions of m and $n+1$ bits. Some of the transformations that can be applied to homogenize operation types and data representation formats have been described by Molina et al. [2].

14.2.2 Scheduling Techniques

The HW waste present in conventional implementations can be reduced in the scheduling phase if we balance the computational cost of the operations executed per cycle (number of bits of every different operation type) instead of the number of operations. In order to obtain homogeneous distributions, some specification operations must be transformed into a set of new ones, whose types and widths may be different from the original.

Operation fragments inherit the mobility of the original operation and are scheduled separately, in order to balance the computational cost of the operations executed per cycle. Therefore, one original operation may be executed across a set of cycles, not necessarily consecutive (saving the partial results and carry information calculated in every cycle). This implies that every result bit is available to be used as input operand the cycle it is calculated in, even if the operation finishes at a later cycle. Also each fragment can begin its execution once its input operands are available, even if its predecessors finish at a later cycle. It can also occur that the first operation fragment executed is not the one that uses the LSB of the input operands. This feature does not imply that the MSB of the result can be calculated before its LSB in the case of operations with rippling effect.

The set of different fragmentations to be applied in every case mainly depends on the type and width of the operation. In the following sections a detailed analysis of the different ways to fragment additions and multiplications is presented. Other types of operations could also be fragmented in a similar one. However, its description has been omitted as most of them can be reduced to multiplications, additions, and logic operations and the remaining ones are less common in behavioural specifications.

14.2.2.1 Fragmentation of Additions

In order to improve the quality of circuits, some specification additions can be fragmented into several smaller additions and executed across several not necessarily consecutive cycles. In this case the new data dependences among operation fragments (propagation of carry signals) obliges to calculate the LSB of the operation in the earliest cycle and the MSB in the latest one. The application of this design technique provides many advantages in comparison to previous ones, such as chaining, bit-level chaining, multicycle, or non-integer multicycle. Some of them are summarized below:

(a) *Area reduction of the required FUs*. The execution of the new operation fragments can be performed using a unique adder. The adder width must equal the width of the biggest addition fragment, which is always narrower than the original operation length. In general, the routing resources needed to share the adder among all the addition fragments do not waste the benefits achieved by the FU reuse. Aside from the remaining additions present in the specification, the best area reduction is obtained when all the fragments have similar widths due to the HW waste minimization, which is also applicable to the routing resources.

    In the particular case that the addition fragments are scheduled in consecutive cycles, this design technique may seem similar to the multicycle technique. However, the main difference resides on the required HW, as the width of the multicycle FU must match the operation width, meanwhile in our case must equal the widest fragment.

(b) *Area reduction of the required storage units*. As well as all the input operand bits of a fragmented addition are not required in the first of its execution cycles, the already summed up operand bits are not longer necessary in the later ones. That is, every input operand bit is needed just in one cycle, and it has to be stored only until that cycle. The storage requirements are quite smaller than when the entire operation is executed at a later cycle, because this implies to save the complete operands during several cycles. In the case of multicycle operators, the input operands are necessary all the cycles that last its execution.

(c) *Cycle length reduction*. The reduction of the addition widths, consequence of the operation fragmentations, directly involves a reduction of their delays, and in many cases, it also results in considerable cycle length reductions. However, this reduction is not achieved in a forthright manner, because it also depends on both the schedule of the remaining specification operations and the set of operations chained in every cycle.

(d) *Chaining of uncompleted operations*. Every result fragment calculated can be used as input operand in the cycle it has been computed in. This allows the beginning of a successor operation in the same cycle. This way, the calculus of addition fragments is less restrictive than the calculus of complete operations, and therefore extends the design space explored by synthesis algorithms.

Figure 14.3 shows graphically the schedule of one n bits addition in two inconsecutive cycles. The m LSB are calculated in cycle i, and the remaining ones in cycle $i+j$. The input operand bits computed in cycle i are not required in the following ones, and thus the only saved operands are $A_{n-1\ldots m}$ and $B_{n-1\ldots m}$. A successor operation requiring $A+B_{m-1\ldots 0}$ can begin as soon as cycle i. Although this example corresponds to the most basic fragmentation of one addition in two smaller ones, many others are also possible, including the ones that produce a bigger number of new operations.

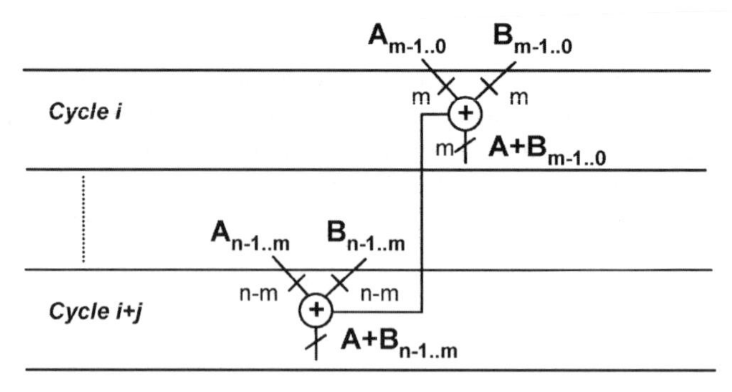

Fig. 14.3 Execution of one addition in two inconsecutive cycles

14.2.2.2 Fragmentation of Multiplications

The multiplications can also be fragmented in smaller operations to be executed in several not necessarily consecutive cycles, with the same aim of reducing the circuit area. In this case, two important differences with the addition fragmentation arise:

- The fragmentation of one multiplication not only produces smaller multiplications but also additions, needed to compose the final result from the multiplication fragments.
- The new data dependences among the operation fragments also oblige to calculate the LSB before the MSB. However, this need does not imply that the LSB must be calculated in the first cycle in this case. In fact, the fragments can be computed in any order as there are not data dependences among them. Only the addition fragments that produce the result bits of the original multiplication must keep their order to propagate adequately the carry signals.

The advantages of this design technique are quite similar to the ones mentioned in the above section dealing with additions. For this reason just the differences between them are cited in the following paragraphs.

(a) *Area reduction in the required FUs*. The execution of the new operation fragments can be performed using a multiplier and an adder, whose widths must equal the widths of the biggest multiplication and addition fragments, respectively. The total area of these two FUs and the required routing resources to share them across its execution cycles is usually quite smaller than the area of one multiplier of the original multiplication width.
(b) *Area reduction in the required storage units*. Although several multiplication fragments may require the same input operands, they are not needed all along every cycle of its execution. Every input operand bit may be available in several cycles, but it only needs to be stored until the last cycle it is used as input operand. The storage requirements are reduced compared to the execution of the entire operation at a later cycle, or the use of a multicycle multiplier, where the operand bits must be kept stable during the cycles needed to complete the operation.

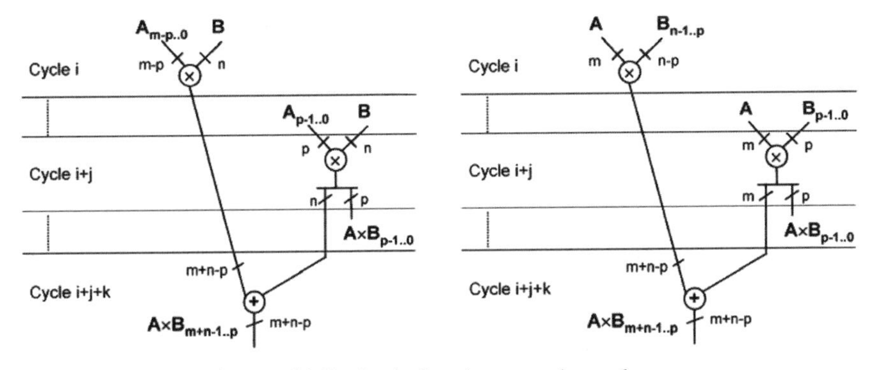

Fig. 14.4 Executions of one multiplication in three inconsecutive cycles

Figure 14.4 illustrates two different schedules of one $m \times n$ bits multiplication in three inconsecutive cycles. The LSB are calculated in cycle $i+j$, and the remaining ones in cycle $i+j+k$. Note that in both cases the multiplication fragment scheduled in cycle i does not produce any result bit. However, if the execution cycles of the fragments assigned to cycles i and $i+j$ were changed, the schedules would produce result bits in cycles i and $i+j+k$. In both schedules some of the input operand bits computed in cycle i are not required in cycle $i+j$, and none of them are required in cycle $i+j+k$. Although these examples correspond to basic fragmentations of one multiplication into two smaller ones and one addition, other more complex ones (that produce more fragments) are also possible.

14.2.3 Allocation and Binding Techniques

The allocation techniques to reduce the HW waste are focused on increasing the bit level reuse of FUs. The maximum reuse of functional resources occurs when all the datapath FUs execute an operation of its same type and width in every cycle (no datapath resource is wasted). If the allocation phase takes place after the scheduling, then the maximum degree of FUs reuse achievable depends on the given schedule. Note that the maximum bit level reuse of FUs can only be reached once a totally homogeneous distribution of operation bits among cycles is provided.

In order to get the maximum FUs reuse for a given schedule, the specification operations must be fragmented to obtain the same number of operations of equal type and width in every cycle. These fragmentations include the transformations of operations into several simpler ones whose types, representations, and widths may be different from those of the original. These transformations imply that one original operation will be executed distributed over several linked FUs. Its operands will also be transmitted and stored distributed over several routing and storage resources, respectively. The type of fragmentation performed is different for every operation type.

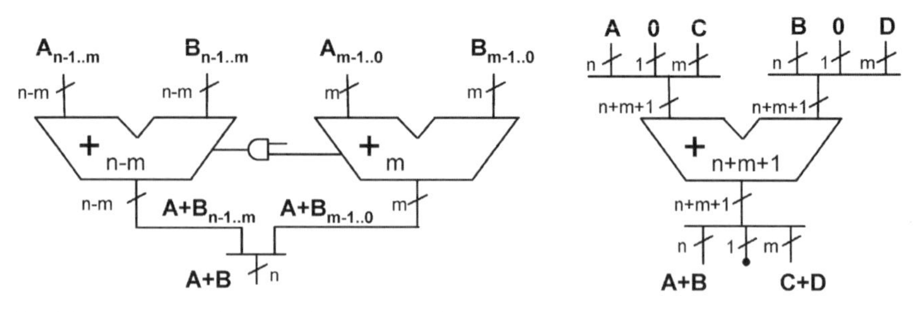

Fig. 14.5 Execution of one addition in two adders, and two additions in one adder

14.2.3.1 Fragmentation of Additions

The minimum HW waste in datapath adders occurs when all of them execute additions of their same width at least in one cycle. For every different schedule there may exist several implementations with the minimum HW waste, and thus similar adders area. The main difference resides on the fragmentation degree of the implementations. These circuits may present some of the following features:

(a) *Execution of one addition over several adders*. The set of adders used to execute every addition must be linked to propagate the carry signals.
(b) *Execution of one addition over a wider adder*. This requires the extension of the input operands and the discard of some result bits. This technique is used to avoid the excessive fragmentation in cycles with smaller computational cost.
(c) *Execution of several additions in the same adder*. Input operands of different operations must be separated with zeros to avoid the carry propagation, and the corresponding bits of the adder result discarded. At least one zero must be used to separate every two operations, such that the minimum adder width equals the sum of the additions widths to the number of operations minus one.

Figure 14.5 shows the execution of one addition over two adders as well as the execution of two additions over the same adder. These trivial examples provide the insight of how these two design techniques can be applied to datapaths with more adders and operations.

14.2.3.2 Fragmentation of Multiplications

Like additions, the minimum HW waste in datapath multipliers occurs when all of them calculate multiplications of their same width at least in one clock cycle. Some of the desirable features of one implementation with minimum HW waste are:

(a) *Execution of one multiplication over several multipliers and adders*. The set of multipliers used to execute every multiplication must be connected to adders

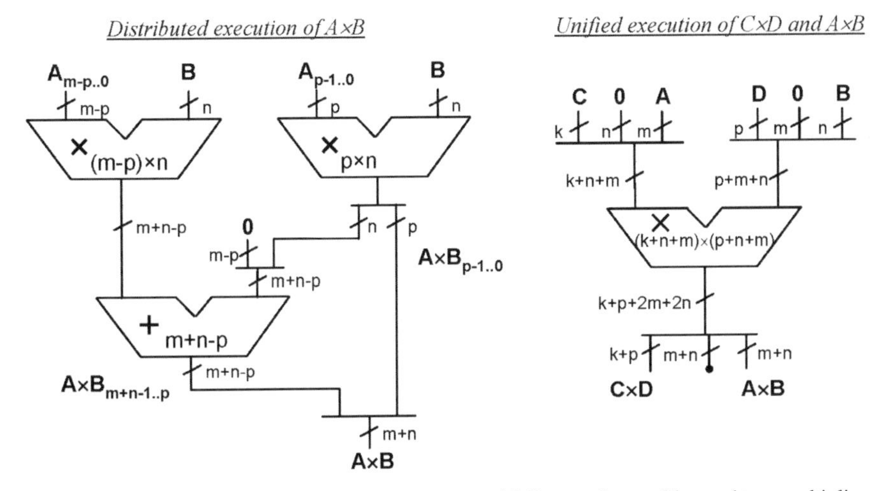

Fig. 14.6 Execution of one multiplication over two multipliers and one adder, and two multiplications in one multiplier

in some way (either chained or via intermediate registers) in order to sum the partial results and get the final one.

(b) *Execution of one multiplication over a wider multiplier.* This technique is used to avoid the excessive fragmentation of multiplications in cycles with smaller computational cost.

(c) *Execution of several multiplications in the same multiplier.* The input operands of different operations must be separated with several zeros to avoid interferences in the calculus matrix of the different multiplications. As the amount of zeros needed equals the operands widths, this technique seems only appropriate when the computational costs of the multiplications scheduled in every cycle are quite different.

Figure 14.6 shows the execution of one multiplication over two multipliers and one adder as well as the execution of two multiplications over the same multiplier. Note here that the unification of operations to be executed in the same FU produces an excessive internal waste of the functional resource, what makes this technique unviable in most designs. By contrast, the fragmentation does not produce any additional HW waste, and can be used to reduce the area of circuits with some HW waste in all the cycles.

14.3 Applications to Scheduling Algorithms

The scheduling techniques proposed to reduce the HW waste can be easily implemented in most HLS algorithms. This requires the common kernel extraction of specification operations and its successive transformations into several simpler ones to obtain balanced distributions in the number of bits computed per cycle.

This section presents a variant of the classical force-directed scheduling algorithm proposed by Paulin and Knight [4] that includes some bit-level design techniques to reduce the HW waste [3]. The intent of the original method is to minimize the HW cost subject to a given time constraint by balancing the number of operations executed per cycle. For every different type of operations the algorithm successively selects, among all operations and all execution cycles, an (operation, cycle) pair according to an estimate of the circuit cost called force.

By contrast, the intent of the proposed variant is to minimize HW cost by balancing the number of bits of every different operation type calculated per cycle. This method successively selects, among all operations (multiplications first and additions afterwards) and all execution cycles, a pair formed by an operation (or an operation fragment) and an execution cycle, according to a new *force* definition that takes into account the widths of operations. It also uses a new parameter, called *bound*, to decide whether an operation should be either scheduled complete, or fragmented and just one fragment scheduled in the selected cycle.

14.3.1 Force Calculation

Our redefinition of the force measure used in the classical force-directed algorithm does not consider all operations equally. Instead, it gives a different weight to every operation in function of its computational cost, leading to a more uniform distribution of the computational costs of operations among cycles.

The proposed algorithm calculates the forces using a set of new distribution graphs (DGs) that represent the distribution of the computational cost of operations. For each DG, the distribution in clock cycle c is given by:

$$\mathrm{DG}(\tau,c) = \sum_{op \in \mathrm{EOP}_c^{\tau}} (\mathrm{COST}(op) \cdot \mathrm{P}(op,c)),$$

where

- $\mathrm{COST}(op)$: computational cost of operation op. For additions it is defined as the width of the widest operand, and for multiplications as the product of the widths of its operands.

$$\mathrm{COST}(op) = \begin{cases} \mathrm{Max}\,(\mathrm{width}(x), \mathrm{width}(y)) & op \equiv x + y \\ \mathrm{width}(x) \cdot \mathrm{width}(y) & op \equiv x \times y \end{cases}$$

- $\mathrm{P}(op,c)$: probability of scheduling operation op in cycle c. It is similar to the classical method.
- EOP_c^{τ} (*estimated operations* of type τ in cycle c): set of operations of type τ whose mobility makes their scheduling possible in cycle c.

Except for the DG redefinition, the force associated with an operation to cycle binding is calculated like the classical method. The smaller or more negative the *force* expression is, the more desirable an operation to cycle binding becomes. In each iteration, the algorithm selects the operation and cycle with the lowest *force*, thus scheduling first operations with less mobility in cycles with smaller sum of computational costs, i.e. with smaller $\mathrm{DG}(\tau, c)$.

14.3.2 Bound Definition

Our algorithm binds operations (or fragments) to cycles subject to an upper bound that represents the most uniform distribution of operation bits of a certain type among cycles reachable in every moment. It is used to decide if an operation should be fragmented and how. Its initial value corresponds to the most uniform distribution of operation bits of type τ among cycles, without data dependencies:

$$bound = \frac{\sum\limits_{op \in \mathrm{OP}^{\tau}} \mathrm{COST}(op)}{\lambda},$$

where

OP^{τ}: set of specification operations of type τ.
λ: circuit latency.

The perfect distribution defined by the bound is not always reachable due to data dependencies among operations and given time constraints. Therefore it is updated during the scheduling in the way explained below.

If the number of bits already scheduled in the selected cycle c plus the computational cost of the selected operation op is equal or lower than the bound, then the operation is scheduled there.

$$\mathrm{CCS}(\tau, c) + \mathrm{COST}(op) \leq bound,$$

where

- $\mathrm{CCS}(\tau, c)$: sum of the computational costs of type τ operations scheduled in cycle c

$$\mathrm{CCS}(\tau, c) = \sum_{op \in \mathrm{SOP}_c^{\tau}} \mathrm{COST}(op).$$

- SOP_c^{τ}: set of operations of type τ scheduled in cycle c.

Otherwise the operation is fragmented and one fragment scheduled in the selected cycle (the remaining fragments continue unscheduled). The computational cost of the fragment to be scheduled *CCostFrag* equals the value needed to reach the bound in that cycle.

$$CCostFrag = bound - \mathrm{CCS}(\tau, c).$$

In order to avoid a reduction in the mobility of the predecessors and successors of the fragmented operation, they are also fragmented.

14.3.3 Bound Update

Once an operation (or a fragment) has been scheduled in a cycle c, it is checked if the distribution defined by the actual value of the bound is still reachable. Otherwise the value of the bound is updated with the next most uniform distribution still reachable. This occurs when:

- The sum of the computational costs of operations scheduled in cycle c does not reach the *bound* and there are not new operations left that could be scheduled in it, either because they are already scheduled, or their mobilities have changed.

  $$(\mathrm{CCS}(\tau,c) < bound) \wedge (\mathrm{UOP}_c^\tau = \phi),$$

  where
  UOP_c^τ: set of unscheduled operations of type τ whose mobility makes their scheduling possible in cycle c.

  The new bound value is the previous one plus the value needed to reach the bound in cycle c divided by the number of open cycles (included in the mobility of the unscheduled operations).

  $$NewBound = bound + \frac{bound - \mathrm{CCS}(\tau,c)}{\|\mathrm{OC}\|} \text{ where, } \mathrm{OC} = \{c \in \mathrm{N} \,|\, \mathrm{UOP}_c^\tau \neq \phi\}.$$

- The sum of the computational costs of the operations scheduled in cycle c equals the bound and there exists at least one unscheduled operation whose mobility includes cycle c, but even fragmented cannot be scheduled in its mobility cycles.

$$(\mathrm{CCS}(\tau,\mathrm{c}) = \mathrm{bound}) \wedge \left(\exists op \in \mathrm{UOP}_c^\tau | \sum_{c \in \mu op} (bound - \mathrm{CCS}(\tau,\mathrm{c})) < \mathrm{width}(op) \right),$$

where
μop: set of cycles included in the mobility of operation op.

The new bound value is the old one plus, for every operation satisfying the above condition, the computational cost of the operation fragment that cannot be scheduled divided by the number of cycles of its mobility.

$$NewBound = bound + \frac{\mathrm{COST}(op) - \sum_{c \in \mu op} (bound - \mathrm{CCS}(\tau,c))}{\|\mu_{op}\|}.$$

14.3.4 Operation Fragmentation

In order to schedule an addition fragment in a certain cycle, it is not necessary to define the portion of the addition to be calculated in that cycle. It will be fixed once the operation has been completely scheduled, i.e. when all the addition fragments have been scheduled. Then the algorithm selects the LSB of the operation to be executed in the earliest of its execution cycles, and so on until the MSB are calculated in the last cycle. Due to carry propagations among addition fragments, any other arrangement of the addition bits would require more computations to produce the correct result. The number of bits executed in every cycle coincides with the width of the addition fragment scheduled in that cycle

Unlike additions, the algorithm must select the exact portion of the multiplication that will be executed in the selected cycle. To do so, it transforms the operation into a set of smaller multiplications and additions. One of these new multiplications corresponds to the fragment to be scheduled there, and the other fragments continue unscheduled. The selection of every fragment type and width is required to calculate the mobility of the unscheduled part of the multiplication, and of the predecessors and successors of the original operation as well. Thus, it must be done immediately after scheduling a multiplication fragment in order to avoid reductions in the mobility of all the affected operations.

Many different ways can be found to transform one multiplication into several multiplications and additions. However, it is not always possible to obtain a multiplication fragment of a certain computational cost. In these cases, the multiplication is transformed in order to obtain several multiplication fragments whose sum of computational costs equals the desired cost.

In order to avoid reductions in the mobility of the successors and predecessors of fragmented operations, these must be fragmented too. In the case of additions, every predecessor and successor is fragmented into two new operations, one of them as wide as the scheduled fragment. The mobility of each immediate predecessor ends just before where the addition fragment is scheduled, and the mobility of each immediate successor begins in the next cycle. The remaining fragments of its predecessors and successors inherit the mobility of their original operations. These fragmentations divide the computational path into two new independent ones, where the two fragments of a same operation have different mobility.

In the case of multiplications, their immediate successors and predecessors may not become immediate successors and predecessors of the new operations. Data dependencies among operations are not directly inherited during the fragmentation. Instead, the immediate predecessors and successors of every fragment must be calculated after each fragmentation.

14.4 Applications to Allocation Algorithms

The proposed techniques to reduce the HW waste during the allocation phase can be easily implemented in most algorithms. This chapter presents a heuristic algorithm that includes most of the proposed techniques [2]. First it calculates the minimum set of functional, storage, and routing units needed to allocate the operations of the given schedule, and afterwards, it successively transforms the specification operations to allocate them to the set of FUs. The set of datapath resources can also be modified in the allocation to avoid the HW waste. These modifications consist basically on the substitution of functional, storage, or routing resources for several smaller ones, but do not represent an increment of the datapath area.

This algorithm also exploits the proposed allocation techniques to guarantee the maximum bit-level reuse of storage and routing units. In order to minimize the storage area, some variables may be stored simultaneously in the same register (wider than or equal to the sum of the variables widths), and some variables may be fragmented and every fragment stored in a different register (the sum of the registers widths must be greater than or equal to the sum of the variables widths). And to achieve the minimal routing area, some variables may be transmitted through the same multiplexer, and some variables may be fragmented and every fragment transmitted through a different multiplexer.

The proposed algorithm takes as input one scheduled behavioural specification and outputs one controller and one datapath formed by a set of adders, a set of multipliers, a set of other types of FUs, some glue logic needed to execute additive and multiplicative operations over adders and multipliers, a set of registers, and a set of multiplexers. The algorithm is executed in two phases:

(1) *Multiplier selection and binding*. A set of multipliers is selected and some specification multiplications are bound to them. Some other multiplications are transformed into smaller multiplications and some additions in order to increase the multipliers reuse, and the remaining ones are converted into additions to be allocated during the next phase.
(2) *Adder selection and binding*. A set of adders is selected and every addition bound to it. These additions may come from the original specification, the transformation of additive operations, or the transformation of multiplications into smaller ones or directly into additions.

The next sections explain the central phases of the algorithm proposed, but first some concepts are introduced to ease their understanding.

14.4.1 Definitions

- *Internal Wastage* (IW) of a FU in a cycle: percentage of bits discarded from the result in that cycle (due to the execution of one operation over a wider FU).
- *Maximum Internal Wastage Allowed* (MIWA): Maximum average IW of every multiplier in the datapath allowed by the designer. A MIWA value of 0% means that no HW waste is permitted (i.e. every multiplier in the datapath must execute one operation of its same width in every cycle).
- *Multiplication order*: One multiplication of width $m \times n$ (being $m \geq n$) is bigger than other one of width $k \times l$ (being $k \geq l$) if either ($m > k$) or ($m = k$ and $n > l$).
- *Occurrence* of width n in cycle c: number of operations of width n scheduled in cycle c.

- *Candidate*: set of operations of the same type which satisfy the following conditions:
  - all of them are scheduled in different cycles
  - $(m \geq n)$ for every width n of the candidate operations, where m is the width of the biggest operation of the candidate

  There exist many different bit alignments of the operations comprised in a candidate. In order to reduce the algorithm complexity, only those candidates with the LSB and the MSB aligned are considered. Thus, if one operation is executed over a wider FU the MSB or the LSB of the result produced are discarded.
- *Interconnection saving* of candidate C (IS): sum of the number of bits of the operands of C candidate operations that may come from the same sources, and the number of bits of the results of C candidate operations which may be stored in the same registers.

$$\mathrm{IS}(C) = \mathrm{BitsOpe}(C) + \mathrm{BitsRes}(C),$$

  where
  BitsOpe(C): number of bits of the left and right operands that may come from the same sources.
  BitsRes(C): number of bits of the C candidate results that may be stored in the same set of storage units.
- *Maximum Computed Additions Allowed per Cycle* (MCAAC): maximum number of addition bits computed per cycle. This parameter is calculated once there are not unallocated multiplications left, and it is obtained as the maximal sum of the addition widths in every cycle.

14.4.2 Multiplier Selection and Binding

In order to avoid excessive multiplication transformations, and thus obtain more structured datapaths, this algorithm allows some HW waste in the instanced multipliers. The maximum HW waste allowed by the designer in every circuit is defined by the MIWA parameter. This phase is divided into the following four steps, and finishes when either there are not remaining unallocated multiplications left, or when it is not possible to instance a new multiplier without exceeding MIWA (due to the given scheduling). This check is performed after the completion of every step. The steps 1–3 are executed until it is not possible to instance a new multiplier with a valid MIWA. Then, step 4 is executed followed by the adder selection and binding phase.

14.4.2.1 Instantiation and Binding of Multipliers Without IW

For every different width $m \times n$ of multiplications, the algorithm instances as many multipliers of that width as the minimum occurrence of multiplications of that width per cycle. Next, the algorithm allocates operations to them. For every instanced multiplier of width $m \times n$, it calculates the candidates formed by as many multiplications of the selected width as the circuit latency, and the IS of every candidate. The algorithm allocates to every multiplier the operations of the candidate with the highest IS. Multipliers instanced in this step execute one operation of its same width per cycle, and therefore their IW is zero in all cycles.

14.4.2.2 Instantiation and Binding of Multipliers with Some IW

The set of multiplications considered in this step may come from either the original specification, or the transformation of multiplications (performed in the next step). For every different width $m \times n$ of multiplications, and from the biggest, the algorithm checks if it is possible to instance one $m \times n$ multiplier without exceeding MIWA. It considers in every cycle the operation (able to be executed over an $m \times n$ multiplier) that produces the lowest IW of an $m \times n$ multiplier. After every successful check the algorithm instances one multiplier of the checked width, and allocates operations to it. Now the candidates are formed by as many operations as the number of cycles in which at least there is one operation that may be executed over one $m \times n$ multiplier. The width of the candidate operation scheduled in cycle c equals the width of the operation used in cycle c to perform the check, such that each candidate has the same number of operations of equal width. Once all candidates have been calculated, the algorithm computes their corresponding IS, and allocates the operations of the candidate with the highest IS. Multipliers instanced in this step may be unused during several cycles, and may also be used to execute narrower operations (being the IW average of these multipliers in compliance with MIWA).

14.4.2.3 Transformation of Multiplications into Several Smaller Multiplications

This step is only performed when it is not possible to instance a new multiplier of the same width as any of the yet unallocated multiplications without exceeding MIWA. It transforms some multiplications to obtain one multiplication fragment of width $k \times l$ from each of them. These transformations increase the number of $k \times l$ multiplications, which may result in the final instance of a multiplier of that width (during previous steps). First the algorithm selects both the width of the operations to be transformed and the fragment width, and afterwards a set of multiplications of the selected width, which are finally fragmented.

The following criteria are used to select the multiplication and fragment widths:

(1) The algorithm selects as $m \times n$ (width of the operations to be transformed) and $k \times l$ (fragment width), the widths of the two biggest multiplications that satisfy the following two conditions:

- There is at least one $k \times l$ multiplication, being $k \times l < m \times n$, that can be executed over one $m \times n$ multiplier (i.e. $m \geq k$ and $n \geq l$).
- At least in one cycle there is one $m \times n$ multiplication scheduled and there are not $k \times l$ multiplications scheduled.

(2) The algorithm selects two different widths as the widths of the operations to be fragmented, and a fragment width independent of the remaining unallocated multiplications. The widths selected of the operations to be fragmented $m \times n$ and $k \times l$, are those of the biggest multiplications that satisfy the following conditions:

- $m \times n \neq k \times l$
- At least in one cycle there is one $m \times n$ multiplication scheduled and there are not $k \times l$ multiplications scheduled.
- At least in one cycle there is one $k \times l$ multiplication scheduled and there are not $m \times n$ multiplications scheduled.

In this case the fragment width equals the maximum common multiplicative kernel of $m \times n$ and $k \times l$ multiplications, i.e. $\min(m,k) \times \min(n,l)$.

Next the algorithm selects the set of operations to be fragmented. In the first case it is formed by one $m \times n$ multiplication per every cycle where there are not $k \times l$ multiplications scheduled. And in the second one, it is formed by either one $m \times n$ or one $k \times l$ multiplication per cycle. In the cycles where there exist operations of both widths scheduled, only one multiplication of the largest width is selected. Once the set of operations to be fragmented and the desired fragment width are selected, the algorithm decides which one out of the eight different possible fragmentations is selected, according to the following criteria:

- The best fragmentations are the ones that obtain, in addition to one multiplication fragment of the desired width, other multiplication fragments of the same width as any of the yet unallocated multiplications.
- Among the fragmentations with identical multiplication fragments, the one that requires the lowest cost in adders is preferable.

Figure 14.7 illustrates the eight different fragmentations of one $m \times n$ multiplication explored by the algorithm to obtain one $k \times p$ multiplication fragment.

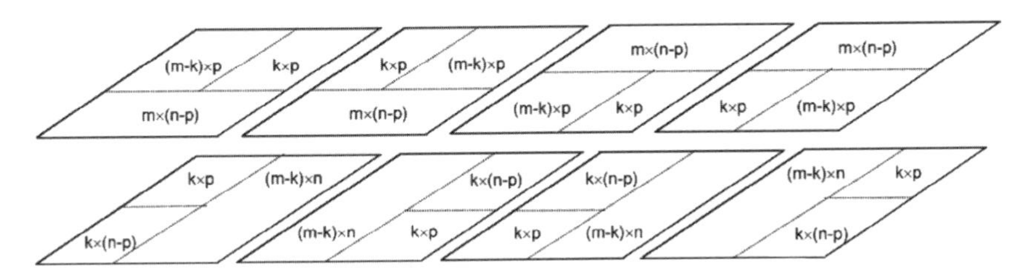

Fig. 14.7 Multiplication fragmentations explored by the algorithm

14.4.2.4 Transformation of Multiplications into Additions

Due to the given schedule it is not always possible to instance a new multiplier without exceeding MIWA. Therefore, unallocated multiplications are transformed into several additions.

14.4.3 Adder Selection and Binding

14.4.3.1 Instantiation and Binding of Adders Without IW

The set of additions considered here may come from the original specification, the transformation of multiplications (performed in the previous phase), or the transformation of additions (step 4.3.3). For every different width n of unallocated additions, the algorithm instances as many adders of that width as the minimum occurrence of additions of that width per cycle. Next, operations are allocated to them. For every instanced adder of width n, it calculates the candidates formed by as many additions of the selected width as the circuit latency, and the IS of every candidate. The algorithm allocates to every adder the operations of the candidate with the highest IS. The IW of the adders instanced here is zero in all the cycles.

14.4.3.2 Instantiation and Binding of Adders with Some IW

For every different width n of unallocated additions, and from the biggest, the algorithm checks if it is possible to instance one n adder without exceeding MCAAC. It considers in every cycle the operation (able to be executed over an n adder) that produces the lowest IW of an n bits adder. After every successful check, the algorithm instances one adder of the checked width, and allocates operations to it. Now the candidates are formed by as many operations as the number of cycles where there is at least one operation that may be executed over one n bits adder. The width of the candidate operation scheduled in cycle c equals the width of the operation used in cycle c to perform the check. Once all candidates are calculated, their corresponding IS are computed, and the additions of the candidate with the highest IS allocated. Adders instanced in this step may be unused during several cycles, and may also be used to execute narrower operations (being the IW of these adders in compliance with MCAAC).

14.4.3.3 Transformation of Additions

This step is only performed when it is not possible to instance a new adder of the same width as any of the yet unallocated additions without exceeding MCAAC. Some additions are transformed to obtain one addition fragment of width m from each of them. These transformations increase the number of m bits additions, which may result in the final instance of an adder of that width (during previous steps).

First the algorithm selects both the set of the operations to be transformed and the fragment width, and afterwards it performs the fragmentation of the selected additions. The fragment size is the minimum width of the widest unallocated operation scheduled in every cycle. A maximum of one operation per cycle is fragmented each time, but only in cycles without unallocated operations of the selected width. The set of fragmented operations is formed by the widest unallocated addition scheduled in every cycle without operations of the selected width. Every selected addition is decomposed into two smaller ones, being one of fragments of the desired width. These fragmentations produce the allocation of at least one new adder of the selected width during the execution of the previous steps, and may also contribute to the allocation of additional adders.

14.5 Analysis of the Implementations Synthesized Using the Proposed Techniques

This section presents some of the synthesis results obtained by the algorithms described previously which include some of the bit level design techniques proposed in this chapter. These results have been compared to those obtained by a HLS commercial tool, *Synopsys Behavioral Compiler* (BC) version 2001.08, to evaluate the quality of the proposed methods and their implementations in HLS algorithms.

The area of the implementations synthesized is measured in number of inverters, and includes the area of the FUs, storage and routing units, glue logic, and controller. The clock cycle length is measured in nanoseconds. The RT-level implementations produced have been translated into VHDL descriptions to be processed by *Synopsys Design Compiler* (DC) to obtain the area and time reports. The design library used in all the experiments is VTVTLIB25 by *Virginia Tech.* based on 0.25 μm TSMC technology.

14.5.1 Implementation Quality: Influential Factors

The main difference between conventional synthesis algorithms and our approach is the number of factors that influence the quality of the implementations obtained. The implementations proposed by conventional algorithms depend on the specification size, the operation mobility, and the specification heterogeneity, measured as the number of different triplets (type, data format, width) present in the original specification divided by the number of operations. Otherwise, our algorithms minimize the influence of data dependencies and get implementations totally independent from the specification heterogeneity, i.e. from the number, type, data format, and width of the operations used to describe behaviours.

Just to illustrate these influences we have synthesized different descriptions of the same behaviour, shown in Table 14.3, first with the proposed algorithms, and afterwards with BC. These descriptions have been created by progressively transforming (from circuit *A* to *G*) some of the specification operations into several smaller ones, in order to increase the number of operations of every different type and width, such that the specification heterogeneities of these descriptions have been progressively reduced from *A* to *G*. Circuit *A* is the original specification formed by 30 operations with six different operation types (MAC, multiplication, addition, subtraction, comparison, and maximum), two different data formats (unsigned and two's complement), and eight different operation widths (4, 8, 12, 16, 24, 32, 48, and 64 bits). And circuit *G* is the specification obtained after the last transformation. It consists of 86 unsigned multiplications and additions of 26 different widths, thus being the description with the smallest heterogeneity. Table 14.3 shows the number of operations (# Operations), the number of different operation types (# Types), the number of different data formats (# Formats), and the number of different operation widths (# Widths) present in each synthesized specification. The latency in all cases is equal to ten cycles. The amount of area saved by the algorithm grows, in general, with the specification heterogeneity. In the circuits synthesized, the homogeneous distribution of the computational costs among cycles achieved by our algorithm has also resulted in substantial clock cycle length reductions. Figure 14.8 shows in this set of examples the amount of area and clock cycle length saved by our approach in function of the heterogeneity.

Table 14.3 Features of the synthesized descriptions

Circuit	# Operations	# Types	# Formats	# Widths	Heterogeneity
A	30	6	2	8	0.7
B	36	6	2	10	0.55
C	44	5	2	12	0.45
D	52	4	1	15	0.42
E	65	3	1	19	0.35
F	73	3	1	22	0.32

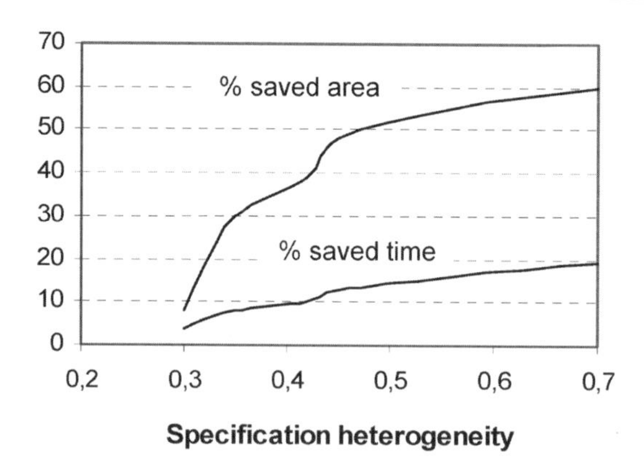

Fig. 14.8 Experimental area and execution time of different descriptions of one specification

Table 14.4 Area results of the synthesis of some modules of the ADPCM decoder algorithm

ADPCM decoder module	Datapath resources	Commercial tool (# inverters)	Fragmentation techniques (# inverters)
IAQ	FUs	388	234
IAQ	Controller	60	62
IAQ	Multiplexers	158	166
IAQ	Registers	189	192
IAQ	Total area	798	664 (16.7% saved)
OPFC + SCA	FUs	928	478
OPFC + SCA	Controller	62	66
OPFC + SCA	Multiplexers	418	470
OPFC + SCA	Registers	461	475
OPFC + SCA	Total area	1,873	1,503 gates (19.7% saved)
TTD	FUs	681	285
TTD	Controller	60	63
TTD	Multiplexers	220	232
TTD	Registers	261	273
TTD	Total area	1,226	865 (29.4% saved)
	Total all modules	3,897	3,032 (22.2% saved)

14.5.2 An Application Example

As an example of a real circuit, we have synthesized the following modules of the ADPCM decoding algorithm described in the Recommendation G.721 of CCITT:

- *Inverse Adaptative Quantizer (IAQ)*
- *Output PCM Format Conversion (OPFC)*
- *Synchronous Coding Adjustment (SCA)*
- *Tone and Transition Detector (TTD)*

Table 14.4 compares the area of the modules synthesized by our approach and BC for a fixed value of the circuit latency. The amount of area saved by our algorithm averages 22%.

OPFC and SCA modules have been synthesized together, and IAQ and TTD independently. Better results could be obtained if all modules were synthesized together, because it would increase the number of operations that could be executed over the same FU, and the number of variables that could be stored in the same register.

14.5.3 Synthesis of Non Heterogeneous Specifications

As shown in previous sections, our algorithm substantially reduces the area of circuits synthesized from heterogeneous specifications, as compared with commercial tools and previous known approaches. But the application method is not just limited to heterogeneous specifications. Important area reductions can also be achieved when specifications formed by operations with the same data formats and widths are synthesized. Indeed, our algorithm becomes the best choice for non heterogeneous specifications where the latency, number of operations, and data dependencies prevent reaching homogeneous distributions of operations among cycles. The areas of conventional implementations synthesized from non heterogeneous specifications may be slightly smaller than ours, but only where conventional algorithms are able to find nearly homogeneous distributions of the number of operations of every different type and width executed per cycle, and for a similar reason as for heterogeneous specifications. The implementations obtained synthesizing non heterogeneous specifications satisfy the following features:

- The amount of cycle length saved increases in inverse ratio to the latency. As latency decreases the number of chained operations that have to be executed in a cycle grows, as well as the potential benefit from distributing over several cycles the execution of certain operations.
- The amount of area saved increases in direct proportion to the circuit latency. As the number of cycles grows, more uniform distributions in the computational costs of operations may be found among them by our algorithm.

In order to illustrate the effectiveness of our method with non heterogeneous specifications, we have synthesized the fifth order elliptic wave filter formed by 34 unsigned operations (26 additions and 8 multiplications). In this specification all variables, input and output ports are 16 bits wide. The implementations obtained have been compared to the ones produced by BC. Table 14.5 shows the area and cycle length of the implementations obtained for three different latencies: 8, 11 and 16 cycles. Our algorithm saves up to 36% of cycle length and 27% of area for 8 and 16 clock cycles, respectively.

14.6 Further Applications of the Proposed Techniques

The proposed design techniques have been implemented in HLS algorithms. However, they can also be applied before or after the synthesis process to optimize behavioural descriptions or RT implementations, respectively. In these cases, conventional HLS algorithms could be used to synthesize the specifications, taking advantage of further improvements in HLS. The transformation of RT implementations usually results more complex than the behavioural optimization, as some design decisions taken during the HLS process might need to be undone. However, the optimization of the behavioural descriptions may produce some different implementations in function of the diverse HLS algorithms used. In order to take advantage of the behavioural optimization, the transformations performed should be in concordance with the design strategies implemented in the HLS algorithms, what requires a previous analysis of the algorithms used to perform the synthesis process.

Circuit area is the optimization parameter discussed along this chapter, but these design techniques can be used to optimize the execution time or power consumption as well.

Table 14.5 Area and time results of the synthesis of the fifth order elliptic wave filter

Circuit latency	Datapath resources	Commercial tool	Fragmentation techniques
8	FUs	3,876 inverters	3,530 inverters
8	Controller	135 inverters	138 inverters
8	Multiplexers	1,696 inverters	1,732 inverters
8	Registers	1,932 inverters	1,974 inverters
8	Total area	7,654 inverters	7,398 inverters (4% saved)
8	Cycle length	58, 63 ns	37, 27 ns (36% saved)
11	FUs	3,552 inverters	2,893 inverters
11	Controller	179 inverters	192 inverters
11	Multiplexers	1,552 inverters	1,632 inverters
11	Registers	1,771 inverters	1,693 inverters
11	Total area	7,065 inverters	6,438 inverters (19% saved)
11	Cycle length	51, 59 ns	41, 81 ns (9% saved)
16	FUs	3,390 inverters	1,937 inverters
16	Controller	194 inverters	208 inverters
16	Multiplexers	1,752 inverters	1,680 inverters
16	Registers	1,449 inverters	1,098 inverters
16	Total area	6,794 inverters	4,953 inverters (27% saved)
16	Cycle length	32, 27 ns	31, 13 ns (4% saved)

Conventional HLS scheduling synthesis algorithms are very conservative when dealing with Read-After-Write dependences, as the execution of one operation is allowed once all its predecessors have been calculated. However, in the execution of arithmetic operations some bits are required later than others, and also some bits are produced earlier than others. The design methods exposed in this chapter may be adapted to ease Read-After-Write dependences in order to improve the circuit performance as has been recently shown by Ruiz-Sautua et al. [5]. A previous analysis of the critical path at bit-granularity must be performed to estimate the most appropriate values of both the cycle length and latency, in order to minimize the slack times wasted in cycles where the results calculated have smaller arrival times than the cycle length. These estimations result quite appropriate to guide the decompositions of operations into sub-words fragments, allowing their execution in different cycles to speed up the circuit execution times. This way the execution of one operation may begin before the calculus of its predecessors has been completed. This becomes feasible when the execution of the predecessor has begun in the selected cycle or in a previous one, and even if it will finish in a posterior cycle. These schedules are out of the current HLS boundaries. The state of the art scheduling techniques (pipelining, chaining, bit-level chaining, multicycle, and non-integer multicycle) cannot achieve designs with these features.

The application of these techniques to reduce the power consumption includes the minimization of both static and dynamic consumptions. On one hand, the static consumption optimization is directly obtained from the circuit area reduction. On the other hand, the minimization of the dynamic dissipation requires the previous data profiling of the circuit input signals. It is obtained by means of simulations of the behavioural description, provided normal operation mode. The analysis of the switching activity information at the bit level become the appropriate parameter to guide the fragmentation of specification operations, in order to reduce the number of commutations occurred in datapath resources. Fragmentation allows the partial application of arithmetic properties, different bit alignments in the execution of operation fragments, and the distributed execution of operations over different FUs. Furthermore, this last feature lets different fragments of the same operation share their functional, storage and routing resources with different specification operations. All these features significantly expand the design space explored by conventional algorithms, resulting in substantial power consumptions savings.

14.7 Conclusions

Several bit-level design techniques have been proposed to improve the quality of the circuits resulting from behavioural synthesis. These techniques are non-compliant with the assertion assumed by conventional HLS algorithms that states the indivisibility of operations. Otherwise, the fragmentation of operations is the method used to expand the design space explored in HLS. These techniques provide several challenges to improve the circuit area, execution time, or power consumption, thanks to some design features infeasible with previous approaches, like the execution of one operation across several inconsecutive cycles, the ease of Read-After-Write dependences, the distributed execution of operations among several functional, storage and routing resources, the reuse of FUs to execute compatible operations, and the partial application of arithmetic properties.

The proposed design methods can be efficiently applied either during architectural synthesis, or to optimize behavioural specifications or RT-level implementations. In this chapter, some of these techniques have been applied during the synthesis process to reduce the circuit area. In particular, the operation fragmentation has been used during the scheduling phase to balance the computational cost of the operations executed in every cycle, and during the HW allocation and binding phase to minimize the HW waste of instanced resources. The set of experiments performed show great area savings in comparison to conventional algorithms, as well as additional reductions in the execution time. Finally, they also demonstrate the independency from the design style used in the specification achieved by the use of these design methods. Therefore, the designer skills become no longer a decisive factor on the quality of the synthesized circuits.

References

1. C.R. Baugh and B.A. Wooley. "A Two's Complement Parallel Array Multiplication Algorithm", IEEE Transactions on Computers, Vol. 22 (12) (1973), pp. 1045–1047
2. M.C. Molina, J.M. Mendías, R. Hermida, "Behavioural Specifications Allocation to Minimise Bit Level Waste of Functional Units", IEE Proceedings-Computers & Digital Techniques, Vol. 150 (5) (2003), pp. 321–329

3. M.C. Molina, R. Ruiz-Sautua, J.M. Mendías, R. Hermida, "Bitwise Scheduling to Balance the Computational Cost of Behavioural Specifications", IEEE Transactions on Computer Aided Design of Integrated Circuits and Systems, Vol. 25 (1) (2006), pp. 31–46
4. P.G. Paulin and J.P. Knight, "Force-Directed Scheduling for the Behavioral Synthesis of ASICS", IEEE Transactions on Computer Aided Design of Integrated Circuits and Systems, Vol. 8 (6) (1989), pp. 661–679
5. R. Ruiz-Sautua, M.C. Molina, J.M. Mendías "Exploiting Bit-Level Delay Calculations in Behavioural Synthesis", IEEE Transactions on Computer Aided Design of Integrated Circuits and Systems, Vol. 26 (9) (2007), pp. 1589–1601

Chapter 15
High-Level Synthesis Algorithms for Power and Temperature Minimization

Li Shang, Robert P. Dick, and Niraj K. Jha

Abstract Increasing digital system complexity and integration density motivate automation of the integrated circuit design process. High-level synthesis is a promising method of increasing designer productivity. Continued process scaling and increasing integration density result in increased power consumption, power density, and temperature. High-level synthesis for integrated circuit (IC) power and thermal optimization has been an active research area in the recent past. This chapter explains the challenges power and temperature optimization pose for high-level synthesis researchers and summarizes research progress to date.

Keywords: Behavioral synthesis, High-level synthesis, Power, Temperature, Thermal modeling, Reliability

15.1 Power and Temperature Optimization

In this section, we give an overview of the key motivations for, and challenges of, optimizing power consumption and temperature during high-level synthesis.

15.1.1 Brief Introduction to High-Level Synthesis

High-level synthesis [1–4] is the process of automatically converting a behavioral, algorithmic, specification to an optimized register-transfer level digital design. The specification indicates the behavior of an algorithm and available hardware resources such as multipliers and multiplexers, but does not indicate the manner in which the algorithm should be implemented. A high-level synthesis algorithm automatically selects the set of hardware resources to use, determines the connections between them, binds operations to functional units such as multipliers, determines a clock frequency, and produces a schedule of operations. High-level synthesis can

P. Coussy and A. Morawiec (eds.) *High-Level Synthesis.*

therefore be formulated as an optimization problem with functionality constraints. Performance, power consumption, temperature, IC area, reliability, or other metrics may be optimized or constrained [5–15].

15.1.2 Importance of Power Consumption and Temperature

Power is the source of the greatest problems facing IC designers. High-power ICs rapidly deplete battery energy. Rapid changes in power consumption result in on-chip voltage fluctuations that lead to transient errors. High spatial and temporal power densities lead to high temperatures, which result in decreased lifetime reliability. High temperatures also increase leakage power consumption, thereby closing a self-reinforcing power–temperature feedback loop. The effects of increasing power consumption, power variation, and power density are expensive to handle. The wages of power are bulky short-lived batteries, huge heatsinks, large on-die capacitors, high server electric bills, and unreliable ICs. The only alternative is optimizing IC power consumption, temperature, and reliability. Power optimization within high-level synthesis has a long history, which we will review in this chapter. In contrast, temperature optimization during high-level synthesis began to receive widespread attention fairly recently, although some researchers foresaw the coming importance of the problem a decade ago.

Temperature is increased by both IC dynamic and leakage power. In addition, IC on-die temperature profiles depend on the temporal and spatial distribution of IC power as well as the packaging and cooling solution. Increasing IC power consumption increases IC peak temperature as well as on-die spatial and temporal thermal variation, which have significant impact on IC power consumption, temperature, reliability, cooling cost, and performance. A high IC temperature increases charge carrier concentrations, resulting in increased subthreshold leakage power consumption. In addition, it decreases charge carrier mobility, decreasing transistor and interconnect performance, and decreases threshold voltage, increasing transistor performance. Moreover, temperature heavily influences the fault processes, i.e., electromigration, dielectric breakdown, and power–thermal cycling, that lead to a large number of IC permanent faults. Finally, increasing IC power density requires the use of more effective cooling and packaging solutions to ensure IC reliable run-time operation, resulting in a significant increase in cooling and packaging cost. In summary, thermal issues have become a major concern in IC design. Modeling and optimizing IC thermal properties is thus essential for reliability, power consumption, and performance.

15.1.3 Power Analysis and Optimization

IC power analysis and optimization have been an active research areas for decades. Researchers developed power modeling techniques at all levels of the IC design hierarchy. High-level synthesis poses unique challenges for IC power modeling and analysis. During behavioral synthesis, the lack of low-level implementation details, such as interconnect length and timing information permitting estimation of transient glitches, makes accurate power analysis challenging. In addition, power optimization during high-level synthesis typically involves the evaluation of numerous optimization decisions, requiring highly-efficient power analysis techniques. Most existing power-aware high-level synthesis systems use microarchitectural or structural power modeling methods to permit fast power estimation. These modeling methods are capable of approximately estimating the relative power savings of behavioral optimization decisions, but unable to characterize the accurate IC power profile.

Power optimization has been a primary focus of high-level synthesis for more than a decade. A variety of power optimization techniques have been proposed to tackle IC dynamic and leakage power consumption during high-level synthesis. IC dynamic power consumption can be reduced by attacking supply voltage, capacitance, switching activity, and frequency. Among these, voltage scaling is the most promising technique for reducing IC dynamic power consumption, due to the fact that IC dynamic power is quadratically proportional to supply voltage. Techniques, such as voltage and frequency scaling, multi-V_{dd}, and voltage islands, have been widely adopted by recently-developed low-power high-level synthesis systems. However, voltage reduction has a negative impact on circuit performance. Moreover, the effectiveness of voltage scaling diminishes as the supply voltage of nanometer-scale ICs approaches the sub-volt range. IC leakage power consumption was once a second-order consideration. However, it is becoming increasingly significant as a result of continued IC process scaling. Leakage accounts for 40% of the power consumption of today's high-performance microprocessors [16]. Leakage power can be the primary limitation on the lifetime of battery-powered systems. Leakage power optimization techniques, such as body biasing and transistor sizing, have been used in several high-level synthesis systems [17–20]. IC subthreshold leakage increases superlinearly with temperature. Due to the increase of IC power density and thermal effects, thermal-aware leakage analysis has gained prominence in high-level synthesis [21, 22].

15.1.4 Thermal Analysis and Optimization

An IC's thermal profile is a complex, time-varying function of its power consumption profile. The chip average temperature is determined by IC average power density and cooling package efficiency. The run-time chip thermal profile, on the other hand, depends on IC spatial and temporal power variation. The occurrence of on-die hotspots is often the result of transient activation of functional units with a high power density.

Behavioral design changes alone cannot effectively solve the IC temperature optimization problem. IC thermal analysis requires detailed physical information, i.e., IC floorplan, interconnect, and chip-package configuration. IC thermal optimization requires the use of behavioral power optimization techniques to minimize IC average power density and temperature-aware physical design to balance and optimize the chip thermal profile. A unified high-level and physical analysis and optimization flow is critical for IC thermal optimization.

One primary challenge of IC thermal optimization comes from the high computational complexity of IC thermal analysis. IC thermal analysis is the process of characterizing the three-dimensional temperature profile of IC chip and cooling package. It requires a detailed simulation of heat conduction from an IC's power sources, i.e., transistors and interconnects, through cooling package layers, to the ambient environment, which can be described using the following equation:

$$\rho c \frac{\partial T(\mathbf{r},t)}{\partial t} = \nabla \cdot (k(\mathbf{r}) \nabla T(\mathbf{r},t)) + p(\mathbf{r},t), \tag{15.1}$$

where ρ is the material density, c is the mass heat capacity, $T(\mathbf{r},t)$ and $k(\mathbf{r})$ are the temperature and thermal conductivity of the material at position $\mathbf{r}$ and time t, and $p(\mathbf{r},t)$ is the power density of the heat source. Steady-state thermal analysis characterizes the chip temperature distribution when the IC power consumption does not vary with time, i.e., when the heat capacity, c, is neglected. Dynamic thermal analysis is used to characterize the temporal variations of the IC thermal profile. This problem is analogous to transient analysis of an electrical circuit [23], with electrical resistance and capacitance replaced with thermal resistance and heat capacity. The rate of temperature change in response to a change in power density is related to the thermal RC time constant of the IC region of interest. The major challenges of numerical IC thermal analysis are high computational complexity and memory usage. For steady-state thermal analysis, high modeling accuracy requires fine-grain modeling of IC chip and cooling package, resulting in high memory usage and long analysis time. For dynamic thermal analysis using time-domain methods, such as the fourth-order Runge-Kutta method, higher modeling accuracy requires fine spatial and temporal discretization granularity, increasing computational overhead and memory usage. Recent IC thermal analysis techniques use spatially and temporally adaptive numerical modeling methods to control the computational complexity and memory usage of IC thermal analysis while maintaining high accuracy [24].

15.2 High-Level Synthesis Algorithms for Power Optimization

Research on power-aware high-level synthesis can be traced back to the early 1990s. This section reviews existing low-power high-level design methodologies and synthesis tools.

15.2.1 Dynamic Power Optimization in High-Level Synthesis

In the past, IC power consumption was dominated by dynamic power. Therefore, early research on low-power synthesis focused on dynamic power optimization. IC dynamic power consumption is a quadratic function of supply voltage. Voltage scaling is therefore the most effective dynamic power optimization technique. However, voltage scaling may have a negative impact on circuit performance. Therefore, the tradeoff between power and performance has been a central theme in power-aware high-level synthesis. Johnson and Roy developed MESVS, a behavioral scheduling algorithm, that minimizes IC power consumption by using multiple supply voltages [25]. This work uses integer linear programming to produce an optimal schedule with discrete voltage-level assignment under timing constraints. Unfortunately, optimal integer linear programming formulations generally cannot be used for large problem instances due to high computational complexity. Raje and Sarrafzadeh proposed a heuristic to solve the voltage assignment problem [26]. The computational complexity of this method is $\mathcal{O}(N^2)$. Chang and Pedram developed a dynamic programming technique to solve the multi-voltage scheduling problem [27]. This technique reduces supply voltages along non-critical paths to optimize IC power consumption and minimize performance impact. Hong et al. designed a multi-voltage scheduling algorithm to minimize the power consumption of core-based systems-on-a-chip [28]. Helms et al. propose a behavioral synthesis system which uses multi-voltage assignment and adaptive body biasing to minimize IC power consumption [29]. These studies demonstrate that voltage scaling can reduce IC power consumption. However, the extra power saving decreases with the number of voltage levels. Recently, Liu et al. propose an approximation algorithm for IC power optimization using multiple supply voltages [30]. The computational complexity of the proposed approximation algorithm is $\mathcal{O}(dkN)$, where d and k are small constants. This work shows significant runtime advantage over the past work.

IC dynamic power consumption can be reduced by minimizing circuit capacitance and run-time switching activity. Chatterjee and Roy designed a behavioral synthesis system, which uses architectural transformation to minimize circuit switching activity [31]. Raghunathan and Jha developed the first optimal, ILP-based formulation of high-level synthesis for switching power minimization [32]. Chandrakasan et al. developed HYPER-LP, a high-level synthesis system using algorithmic transformation to reduce circuit capacitance, thereby reducing IC power consumption [9]. Chang and Pedram developed an low-power allocation and resource binding technique to minimize the switching activity in registers [11] and datapath functional components [33]. In this work, the power-optimal register and functional component assignment problem is formulated as a max-cost flow problem. Dasgupta and Karri developed binding and scheduling techniques to minimize the switching activity of buses [6]. Musoll and Cortadella developed a high-level synthesis system, which uses loop interchange, operand reordering, operand sharing, idle units, and operand correlation, for reducing the activities of IC functional units [34]. Raghunathan and Jha designed SCALP, an iterative-improvement-based high-level synthesis system [13], which integrates a variety of power optimization techniques, including architectural transformation, scheduling, clock selection, module selection, and hardware allocation and assignment. Lakshminarayana et al. proposed a power-aware register binding technique for high-level synthesis, which provides the first formulation of a perfect power management philosophy, i.e., no functional unit that does not need to be active in a given cycle should consume any switching power in that cycle [35]. Dasgupta and Karri developed a high-level synthesis system for IC energy and reliability optimization [36]. They proposed a resource binding and scheduling algorithm to minimize circuit switching activity, thereby optimizing IC power consumption and minimizing electromigration-induced failure effects in on-chip buses. Ercegovac et al. proposed a behavioral synthesis system [37] that uses multi-gradient search for system resource allocation using multiple-precision arithmetic units. Karmarkar-Karp's number partitioning heuristic is used to determine task assignment. Lakshminarayana et al. proposed a high-level power optimization technique which extracts common-case behavior from the given behavioral description and then synthesizes an RTL implementation of the common-case circuit, which is a much smaller than the circuit that implements the complete behavior and runs most of the time [38]. Wang et al. proposed a high-level design methodology for IC energy and performance optimization [39] called input space adaptive design. This technique identifies the behavioral equivalence among sub-circuits and eliminates redundant logical operations, thereby optimizing IC energy and performance.

15.2.2 Leakage Power Optimization in High-Level Synthesis

IC leakage power consumption is becoming increasingly significant as a result of technology scaling. Therefore, leakage power optimization during high-level synthesis has drawn significant attention. Khouri and Jha [17] developed a behavioral, iterative algorithm to minimize IC leakage power consumption using dual-V_{th} technology. The proposed algorithm is a greedy approach that iteratively identifies the operation with the maximum leakage power reduction potential and binds it with a high-V_{th} implementation. Gopalakrishnan and Katkoori developed a leakage-aware resource allocation and binding algorithm using multi-V_{th} technology [18]. This algorithm seeks to maximize the idle time slots of datapath components. Idle functional modules are scheduled to enter the sleep mode at runtime to minimize the IC leakage power consumption. Tang et al. formulated the leakage optimization problem as the maximum weight independent set problem [19]. A heuristic was proposed to identify the datapath components with maximum or near-maximum leakage reduction potentials, which are then replaced with low-leakage alternatives. Dal et al. developed a low-power high-level synthesis algorithm using power islands [20]. The supply voltage of each power island can be controlled independently. The proposed algorithm conducts circuit partitioning and assigns circuit components with overlapping idle times to the same power island. Idle power islands are then scheduled to be power-gated to minimize leakage power consumption. IC sub-threshold leakage power is a strong function of chip temperature. Therefore, thermal effects must be considered during leakage power optimization. We will later survey thermal-aware leakage optimization techniques.

15.2.3 Importance of Incorporating Physical Design Within High-Level Synthesis

It is becoming increasingly important to consider physical design decisions within high-level synthesis. Interconnect power consumption and delay are increasing relative to logic delay. Increasing power densities are making it necessary to determine and optimize the IC thermal profile at design time; computing a thermal profile requires a power profile. Determining the interconnect structure and power profile depends on the knowledge of the IC floorplan. As a result, a number of researchers have considered the impact of physical details, e.g., floorplanning information, on high-level synthesis [40–46].

Taking interconnect power consumption and delay into consideration during high-level synthesis has attracted significant attention. In previous work [47–51], the number of interconnects or multiplexers was used to estimate the interconnect cost. The performance and power impact of the interconnect and interconnect buffers are now first-order considerations [52]. It is no longer possible to accurately predict the power consumption and performance of a design without first knowing enough about its floorplan to predict the structure of its interconnect. This change has complicated both design and synthesis. For this reason, a number of researchers have worked on interconnect-aware high-level synthesis algorithms [53–55]. These approaches typically use a loosely coupled independent floorplanner for physical estimation. This technique has the advantage of allowing estimation of physical properties but has a drawback. Creating a floorplan from scratch for each high-level synthesis move is inefficient, given the fact that the new floorplan frequently has only small differences with the previous one. The constructive approach works for small problem instances but is unlikely to scale to large designs. New techniques for tightly coupling behavioral and physical synthesis that dramatically improve their combined performance and quality are now necessary.

Incremental automated design promises to build tighter relationship between high-level synthesis and physical design, improving the quality of each [56, 57]. A number of high-level synthesis algorithms are based on incremental optimization and are therefore amenable to integration with incremental physical design algorithms. This has the potential of improving both quality and performance. Incremental methods improve quality of results by maintaining important properties across consecutive physical estimations during synthesis. Moreover, they shorten CPU time by reusing and building upon previous high-quality physical design solutions that required a huge amount of effort to produce. Recent work has proposed unified incremental behavioral synthesis and floorplanning to permit more accurate communication delay, communication power consumption, and power profile estimation [58].

15.3 Modeling and Optimizing Temperature in High-Level Synthesis

This section introduces the main challenges of temperature-aware high-level synthesis and describes a number of recent techniques to overcome them.

15.3.1 Thermal Model Selection for Use in High-Level Synthesis

It is important to select appropriate thermal modeling and analysis techniques for use in temperature-aware high-level synthesis. In reality, ICs experience temporal and spatial temperature variation. However, accurately modeling spatial and temporal variation during thermal analysis can be the most time consuming part of high-level synthesis. Given a fixed amount of time for synthesis, there is a trade-off between the amount of time spent on thermal analysis and the number of tentative behavioral synthesis solutions that can be considered. Therefore, it is important to model temporal and spatial temperature variation with as much detail as necessary for accuracy, but no more.

A number of high-level synthesis formulations consider energy consumption or average power consumption. This is equivalent to optimizing temperature while neglecting temporal and spatial variation in temperature. In some applications, this is legitimate. In others, it can result in extremely large errors. Let us now consider the circumstances in which it is necessary to model spatial and temporal variation in temperature.

IC packaging has a strong influence on heat flow, and therefore on the importance of modeling spatial temperature variation. Packaging and cooling solutions that more efficiently remove heat tend to be more expensive. In order to minimize cost, it is reasonable to select a cooling solution that permits the temperature to approach its constraint under worst-case or average-case conditions. As a result, in low power density designs the package will have poor thermal conductance, e.g., a plastic package without heatsink. Is this case, the conductance between different points on the silicon die is high relative to the conductance between a point on the die, through the package, to the ambient. As a result, the temperature of the active layer will generally be fairly uniform despite spatial variation in power density. For this reason, a simple thermal model is sufficient for low power density ICs using low thermal conductance packages and cooling solutions [59,60]. High power density designs require more efficient packaging and cooling solutions to maintain safe temperatures. As a result, the thermal conductance between different points on the silicon die can decrease relative to the thermal conductance to the ambient. In this case, spatial variation in the power profile will result in spatial variation in temperature.

The properties of temporal variation in IC power consumption have a strong influence on the thermal modeling requirements. Most existing work on temperature-aware high-level synthesis assumes that power density does not vary with time and uses steady-state thermal analysis based on the temporal averages of power density. This is legitimate when the temporal variation of power densities occurs in a much shorter timescale than the IC thermal RC time constants, e.g., a high-frequency periodic system in which power density does not change on long time scales due to changing input data. However, it is not legitimate when there are long time scale changes in power density. If the interval of change in power density is long relative to the thermal RC time constants, it may be possible to accurately approximate the temperature by conducting steady-state analysis for each power density phase. However, in general, accurately modeling the thermal impact of time-varying power profiles requires dynamic thermal analysis, which is generally much more time-consuming than steady-state analysis.

Thus far, we have considered the conditions in which spatial and temporal thermal variation can be entirely neglected. However, once the decision is made to model spatial and/or temporal variation, it is still necessary to determine the required modeling resolution. Increasing the number of thermal elements or temperature evaluation time instants can dramatically increase the run-time of thermal analysis.

The required thermal model spatial resolution depends on material properties, cooling environment, and power density variation. During thermal analysis, it is common for an IC to be partitioned into multiple elements, each of which is assumed to be isothermal, i.e., to have internally-uniform temperature. To minimize analysis time, thermal elements should generally be as large as possible while still honoring the isothermal assumption. Note that an element with uniform power density does not necessarily honor the isothermal assumption because its neighboring thermal elements may have different temperatures, resulting in a substantial temperature gradient. The architectural thermal analysis tools commonly used in high-level synthesis thermal analysis support manual [61] or automatic [60] adaptation of spatial modeling granularity.

Dynamic thermal analysis is frequently formulated as a time-domain initial value problem in which the thermal profile is iteratively updated at increasing time instants. There is a tradeoff between the number of time instants, at which the temperature is explicitly evaluated, and accuracy. Assuming a constant error bound, the duration between explicit temperature evaluations depends on the rate and complexity of changes in the power profile. Therefore, dynamic adaptation is required to minimize analysis time under a constraint on maximum error. The thermal analysis tools commonly used in high-level synthesis support dynamic temporal adaptation to varying degrees [60, 61].

15.3.2 High-Level Synthesis Algorithms for Temperature Optimization

Temperature-aware high-level synthesis is currently a thriving research area, with new work appearing monthly in top conferences and journals. Ten years ago, Weng and Parker were the first to address the problem by moving high power density functional units away from high-temperature areas to reduce the spatial power density and introducing redundant operators to reduce the temporal power density [62]. It is interesting to note that Prakash and Parker were also the first to formulate the system-level heterogeneous distributed system synthesis problem, also 10 years before it became a highly-active research area [63]. Mukherjee et al. proposed to incrementally improve binding decisions to reduce the temperature of the hottest functional unit, thereby reducing both dynamic and leakage power consumption [21]. Gu et al. designed TAPHS, a temperature-aware unified physical and behavioral synthesis system [64]. TAPHS integrates behavioral and physical thermal optimization techniques, including voltage assignment, voltage island generation, and floorplanning, to optimize chip temperature, power, performance and area. Lim and Kim propose a network flow based method for temperature-aware binding that minimizes both peak and average switched capacitance [65]. Ni and Öğrenci Memik proposed a technique to reduce leakage power consumption using selective resource redundancy [22].

15.4 Conclusions

This chapter has described the current state-of-the-art in high-level synthesis algorithms that optimize power consumption and temperature. International Technology Roadmap for Semiconductors imply that power consumption will continue to be a primary concern for IC designers. Emerging power and power-induced problems, such as process variation influenced IC leakage power consumption, IC leakage-thermal coupling, and power–thermal dependent IC lifetime reliability problems, further exacerbate the challenges for high-level synthesis algorithms. On the other hand, power optimization techniques that were widely used in the past, such as voltage scaling and body biasing, will soon start running out of steam as a result of continued process scaling. Moreover, as power-aware unified architectural–physical design flows cease to be luxuries and become necessities, it will become necessary to cooperatively solve many problems that were once orthogonal to high-level synthesis. These challenges may require fundamental changes to existing high-level synthesis flows.

References

1. R. Camposano and W. Wolf, *High Level VLSI Synthesis*. Kluwer, MA, 1991
2. D. C. Ku and G. D. Micheli, *High Level Synthesis of ASICs Under Timing and Synchronization Constraints*. Kluwer, MA, 1992
3. D. Gajski, N. Dutt, A. Wu and S. Lin, *High-Level Synthesis: Introduction to Chip and System Design*. Kluwer, MA, 1992
4. A. Raghunathan, N. K. Jha and S. Dey, *High-Level Power Analysis and Optimization*. Kluwer, MA, 1998

5. R. Mehra and J. Rabaey, "Behavioral level power estimation and exploration," in *Proc. Int. Wkshp Low Power Design*, Apr. 1994, pp. 197–202
6. A. Dasgupta and R. Karri, "Simultaneous scheduling and binding for power minimization during microarchitecture synthesis," in *Proc. Int. Symp. Low-Power Design*, Apr. 1994
7. L. Goodby, A. Orailoglu and P. M. Chau, "Microarchitecture synthesis of performance-constrained, low-power VLSI designs," in *Proc. Int. Conf. Computer Design*, Oct. 1994
8. A. Raghunathan and N. K. Jha, "Behavioral synthesis for low power," in *Proc. Int. Conf. Computer Design*, Oct. 1994, pp. 318–322
9. A. P. Chandrakasan, M. Potkonjak, R. Mehra, J. Rabaey and R. Brodersen, "Optimizing power using transformations," *IEEE Transactions on Computer-Aided Design of Integrated Circuits and Systems*, vol. 14, no. 1, pp. 12–31, 1995
10. R. S. Martin and J. P. Knight, "Power profiler: Optimizing ASICs power consumption at the behavioral level," in *Proc. Design Automation Conf.*, June 1995
11. J. M. Chang and M. Pedram, "Register allocation and binding for low power," in *Proc. Design Automation Conf.*, June 1995
12. N. Kumar, S. Katkoori, L. Rader and R. Vemuri, "Profile-driven behavioral synthesis for low-power VLSI systems," *IEEE Design Test*, vol. 12, no. 3, pp. 70–84, 1995
13. A. Raghunathan and N. K. Jha, "SCALP: An iterative-improvement-based low-power data path synthesis system," *IEEE Transactions on Computer-Aided Design of Integrated Circuits and Systems*, vol. 16, no. 11, pp. 1260–1277, 1997
14. K. S. Khouri, G. Lakshminarayana and N. K. Jha, "High-level synthesis of low power control-flow intensive circuits," *IEEE Transactions on Computer-Aided Design of Integrated Circuits and Systems*, vol. 18, no. 12, pp. 1715–1729, 1999
15. H. P. Peixoto and M. F. Jacome, "A new technique for estimating lower bounds on latency for high level synthesis," in *Proc. Great Lakes Symp. VLSI*, Mar. 2000, pp. 129–132
16. S. Naffziger et al., "The implementation of a 2-core, multi-threaded Itanium family processor," *IEEE Journal of Solid-State Circuits*, vol. 41, no. 1, pp. 197–209, 2006
17. K. S. Khouri and N. K. Jha, "Leakage power analysis and reduction during behavioral synthesis," *IEEE Transactions on Computer-Aided Design of Integrated Circuits and Systems*, vol. 10, no. 6, pp. 876–885, 2002
18. C. Gopalakrishnan and S. Katkoori, "KnapBind: An area-efficient binding algorithm for low-leakage datapaths," in *Proc. Int. Conf. Computer Design*, Oct. 2003, pp. 430–435
19. X. Tang, H. Zhou and P. Banerjee, "Leakage power optimization with dual-Vth library in high-level synthesis," in *Proc. Design Automation Conf.*, June 2005, pp. 202–207
20. D. Dal, A. Nunez and N. Mansouri, "Power islands: A high-level technique for counteracting leakage in deep sub-micron," in *Proc. Int. Symp. Quality of Electronic Design*, Mar. 2006, pp. 165–170
21. R. Mukherjee, S. Öğrenci Memik and G. Memik, "Temperature-aware resource allocation and binding in high-level synthesis," in *Proc. Design Automation Conf.*, June 2005
22. M. Ni and S. Öğrenci Memik, "Thermal-induced leakage power optimization by redundant resource allocation," in *Proc. Int. Conf. Computer-Aided Design*, Nov. 2006, pp. 297–302
23. G. S. Ohm, "The Galvanic circuit investigated mathematically," in *Die galvanische Kette: mathematisch bearbeitet*, 1827
24. Y. Yang, C. Zhu, Z. P. Gu, L. Shang and R. P. Dick, "Adaptive multi-domain thermal modeling and analysis for integrated circuit synthesis and design," in *Proc. Int. Conf. Computer-Aided Design*, Nov. 2006, pp. 575–582
25. M. Johnson and R. K. Roy, "Optimal selection of supply voltages and level conversion during datapath scheduling under resource constraints," in *Proc. Int. Conf. Computer Design*, Oct. 1996, pp. 72–77
26. S. Raje and M. Sarrafzadeh, "Variable voltage scheduling," in *Proc. Int. Symp. Low Power Electronics & Design*, Aug. 1995, pp. 9–14
27. J. Chang and M. Pedram, "Energy minimization using multiple supply voltages," in *Proc. Int. Symp. Low Power Electronics & Design*, Aug. 1996, pp. 157–162

28. I. Hong, D. Kirovski, G. Qu, M. Potkonjak and M. B. Srivastava, "Power optimization of variable voltage core-based systems," *IEEE Transactions on Computer-Aided Design of Integrated Circuits and Systems*, vol. 18, no. 12, pp. 1702–1714, 1999
29. D. Helms, O. Meyer, M. Hoyer and W. Nebel, "Voltage- and ABB-island optimization in high level synthesis," in *Proc. Int. Symp. Low Power Electronics & Design*, Aug. 2007, pp. 153–158
30. H. Liu, W. Lee and Y. Chang, "A provably good approximation algorithm for power optimization using multiple supply voltages," in *Proc. Design Automation Conf.*, June 2007, pp. 887–890
31. A. Chatterjee and R. K. Roy, "Synthesis of low power linear DSP circuits using activity metrics," in *Proc. Int. Conf. VLSI Design*, Jan. 1994, pp. 261–264
32. A. Raghunathan and N. K. Jha, "An ILP formulation for low power based on minimizing switched capacitance during datapath allocation," in *Proc. Int. Symp. Circuits & Systems*, May 1995, pp. 1069–1073
33. J. Chang and M. Pedram, "Module assignment for low power," in *Proc. European Design Automation Conf.*, Sept. 1996, pp. 376–381
34. E. Musoll and J. Cortadella, "High-level synthesis techniques for reducing the activity of functional units," in *Proc. Int. Symp. Low Power Electronics & Design*, Aug. 1995, pp. 99–104
35. G. Lakshminarayana, A. Raghunathan, N. K. Jha and S. Dey, "A power management methodology for high-level synthesis," in *Proc. Int. Conf. VLSI Design*, Jan. 1998
36. A. Dasgupta and R. Karri, "High-reliability, low-energy microrchitecture synthesis," *IEEE Transactions on Computer-Aided Design of Integrated Circuits and Systems*, vol. 17, pp. 1273–1280, 1998
37. M. Ercegovac, D. Kirovski and M. Potkonjak, "Low-power behavioral synthesis optimization using multiple precision arithmetic," in *Proc. Design Automation Conf.*, June 1999, pp. 568–573
38. G. Lakshminarayana, A. Raghunathan, K. S. Khouri, N. K. Jha and S. Dey, "Common case computation: A high-level power-optimizing technique," in *Proc. Design Automation Conf.*, June 1999
39. W. Wang, A. Raghunathan, G. Lakshminarayana and N. K. Jha, "Input space adaptive design: A high-level methodology for energy and performance optimization," in *Proc. Design Automation Conf.*, June 2001, pp. 738–743
40. M. C. McFarland and T. J. Kowalski, "Incorporating bottom-up design into hardware synthesis," *IEEE Transactions on Computer-Aided Design of Integrated Circuits and Systems*, vol. 9, no. 9, pp. 938–950, 1990
41. J.-P. Weng and A. C. Parker, "3D scheduling: High-level synthesis with floorplanning," in *Proc. Design Automation Conf.*, June 1991, pp. 668–673
42. D. W. Knapp, "Fasolt: A program for feedback-driven data-path optimization," *IEEE Transactions on Computer-Aided Design of Integrated Circuits and Systems*, vol. 11, no. 6, pp. 677–695, 1992
43. J. P. Weng and A. C. Parker, "3D scheduling: High-level synthesis with floorplanning," in *Proc. Design Automation Conf.*, June 1992
44. Y. M. Fang and D. F. Wong, "Simultaneous functional-unit binding and floorplanning," in *Proc. Int. Conf. Computer-Aided Design*, Nov. 1994
45. M. Xu and F. J. Kurdahi, "Layout-driven RTL binding techniques for high-level synthesis using accurate estimators," *ACM Transactions on Design Automation Electronic Systems*, vol. 2, no. 4, pp. 312–343, 1997
46. W. E. Dougherty and D. E. Thomas, "Unifying behavioral synthesis and physical design," in *Proc. Design Automation Conf.*, June 2000
47. P. G. Paulin and J. P. Knight, "Scheduling and binding algorithms for high-level synthesis," in *Proc. Design Automation Conf.*, June 1989, pp. 1–6
48. C. A. Papachristou and H. Konuk, "A linear program driven scheduling and allocation method followed by an interconnect optimization algorithm," in *Proc. Design Automation Conf.*, June 1990
49. T. A. Ly, W. L. Elwood and E. F. Girczyc, "A generalized interconnect model for data path synthesis," in *Proc. Design Automation Conf.*, June 1990

50. S. Tarafdar and M. Leeser, "The DT-model: High-level synthesis using data transfer," in *Proc. Design Automation Conf.*, June 1998
51. C. Jego, E. Casseau and E. Martin, "Interconnect cost control during high-level synthesis," in *Proc. Design Circuits & Integration Systems Conf.*, Nov. 2000
52. R. Ho, K. Mai and M. Horowitz, "The future of wires," *Proceedings of the IEEE*, vol. 89, no. 4, pp. 490–504, 2001
53. P. Prabhakaran and P. Banerjee, "Simultaneous scheduling, binding and floorplanning high-level synthesis," in *Proc. Int. Conf. VLSI Design*, Jan. 1998
54. L. Zhong and N. K. Jha, "Interconnect-aware low power high-level synthesis," *IEEE Transactions on Computer-Aided Design of Integrated Circuits and Systems*, vol. 24, no. 3, pp. 336–351, 2005
55. A. Stammermann, D. Helms, M. Schulte, A. Schulz and W. Nebel, "Binding, allocation and floorplanning in low power high-level synthesis," in *Proc. Int. Conf. Computer-Aided Design*, Nov. 2003
56. O. Coudert, J. Cong, S. Malik and M. Sarrafzadeh, "Incremental CAD," in *Proc. Int. Conf. Computer-Aided Design*, Nov. 2000, pp. 236–244
57. W. Choi and K. Bazargan, "Hierarchical global floorplacement using simulated annealing and network flow migration," in *Proc. Design, Automation & Test in Europe Conf.*, Mar. 2003
58. Z. P. Gu, J. Wang, R. P. Dick and H. Zhou, "Unified incremental physical-level and high-level synthesis," *IEEE Transactions on Computer-Aided Design of Integrated Circuits and Systems*, 2007
59. G. Paci, P. Marchal, F. Poletti and L. Benini, "Exploring "temperature-aware design" in low-power MPSoCs," in *Proc. Design, Automation & Test in Europe Conf.*, Mar. 2006
60. Y. Yang, Z. P. Gu, C. Zhu, R. P. Dick and L. Shang, "ISAC: Integrated Space and Time Adaptive Chip-Package Thermal Analysis," *IEEE Transactions on Computer-Aided Design of Integrated Circuits and Systems*, 2007
61. W. Huang et al., "HotSpot: A compact thermal modeling methodology for early-stage VLSI design," *IEEE Transactions on VLSI Systems*, vol. 14, no. 5, pp. 501–524, 2006
62. J.-P. Weng and A. C. Parker, "Taking thermal considerations into account during high-level synthesis," *VLSI Design*, vol. 5, no. 2, pp. 183–193, 1997
63. S. Prakash and A. Parker, "Synthesis of application-specific multiprocessor architectures," in *Proc. Design Automation Conf.*, June 1991
64. Z. P. Gu, Y. Yang, J. Wang, R. P. Dick and L. Shang, "TAPHS: Thermal-aware unified physical-level and high-level synthesis," in *Proc. Asia & South Pacific Design Automation Conf.*, Jan. 2006, pp. 879–885
65. P. Lim and T. Kim, "Thermal-aware high-level synthesis based on network flow method," in *Proc. Int. Conf. Hardware/Software Codesign and System Synthesis*, Oct. 2006